T0182069

Compact Textbooks in Mathematics

Compact Textbooks in Mathematics

This textbook series presents concise introductions to current topics in mathematics and mainly addresses advanced undergraduates and master students. The concept is to offer small books covering subject matter equivalent to 2- or 3-hour lectures or seminars which are also suitable for self-study. The books provide students and teachers with new perspectives and novel approaches. They may feature examples and exercises to illustrate key concepts and applications of the theoretical contents. The series also includes textbooks specifically speaking to the needs of students from other disciplines such as physics, computer science, engineering, life sciences, finance.

- **compact:** small books presenting the relevant knowledge
- **learning made easy:** examples and exercises illustrate the application of the contents
- **useful for lecturers:** each title can serve as basis and guideline for a semester course/lecture/seminar of 2-3 hours per week.

More information about this series at http://www.springer.com/series/11225

Miroslav Josipović

Geometric Multiplication of Vectors

An Introduction to Geometric Algebra in Physics

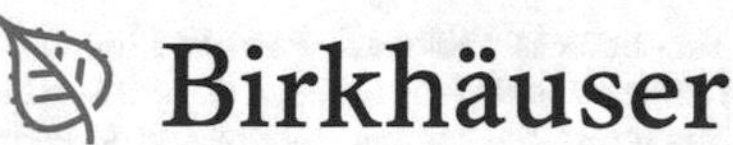

Miroslav Josipović
Zagreb, Croatia

Additional material to this book can be downloaded from http://extras.springer.com.

ISSN 2296-4568 ISSN 2296-455X (electronic)
Compact Textbooks in Mathematics
ISBN 978-3-030-01755-2 ISBN 978-3-030-01756-9 (eBook)
https://doi.org/10.1007/978-3-030-01756-9

Mathematics Subject Classification (2010): 15A66, 51Pxx, 14-XX, 00-01

This book is published under the imprint Birkhäuser, www.birkhauser-science.com, by the registered company Springer Nature Switzerland AG.
The registered company address is: Gewerbestrasse 11, 6330 Cham, Switzerland

So God made the people speak many different languages...

Virus

The ***World Health Organization*** has announced a worldwide epidemic of the ***coordinate virus*** in mathematics and physics courses at all grade levels. Students infected with the virus exhibit compulsive ***vector avoidance*** behavior, unable to conceive of a vector except as a list of numbers, and seizing every opportunity to replace vectors by coordinates. At least two-thirds of physics graduate students are severely infected by the virus, and half of those may be so permanently damaged that they will never recover. The most promising treatment is a strong dose of ***geometric algebra***. (*Hestenes*)

Cat

When the spiritual teacher and his disciples began their evening meditation, the cat who lived in the monastery made so much noise that it distracted them. Therefore, the teacher ordered tying up the cat during the evening practice. Years later, when the teacher died, tying the cat continued during meditation sessions. When the cat died, another cat was brought to the monastery and tied up. Centuries later, a learned descendant of the spiritual teacher wrote a scholarly treatise on the religious significance of tying up a cat during meditation. (Zen story)

Empty Your Cup

A university professor went to visit a famous Zen master. While the master quietly served tea, the professor talked about Zen. The master filled the visitor's cup to the brim, and then kept pouring. The professor watched the overflowing cup until he could no longer restrain himself. It is overfull! No more will go in! *the professor expostulated.* You are like this cup, *the master replied,* how can I show you Zen unless you first empty your cup*? (*Zen story*)*

To my sisters

Preface

The purpose of this text is to introduce the interested reader to the world of geometric algebra. Why?

All right, imagine Neelix and Tuvok (Tuvok is a Vulcan from the starship Voyager) are engaged in conversation. Neelix hopes to sell a new product to Tuvok. His intention is to intrigue Tuvok quickly, giving him as little information as possible, and the ultimate goal is that Tuvok, after using it, will be surprised by the quality of the product and recommend it to others. Do not forget that Tuvok can barely tolerate Neelix's nonstop blathering. Let us begin.

Neelix Mr. Vulcan, would you like to rotate objects without matrices, in any dimension?

Tuvok Mr. Neelix, do you offer me quaternions?

Neelix No, they work in 3D only. I have something much better. In addition, you will be able to calculate with spinors as well.

Tuvok Spinors? Come on, Mr. Neelix, you're not trying to say that I will be able to work with complex numbers, too?

Neelix Yes, Mr. Vulcan, the whole of complex analysis, generalized to higher dimensions. And you will get rid of tensors.

Tuvok Excuse me, what? I'm a physicist... it is impossible...

Neelix It is indeed possible. You do not need coordinates. Moreover, you will be able to do the special theory of relativity and quantum mechanics using the same tool. And all integral theorems that you recognize, including those in the complex field, become a single theorem.

Tuvok Come on... nice idea... I work a lot with Lie algebras and groups...

Neelix In the package...

Tuvok Are you kidding me, Mr. Neelix? Ok, let's suppose that I believe you. How much would that product cost me?

Neelix Pennyworth, Mr. Vulcan; the only price is to multiply vectors differently.

Tuvok That's all? You are offering me all of this for such a small price? What's the trap?

Neelix There isn't one. Except that you will have to spend some time to learn how to use the new tool.

Tuvok Time? I just do not have... Besides, why would I ever give up coordinates? You know, I am quite adept at juggling indices. I have my career...

Neelix Do the physical processes you are studying depend on the coordinate systems you choose?

Tuvok I hope not.

Neelix There. Does a rotation by matrices provide you a clear geometric meaning when you do it?

Tuvok No. I need to make an effort to discover one.

Neelix Now you will not have to for the most part. It will be available to you at each step.

Tuvok Mr. Neelix, I'm curious. Where did you get this new tool?

Neelix Well, Mr. Vulcan, it is an old tool from the Earth, nineteenth century, I think, invented by humans Grassmann and Clifford.

Tuvok What? How is that I am not familiar with it? Isn't that strange?

Neelix Well, I believe that human Gibbs and his followers had a hand in it. Allegedly, human Hestenes was trying to tell the other humans about it, but they did not pay any attention. You will agree, Mr. Vulcan, that humans are really funny sometimes.

Tuvok Mr. Neelix, this is a once in a blue moon case in which I just have to agree with you.

Tuvok buys the product and lives long and prospers. Then, of course, he recommends the new tool to the captain...

This text is motivational, so that the reader can see "what's up." It introduces basic concepts and provides some insights into possible applications. We endeavored to choose simple examples and solved problems; the reader will also find problems to solve. An active reading of the text is suggested, with paper and pencil in hand. Once you achieve some clear insights, computers will help in calculations and visualization. There is a great deal of literature, as well as computer programs, available on the Internet; the reader should do some research in this direction. Some facts are intentionally repeated throughout the text, from slightly different points of view, just to help novices to become comfortable with concepts and relationships. There are grounds to believe that geometric algebra is the mathematics of the future. Paradoxically, although formulated in the mid-nineteenth century, it was mainly ignored, due to a range of (unfortunate) circumstances. This book is intended primarily for young people; those with established careers probably will not easily accept something new. A background in physics and mathematics at the undergraduate level is welcome for some parts of the text. However, it is possible to follow the text using the Internet to find explanations for less familiar terms. Advanced high-school students should be able to read at least parts of the text. A useful source is the book [1]; it can help those who are beginning with algebra and geometry. The book [2] is rather difficult; we recommend it to those who want to do some serious thinking. We also recommend Hestenes's articles.

The reader should adopt the idea that vector multiplication as presented here is natural and justified; what follows are the consequences of such multiplication. The reader can come up independently with arguments to account for the definition of the geometric product. My intention is for the reader to accept that the geometric product is not just a "neat trick." It arises naturally from the concept of a vector. That fact changes a great deal of mathematics. A simple rule that parallel vectors commute, while orthogonal vectors anticommute, produces an incredible amount of mathematics, uniting various mathematical disciplines into the universal language of geometric algebra. Finally, readers well disposed toward geometry may find a great deal of pleasure in geometric algebra. Objects like bivectors are not just new and interesting; they are magical, and they evoke juvenescence and intuitive conceptions about mathematics. In fact, in this book we are trying to explain a new concept of numbers and a special language in which such numbers represent oriented geometric objects. Such oriented numbers are intuitive and can replace unintuitive objects like matrices and tensors, giving us, along the way, universality and great possibilities for the unification of different branches of mathematics.

In the text, we treat for the most part the three-dimensional (3D) Euclidean vector space, and we do so for three main reasons. First, generalizations are usually straightforward; second, the reader can rely on a powerful geometric intuition; and finally, the 3D Euclidean vector space has a lot to offer in the new mathematical language. Its beauty and richness are hidden behind the veil of traditional mathematics.

The Mathematica users can find a special implementation of *Cl*3 at official web address from the Publisher.

You are welcome to send comments or questions to me at miroslav.josipovic@gmail.com

Zagreb, Croatia
2018

Miroslav Josipović

References

1. J. Vince, *Geometric Algebra for Computer Graphics* (Springer, London, 2008)
2. D. Hestenes, G. Sobczyk, *Clifford Algebra to Geometric Calculus* (Reidel, Dordiecht, 1984)

The original version of this book was revised. An correction to this book can be found at https://doi.org/10.1007/978-3-030-01756-9_7

I would like to thank my family and my friends who helped me during the writing of this book in difficult circumstances. Dr. Selim Pašić helped me to solve some technical difficulties with images. Especially, I would like to thank Eckhard Hitzer for his patience, support, and encouragement to start writing this book.

Contents

About the Author

Miroslav Josipović is a Croatian physicist and musician with special interest in revising the language of mathematical physics and in creating new ways of teaching physics. Revision of the language of mathematical physics can be achieved by relying on geometric algebra, a powerful language of physics and mathematics that brings geometric clarity and unseen possibilities of unification and generalizations. Experiments with physics teaching lead the author to the belief that students can understand nontrivial concepts of physics at an early age, but teachers have to give up the teaching of formalism and offer living examples that encourage curiosity. The feeling that it is possible to understand motivates the students greatly, and the job of a teacher is to enable young people to attain such a feeling. If they are successful, elementary school pupils will be able to solve difficult conceptual problems related to Newton's first law, for example, or high-school students will be able to deal with problems of the special theory of relativity that are generally considered to be difficult at that age. Young students are smart. Mathematics has to rely on geometric intuition, and the author firmly believes that geometric algebra can be gradually introduced as early as elementary school, first through games with oriented geometric objects, and later by extending the concept of a vector to oriented surfaces (bivectors) and the oriented volumes (3-vectors).

Mathematical Notation

$x, \vec{x}, \mathbf{x}$	a vector
$\lvert\mathbf{x}\rvert$	the magnitude (length) of the vector $\mathbf{x}$
$\lVert A \rVert$	the norm of the element A, defined by the scalar product
$x \propto y$	x is proportional to y
$\equiv$	equal to (by definition)
AB	the geometric product (GP)
$A \cdot B$	the inner product
$A * B$	the scalar product
$A \bullet B$	the dot product
$A \wedge B$	the outer (wedge) product
$A \vee B$	the regressive product
$a \times b$	the cross product (for vectors in 3D)
$A \rfloor B$	the left contraction (LC)
$[A, B]$	the commutator of A and B, $(AB - BA)/2$
$\{A, B\}$	the anticommutator of A and B, $(AB + BA)/2$
$\mathrm{P}_B(A)$	a projector of A onto the subspace characterized by B
Cln	the Clifford (geometric) algebra in n-dimensional Euclidean space
e_i	a vector of a basis (usually orthonormal, in $Cl3$ we sometimes use σ_i)
e^i	a vector of a reciprocal basis, $e^i \cdot e_j = \delta_{ij}$
$\hat{\sigma}_i$	the ith Pauli matrix
$e_ie_j \equiv e_{ij}$	a unit bivector (in an orthonormal basis e_i)
$j = e_{123}$	the unit pseudoscalar in $Cl3$
k-vector	single graded element of the grade k
$M^\Delta = M \rfloor I^{-1}$	the dual operation, I is the unit pseudoscalar
M^*	complex conjugation, e.g., $(\alpha + \mathbf{n}j + \beta j)^* = \alpha - \mathbf{n}j - \beta j$, $\alpha, \beta \in \mathbb{R}$
$\langle M \rangle_r$	the grade r
$\langle M \rangle = \langle M \rangle_0$	the grade 0
$M^\dagger$	the reverse involution
$\overleftarrow{M}$	the grade involution ($\hat{M}$ is common; however, we use the hat for unit elements)
$\bar{M}$	the Clifford involution (conjugation, main involution)
$\langle M \rangle_R = (M + M^\dagger)/2$	the real part of a multivector (*paravector*) in $Cl3$
$\langle M \rangle_I = (M - M^\dagger)/2$	the imaginary part of a multivector in $Cl3$
$\langle M \rangle_S = (M + \bar{M})/2$	the scalar part of a multivector (complex scalar, $\alpha + \beta j$) in $Cl3$, $\alpha, \beta \in \mathbb{R}$
$\langle M \rangle_V = (M - \bar{M})/2$	the vector part of a multivector (complex vector, $\mathbf{x} + \mathbf{n}j$) in $Cl3$

$\langle M\rangle_+ = \left(M + \overleftarrow{M}\right)/2$	the even part of a multivector (α + $\mathbf{n}j$) in *Cl*3, $\alpha \in \mathbb{R}$
$\langle M\rangle_- = \left(M - \overleftarrow{M}\right)/2$	the odd part of a multivector ($\mathbf{x}$ + βj) in *Cl*3, $\beta \in \mathbb{R}$
$\lvert M\rvert = \sqrt{M\bar{M}}$	the multivector amplitude (MA; sometimes we use it as $M\bar{M}$)
δ_{ij}	the Kronecker delta symbol, $\delta_{ij} = 1$ for $i = j$ and $\delta_{ij} = 0$ for $i \neq j$
$\partial_k \equiv \frac{\partial}{\partial x^k}$	a partial derivative, $\frac{\partial x^i}{\partial x^j} = \delta_{ij}$
$\nabla = e^k \partial_k$	the differential operator *nabla* (summation over $k = 1$, 2, 3)
$\partial = \partial_t + \nabla$	a differential operator as a *paravector*
$\mathfrak{R}^{(p,q,r)}$	a vector space with the *signature* (p, q, r)
$\mathbb{R}$	the set of real numbers (α, β, ...)
i	the imaginary unit, $i = \sqrt{-1}$
$\mathbb{C}$	the set of complex numbers in *Cl*3, such as $\alpha + \beta j$; sometimes $\alpha + \beta i,\ \ i = \sqrt{-1},\ \ \alpha, \beta \in \mathbb{R}$
F	a linear transformation (note the text format)
$\bar{\mathrm{F}}$	adjoint (transpose) of the linear transformation F
$\lvert\psi\rangle$	a vector in a complex vector space, *ket* (a column vector, usually represents spinors)
$\langle\psi\rvert$	the Hermitian conjugate (transposed and complex conjugated) of $\lvert\psi\rangle$, *bra* (a row vector)
$\langle\phi\lvert\psi\rangle$	the inner product in a complex vector space
$\lvert\psi\rangle\langle\phi\rvert$	the outer product in a complex vector space
$\langle\phi\lvert O\rvert\psi\rangle$	the *expectation value* of the *operator* O in a complex vector space

Acronyms

APS	algebra of physical space, Pauli algebra, *Cl*3
CGA	the *conformal model* in geometric algebra
ICS	inertial coordinate system
IRF	inertial reference frame
GA	geometric algebra
GP	geometric product
LC	left contraction
MA	multivector amplitude
STA	spacetime algebra
STR	the special theory of relativity

Reference Labels

E*n*	the *n*th task with a solution in Solutions (Sect. 6.1)
Fig.*n*	Figure *n* from the text
L*n*	the link *n* from the References and Links to specific subjects
n	problem *n* in Problems (Sect. 6.2)
[*n*]	reference *n*
R*n*	reference *n* from the References and Links to specific subjects

1 Basic Concepts

Miroslav Josipović

M. Josipović, *Geometric Multiplication of Vectors*,
Compact Textbooks in Mathematics,
https://doi.org/10.1007/978-3-030-01756-9_1

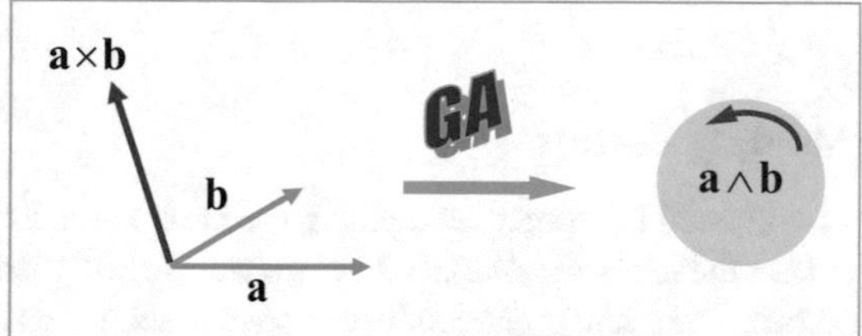

1.1 Geometric Product

We denote vectors by small Latin letters in *italic* format wherever there is no possibility of confusion. If necessary, we will use the **bold** format or arrows. For *multivectors*, we use uppercase *italic* format, and for real numbers we use the Greek alphabet or Latin letters in roman format. If we define an orthonormal basis in a vector space, then the number of unit vectors that square to 1 is denoted by p, that square to -1 by q, and that square to 0 by r. The common designation for such a vector space is $\mathfrak{R}(p, q, r)$ or $\mathfrak{R}^{p,\, q,\, r}$, while the triplet $(p,\ q,\ r)$ defines the *signature*. For the *geometric algebra* of the n-dimensional Euclidean vector space $\mathfrak{R}^n$ (with $q = r = 0$), we use the abbreviation *Cln*, motivated by the surname Clifford.

When we say "vector," we (usually) are not referring to elements of an abstract vector space; we rather are thinking of an "oriented straight line." We use the parallelogram rule to add vectors. Vectors a and b that satisfy the relation $b = \alpha a,\ \ \alpha \in \mathbb{R},\ \ \alpha \neq 0$, are said to be *parallel*. For parallel vectors, we say that they have the same *direction* (*attitude*), but they could have the same or opposite *orientation*. We can resolve any vector b into the component in the direction of the vector a (*projection*) and the component without a component parallel to the vector a (*rejection*):

The original version of this chapter was revised. A correction to this chapter can be found at https://doi.org/10.1007/978-3-030-01756-9_7.

$b = b_{\|} + b_{\perp}, \quad b_{\|} = \alpha a, \quad \alpha \in \mathbb{R}.$

Here we can immediately anticipate objections, such as, "Yes, but if we are talking about orthogonal vectors, we need a scalar product." Although we use the character $\perp$ here for the moment, we are **not** talking about the orthogonality of vectors. Simply, since we can sum vectors, any vector can be expressed as a vector sum of two vectors, in an infinite number of ways. The previous relation articulates one of these possibilities; consequently, we can treat this decomposition as a question of **existence**, not how to implement it practically. Explicitly, for $b_{\perp} = b - b_{\|} = b - \alpha a$, if we assume that the vector $b_{\perp}$ contains a component parallel to a, we can write $b'_{\perp} + \beta a = b - \alpha a$; then the vector $b'_{\perp}$ is our rejection. If there is no $b'_{\perp}$, then vectors a and b are parallel. Eventually, after we have succeed in defining a new product of vectors, we can return to the question of how to find $b_{\perp}$ practically, and that is what the new product of vectors should certainly enable us to do. We will see that these concepts are connected to the commutativity properties.

1.1.1 How to Multiply Vectors

For the moment, we will need to "forget" everything we have learned about the multiplication of vectors (at least the *cross product*). But before we "forget" them, let us analyze a couple of their properties. The scalar product of two vectors is defined as $\mathbf{a} \cdot \mathbf{b} = |\mathbf{a}||\mathbf{b}| \cos \theta$, where $|\mathbf{a}|$ is the length of the vector $\mathbf{a}$, while θ is the angle between the vectors. The cross product is defined as the (*axial*) vector $\mathbf{a} \times \mathbf{b}$ of length $|\mathbf{a} \times \mathbf{b}| = |\mathbf{a}||\mathbf{b}||\sin\theta|$, orthogonal to both vectors $\mathbf{a}$ and $\mathbf{b}$, while its orientation is defined by the *right-hand rule* (see Fig. 1.1, left). Can we uniquely solve the equation $\mathbf{a} \cdot \mathbf{x} = \alpha$? The answer is, clearly, that we cannot, because if $\mathbf{x}$ is a solution, then each vector of the form $\mathbf{x} + \mathbf{b}, \quad \mathbf{b} \cdot \mathbf{x} = 0$, is a solution as well. What about the equation $\mathbf{a} \times \mathbf{x} = \mathbf{b}$? It also cannot be solved uniquely, because if $\mathbf{x}$ is a solution, then each vector of the form $\mathbf{x} + \beta\mathbf{a}$ is also a solution.

However, interestingly, we can find the unique solution if we consider both equations. First, for the existence of a solution, it must be $\mathbf{a} \cdot \mathbf{b} = 0$ and $\alpha \leq |\mathbf{a}||\mathbf{x}|$. From $\mathbf{a} \cdot \mathbf{x} = \alpha = |\mathbf{a}||\mathbf{x}| \cos \theta$ and $|\mathbf{b}| = |\mathbf{a} \times \mathbf{x}| = |\mathbf{a}||\mathbf{x}||\sin\theta|$, it follows that $|\mathbf{x}| = \sqrt{\alpha^2 + |\mathbf{b}|^2}/|\mathbf{a}|$. Both $\mathbf{a}$ and $\mathbf{x}$ are orthogonal to $\mathbf{b}$, giving $\mathbf{x} \cdot \mathbf{b} = 0$, while the orientation of $\mathbf{x}$ is defined by the right-hand rule. For example, using the orthonormal basis e_i and specifying $\mathbf{a} = e_1$, $\mathbf{b} = \sqrt{2}e_3$, and $\alpha = \sqrt{2}$, we get $\mathbf{x} = \sqrt{2}(e_1 + e_2)$. ◆ Try $\mathbf{x} = \sqrt{2}(e_1 + e_2) + \lambda e_3$. ◆

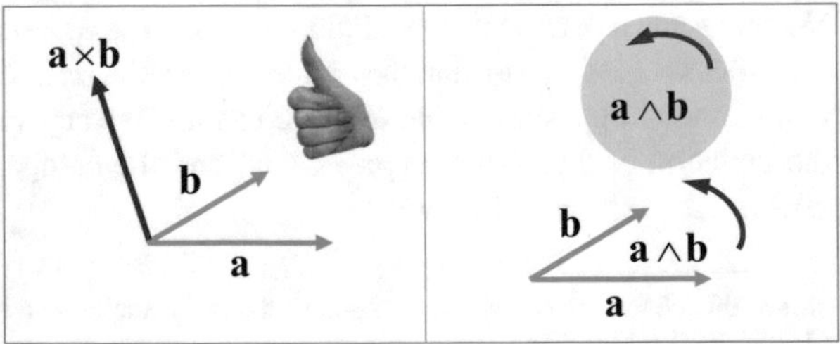

Fig. 1.1 The cross product versus the outer product. (Left) The **cross product** lives in 3D and has several problems. (Right) A **bivector** is an oriented part of a plane, living in all dimensions greater than 1, and it is almost magical

Note that the scalar product is commutative, while the cross product is anti-commutative. For two unit vectors **m** and **n** in $\mathfrak{R}^3$, we have

$$\mathbf{m} \cdot \mathbf{n} = \cos\alpha, \quad \mathbf{m} \times \mathbf{n} \propto \sin\alpha,$$

suggesting that **the two products are related somehow**, as an after effect of

$$\sin^2\alpha + \cos^2\alpha = 1.$$

A relation between the products can be anticipated if we look at the multiplication tables in $\mathfrak{R}^3$ (e_i are orthonormal basis vectors):

$\cdot$	e_1	e_2	e_3
e_1	1	0	0
e_2	0	1	0
e_3	0	0	1

$\times$	e_1	e_2	e_3
e_1	0	e_3	$-e_2$
e_2	$-e_3$	0	e_1
e_3	e_2	$-e_1$	0

We see that the scalar product has nonzero values on the main diagonal, while the cross product has zeros on the main diagonal (due to anticommutativity). The multiplication tables simply invite us to unite them. Furthermore, the form of both products suggests a similarity with complex numbers in the trigonometric form

$$z = |z|(\cos\varphi + i\sin\varphi).$$

To achieve this form, we need a quantity that gives -1 when squared, like the imaginary unit. However, it is not clear how to relate the cross product to an imaginary-unit-like quantity naturally. On the other hand, the cross product is anticommutative, which suggests that it "should" have the feature to give -1 when squared. Specifically, if we imagine any quantities that give a positive real value when squared and whose product is anticommutative and associative, we will have

$$(AB)^2 = ABAB = -ABBA = -A^2B^2 < 0.$$

The cross product is not even associative; it is easy to see that

$$(\mathbf{a} \times \mathbf{b}) \times \mathbf{c} \neq \mathbf{a} \times (\mathbf{b} \times \mathbf{c}).$$

For instance, try vectors e_1, e_2, and $e_1 + e_2 + e_3$. Further, if we make an inversion of space, for vectors we get $\mathbf{a} \to -\mathbf{a}$ (*polar vectors*); however, $\mathbf{a} \times \mathbf{b} \to (-\mathbf{a}) \times (-\mathbf{b}) = \mathbf{a} \times \mathbf{b}$ (*axial* vectors). Moreover, considering an orthonormal basis in 3D, we can say that the vector e_1 is a polar vector, while $e_2 \times e_3 = e_1$ is an axial vector. So what is e_1 like? Of course, we could play with definitions that are more general by invoking tensors, nevertheless, it is strange that in such an elementary example we immediately have a problem. Some mathematicians would argue that the cross product can generally be defined in dimensions greater than 3. However, if we think a little about it, requiring a natural and simple definition, some questions arise immediately.

1.1.2 Grassmann's Great Idea

Consider a 2D world in which flatbed physicists want to define a torque. If they do not wish to look for new dimensions outside their world, they will not even try to define a cross product; there is no vector orthogonal to their world. However, we can appreciate the fact that torque makes sense in the 2D world as well; it is proportional to the magnitude of both a force (**F**) and a force arm (**r**), while the two possible orientations of rotation are possible. Vectors **r** and **F** belong to their flat world. Therefore, how are we to multiply a force arm vector and a force vector to provide the desired torque (Fig. 1.2)? In 4D, we also have a dilemma: should it be $e_1 \times e_2 = e_3$ or $e_1 \times e_2 = e_4$ (or some linear combination of e_3 and e_4)?

The great mathematician Hermann Grassmann, underestimated and neglected by his contemporaries, found the answer to such questions already in the nineteenth century. He defined an anticommutative *exterior* (*outer*, *wedge*) product of vectors and thus obtained a **bivector**, an object **contained in a plane**, with an orientation and a magnitude and ideal for our 2D problem. In addition, it is easy to generalize bivectors to higher dimensions. Grassmann and Clifford instigated the unification of the scalar and the exterior products into one: the *geometric product*, exactly what we are talking about here. The scalar product of vectors is unchanged; however, the outer product substitutes for the cross product. The artificial difference between "axial" and "polar" vectors disappears, since we replace all "axial" vectors by bivectors (magnetic field vector, for example, as discussed in the text).

All right, now "forget" the scalar and the cross products and let us discover how to define a **new** product. It is reasonable to require associativity and distributivity for a new multiplication (as is the case for real numbers, which should be included), that is,

$$a(bc) = (ab)c, \quad (associativity)$$
$$a(\beta b + \gamma c) = \beta ab + \gamma ac, \quad (left\ distributivity)$$
$$(\beta b + \gamma c)a = \beta ba + \gamma ca, \quad \beta, \gamma \in \mathbb{R}. \quad (right\ distributivity)$$

Of course, we do not expect commutativity for our new multiplication, except for scalars (such as real numbers); consequently, we have two distributivity rules. After all, the definition of the cross product was actually motivated by a need for such noncommutative constructs (like a torque or the Lorentz force).

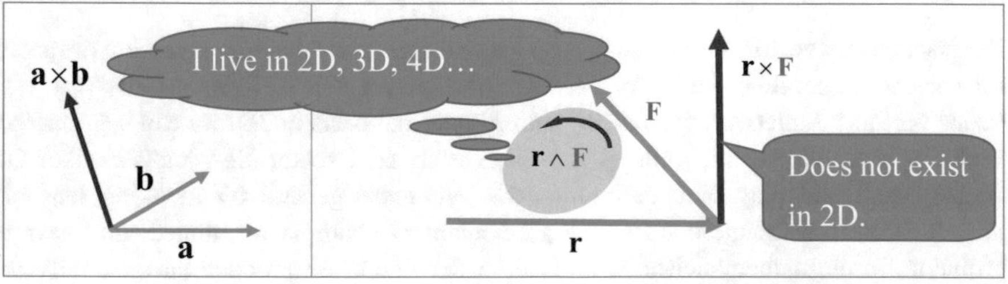

Fig. 1.2 The cross product problems. (Left) The **cross product** lives in 3D and has several problems. (Right) Torque in 2D: the cross product cannot give us a solution, but the **bivector** can

Let us consider the term a^2 first (a is a vector, later we will assume that $a^2 \in \mathbb{R}$). Let us clarify immediately that we do **not** imply that $ab \equiv a \cdot b$, where as usual, the scalar product is denoted by a *dot*. This is essential to note, for it would contribute to confusion otherwise. We expect that the square of a vector does not depend on its direction but that it does depend on its length (we exclude nonzero vectors of length zero for now).

We expect that the multiplication of a vector by a real number is commutative, whence immediately follows that multiplication of parallel vectors ($a \parallel b$) is commutative:

$$\lambda a = a\lambda \Rightarrow ab = a\lambda a = \lambda aa = ba, \quad \lambda \in \mathbb{R}.$$

Actually, we can use principles of symmetry to help us (Fig. 1.3), we immediately see that multiplication of parallel vectors must be commutative, that is, we have no criterion to distinguish which vector is "first" and which is "second." This is obvious if vectors have the same orientation; for vectors with opposite orientations, we can rely on the fact that all orientations in space are equal (*isotropy*). On the other hand, for orthogonal vectors we just expect anticommutativity, since we have the possibility to choose the order of multiplication, with a clear geometric content. Specifically, orthogonal vectors define a part of the oriented plane (parallelogram), where an orientation can be defined by the order of multiplication. In addition, we can rotate the Fig. 1.3, and the products will not change, which is an important symmetry. **Note that** *aa* **has no orientation.**

Due to the independence of the square of the vector from its direction, we have (Fig. 1.4; recall that $b_\perp$ has no component in the direction of a)

$$(b_\perp + a)^2 - (b_\perp - a)^2 = 0 = 2(b_\perp a + ab_\perp),$$

meaning that vectors $b_\perp$ and a anticommute. You can find other arguments. Recall, however, that we do not assume the scalar or the cross product, since we are looking for the properties of a **new** product of vectors "from scratch." The previous example is **not** a proof; it is just an idea of how to think about it. We simply required that the square of a vector be invariant under a change of direction and that product of vectors be generally

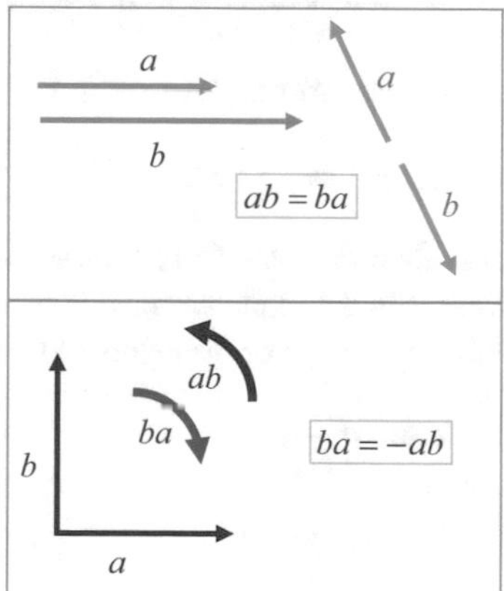

Fig. 1.3 Parallel vectors commute, orthogonal vectors anti-commute

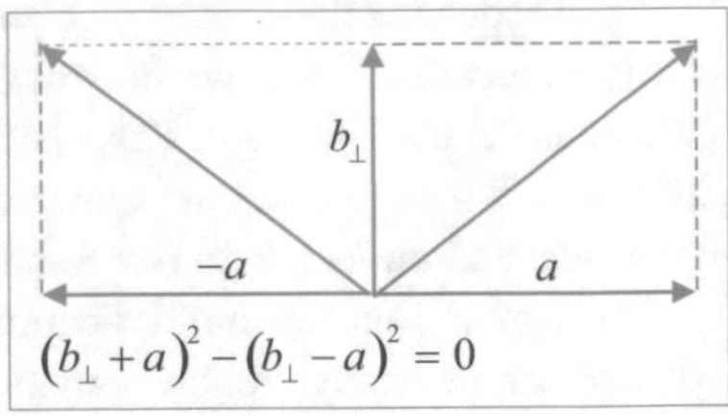

Fig. 1.4 Independence of the square of a vector on its direction

noncommutative. It is clear from Fig. 1.3 that noncommutativity is connected with orientations of geometric objects. Actually, we will see that the elements of geometric algebra are oriented geometric objects.

We could, of course, after we have assumed noncommutative multiplication, just use

$$(a + b_\perp)^2 = a^2 + b_\perp^2 + ab_\perp + b_\perp a$$

and immediately conclude that we must have $ab_\perp + b_\perp a = 0$, since we expect the *Pythagorean theorem* to be true. However, Fig. 1.4 shows us that we have a symmetry here; specifically, vectors a and $-a$ define a "straight line"; here "right" and "left" are not important concepts (go behind, and "left" becomes "right"). Accordingly, we see that the direction of the vector $b_\perp$ suggests the symmetry in accordance with our intuitive concept of orthogonality. Without this symmetry, we enter a "skew land"; however, let pure mathematicians go there if they so choose.

Let us show that a^2 commutes with b (**without any assumption as to what a^2 is; it is enough to use it as** *aa*):

$$a^2 b = a^2 (b_\| + b_\perp) = ab_\| a - ab_\perp a = b_\| a^2 + b_\perp a^2 = ba^2.$$

Note that the commutation occurs due to the fact that a^2 is just aa, so it anticommutes with $b_\perp$ twice; **it is unimportant whether or not a^2 is a real number**. It follows immediately that $ab_\|$ commutes with b, due to $b_\| = \alpha a,\ \ \alpha \in \mathbb{R}$. Now we have

$$ab + ba = ab_\perp + b_\perp a + 2ab_\| = 2ab_\|,$$

so $ab + ba$ commutes with b. We can see this another way:

$$b(ab + ba) = bab + b^2 a = bab + ab^2 = (ab + ba)b.$$

It is clear that $ab + ba$ commutes with a as well, which means that it commutes with every vector in the plane spanned by vectors a and b. However, $ab + ba$ obviously commutes with every vector orthogonal to that plane ($c \perp a$ and $c \perp b$):

$$cab_\| = -acb_\| = ab_\| c.$$

1.1.3 The Symmetric and Antisymmetric Parts of the Product

We can always decompose any noncommutative product into its symmetric and antisymmetric parts (this is a standard procedure):

$$ab = \frac{ab + ba}{2} + \frac{ab - ba}{2} = \{a, b\} + [a, b] = S + A$$

(see Mathematical Notation), where we have

$$S = ab_{\parallel}, \quad A = ab_{\perp}.$$

The last two relations are very handy. For example, we can see that

$$A^2 = ab_{\perp}ab_{\perp} = -a^2 b_{\perp}^2.$$

Note that from $A = 0$ it follows that $ab = ba$ (parallel vectors), while from $S = 0$ it follows that $ab = -ba$ (orthogonal vectors). The symmetric part, as we have seen, commutes with all vectors. This also follows from

$$ab + ba = (a + b)^2 - a^2 - b^2,$$

due to the commutativity of the square of a vector. We have not defined a^2 yet; however, it is obvious that regardless of the explicit value of a^2 (it is just aa), we have for vectors a and $b_{\perp}$ that

$$(a + b_{\perp})^2 = a^2 + b_{\perp}^2 + ab_{\perp} + b_{\perp}a = a^2 + b_{\perp}^2 + ab_{\perp} - ab_{\perp} = a^2 + b_{\perp}^2;$$

that is, we have the Pythagorean theorem, here expressed through the new multiplication of vectors. If we define the term "orthogonal" as the relation between vectors in which the projection of one onto another is zero ($b_{\perp} = a - b_{\parallel}$), we get the Pythagorean theorem, which now applies to orthogonal vectors, regardless of the specific value of a^2, of course, if we accept the previous arguments. Recall that the Pythagorean theorem, as a rule, is expressed over the scalar product of vectors, but we see that it is a consequence of the anticommutativity of orthogonal vectors, without any assumptions about a^2.

For any two vectors, the relation

$$(a + b)^2 = a^2 + b^2 + ab + ba$$

can be taken as the *cosine rule*, since the symmetric part of the new product commutes with all vectors and is thus a "scalar." Note that the vector a (and b as well) anticommutes with the antisymmetric part

$$a(ab - ba) = a^2 b - aba = ba^2 - aba = (ba - ab)a = -(ab - ba)a,$$

or

$$aab_\perp = aab_\perp = -ab_\perp a.$$

We now assume that a^2 is a real number equal to $\pm|a|^2$, where $|a|$ is the magnitude of the vector a. Now we can write for the symmetric part

$$a \cdot b \equiv (ab + ba)/2 = ab_\parallel = |a||b| \cos \measuredangle(a, b).$$

The product $a \cdot b$ we call the *inner product*. We see that it coincides with the usual scalar product of vectors, but here we need a little bit of caution with names. Namely, in geometric algebra, we generally distinguish several types of "scalar" products of elements (not necessarily vectors). One of them is the *scalar product* (defined for all elements of the algebra), but we also have the *dot product, left contraction*, etc. For vectors, all types of "scalar" products coincide, however. Generally, they are slightly different (see Appendix 5.2). Here we will mainly work with the *inner product* and the *left contraction*. Note that if $ab = ba$, then we have

$$ab = ba \Rightarrow (ab + ba)/2 = ba = a \cdot b,$$

and therefore,

$$ab = ba \Rightarrow a^2 b = aba = a(a \cdot b) \Rightarrow a \propto b.$$

We can state this as follows: *vectors commute iff they are proportional (i.e., parallel).*

1.1.4 Orthonormal Basis Vectors

For the unit vectors of an orthonormal basis, we have $e_i^2 = \pm 1$ (null vectors $e_i^2 = 0$ are not included here), which means that

$$e_i e_j + e_j e_i = \pm 2\delta_{ij}.$$

Caution Do not confuse $e_i e_j$ with $e_i \cdot e_j$! If you are wondering what $e_i e_j$ is, the answer could be that it is a **completely new type of object** (*bivector*). We will explain this in the text.

Let us look at 2D examples:

$$\begin{aligned} \Re^2 &: \ e_1^2 = e_2^2 = 1 \Rightarrow (e_1 + e_2)^2 = 1 + 1 + e_1 e_2 + e_2 e_1 = 2 = e_1^2 + e_2^2, \\ \Re^{1,1} &: \ e_1^2 = -e_2^2 = 1 \Rightarrow (e_1 + e_2)^2 = 1 - 1 + e_1 e_2 + e_2 e_1 = 0 = e_1^2 + e_2^2. \end{aligned}$$

We see that with the new multiplication of vectors the Pythagorean theorem is valid in both cases, because of the anticommutativity of orthogonal vectors.

For $\Re^3$, we have

$$e_1^2 = e_2^2 = e_3^2 = 1, \quad e_i e_j + e_j e_i = 2\delta_{ij}.$$

Here, however, we have some magic. There are mathematical objects that meet these relations precisely: the *Pauli matrices*, discovered in the glorious years of the development of quantum mechanics. We can say that the Pauli matrices are the 2×2 *matrix representation* of unit vectors in $\Re^3$. We only need vectors to be multiplied in a new manner, as just described. That is to say, the Pauli matrices (see Sects. 4.1, 4.2, 4.4) have the same multiplication table as the orthonormal basis vectors. Let us make sure of that. The Pauli matrices are defined as

$$\hat{\sigma}_1 = \begin{pmatrix} 0 & 1 \\ 1 & 0 \end{pmatrix}, \quad \hat{\sigma}_2 = \begin{pmatrix} 0 & -i \\ i & 0 \end{pmatrix}, \quad \hat{\sigma}_3 = \begin{pmatrix} 1 & 0 \\ 0 & -1 \end{pmatrix},$$

and thus, for example,

$$\hat{\sigma}_2\hat{\sigma}_2 = \begin{pmatrix} 1 & 0 \\ 0 & 1 \end{pmatrix}, \quad \hat{\sigma}_1\hat{\sigma}_2 + \hat{\sigma}_2\hat{\sigma}_1 = \begin{pmatrix} 0 & 0 \\ 0 & 0 \end{pmatrix}.$$

We use the designation $\hat{\sigma}_i$ for the Pauli matrices. Consequently, we sometimes (as in quantum mechanics) use σ_i for the basis unit vectors in $\Re^3$. The Pauli matrices are important for describing *spin* in quantum mechanics, which means that vectors could serve this purpose as well, though with our new product of vectors. Indeed, quantum mechanics can be formulated by such mathematics nicely, **without matrices and the imaginary unit** (see Sect. 2.10).

Note that by a transposition of the Pauli matrices followed by a complex conjugation (*Hermitian adjoint*), we get the same matrices (*Hermitian matrices*), for example,

$$\hat{\sigma}_2 = \begin{pmatrix} 0 & -i \\ i & 0 \end{pmatrix} \xrightarrow{T} \begin{pmatrix} 0 & i \\ -i & 0 \end{pmatrix} \xrightarrow{*} \begin{pmatrix} 0 & -i \\ i & 0 \end{pmatrix} = \begin{pmatrix} 0 & -i \\ i & 0 \end{pmatrix}^{\dagger},$$

or simply $\hat{\sigma}_2^\dagger = \hat{\sigma}_2$. In addition, we have, for example, $(\hat{\sigma}_2\hat{\sigma}_3)^\dagger = \hat{\sigma}_3^\dagger\hat{\sigma}_2^\dagger = \hat{\sigma}_3\hat{\sigma}_2$ (*antiautomorphism*; ◆ E01: Show this. ◆). This exactly matches the operation *reverse* (see below) on geometric products of vectors, for example $e_1e_2e_3 \xrightarrow{\dagger} e_3e_2e_1$, which is the important fact. Therefore, we can use the character † to denote the *reverse* operation (we will do so).

Here we can immediately spot the important feature of the new multiplication of vectors. The vector is a geometrically clear and intuitive concept, but we will see that the new product of vectors has a clear geometric interpretation as well (see below). For example, we can clearly interpret the product e_1e_3 geometrically as the oriented area; it has the ability to rotate ($e_1e_3e_3 = e_1$), unambiguously defines the plane spanned by the vectors e_1 and e_3, etc. All this we can conclude at a glance. For comparison, consider the matrix representation of vectors e_1 and e_3 with their product

$$\hat{\sigma}_1\hat{\sigma}_3 = \begin{pmatrix} 0 & 1 \\ 1 & 0 \end{pmatrix} \cdot \begin{pmatrix} 1 & 0 \\ 0 & -1 \end{pmatrix} = \begin{pmatrix} 0 & -1 \\ 1 & 0 \end{pmatrix} = -i\hat{\sigma}_2.$$

Can we infer a similar geometric interpretation just by looking at the resultant matrix? Just by looking, certainly not. It would take some effort. However, we will often fail to get a clear geometric interpretation from matrices. Which plane does the resultant matrix define (if any is defined at all)? **The Pauli matrices cannot do all that vectors can do**. In this text, with a bit of luck, we will illuminate such things in order to obtain an idea of the importance of the new multiplication of vectors. ◆ Show that $-i\hat{\sigma}_k$ $(k = 1, 2, 3)$ can represent the quaternion units. ◆

1.1.5 The Inner and the Outer Products

It is time for the new multiplication of vectors to get a name (due to *Clifford, Hestenes*...): *geometric product*. The symmetric and antisymmetric parts of the geometric product of vectors have special notation: $a \cdot b$ and $a \wedge b$, where $a \cdot b$ is the *inner product* and $a \wedge b$ is the *outer (wedge) product*. Consequently, we can write (just for vectors, for the time being)

$$ab = a \cdot b + a \wedge b.$$

An important concept, which we will often use, is the *grade*. Real numbers have grade 0, vectors have grade 1, all elements that are linear combinations of products $e_i \wedge e_j$, $i \neq j$, have grade 2, and so on. Note that the geometric product of two vectors is a combination of grades 0 and 2; it is *even*, since its grades are even. ◆ E02: What grades do the geometric products of three vectors have in general? ◆

A vector space over the real field with the geometric product (GP in the text) becomes an *algebra* (*geometric algebra*, GA in the text). Elements of a GA, obviously, are not scalars and vectors only. For the orthonormal basis vectors we have, for example,

$$e_1 \cdot e_1 = \frac{e_1 e_1 + e_1 e_1}{2} = \pm 1, \quad e_1 \cdot e_2 = \frac{e_1 e_2 + e_2 e_1}{2} = 0 \Rightarrow$$
$$e_1 e_2 = e_1 \cdot e_2 + e_1 \wedge e_2 = e_1 \wedge e_2;$$

consequently, **for orthogonal vectors, the geometric product is the same as the outer product**. For the antisymmetric part, we have

$$e_1 \wedge e_2 = \frac{e_1 e_2 - e_2 e_1}{2} = \frac{e_1 e_2 + e_1 e_2}{2} = e_1 e_2, \quad e_1 \wedge e_1 = \frac{e_1 e_1 - e_1 e_1}{2} = 0.$$

Obviously, $e_1 e_2$ is not a real number, since it does not commute with all vectors; for example,

$$(e_1 \wedge e_2)\mathrm{e}_1 = (e_1 e_2)\mathrm{e}_1 = -e_1 e_1 e_2 = -e_1(e_1 \wedge e_2).$$

However, it is not a vector either; it squares to -1 for $e_i^2 = 1$ and to 1 for $e_2^2 = -1$:

$$(e_1 e_2)^2 = e_1 e_2 e_1 e_2 = -e_1 e_1 e_2 e_2 = \pm 1.$$

Accordingly, we have a **new** type of mathematical object; it is like the **imaginary unit**, except that it is noncommutative. The name for such an object is a *bivector*. Generally, we will define a bivector (2-*vector*) as an element of the form $a \wedge b$ (a and b are vectors) or a linear combination of such products. Let us find some more properties of the bivector $e_1 e_2$. We have

$$(e_1 e_2)e_1 = -e_1 e_1 e_2 = -e_2, \quad (e_1 e_2)e_2 = e_1;$$

consequently, acting on vectors from the left, it rotates them by $-\pi/2$. ◆ E03: How does $e_1 e_2$ rotate vectors acting from the right? ◆

Using the *reverse* operation on geometric product of vectors, such as $X = abc \rightarrow X^\dagger = cba$, we have for $e_1 e_2$,

$$(e_1 e_2)(e_1 e_2)^\dagger = (e_1 e_2)(e_2 e_1) = 1.$$

Therefore, we call it a *unit bivector*. Thus, our bivector has the magnitude and the orientation ($e_1 e_2 \neq e_2 e_1 = -e_1 e_2$). Furthermore, unit bivectors of the form $a \wedge b$, like $e_1 e_2$, except for a magnitude, an orientation, and the ability to rotate vectors, have another very important feature that the ordinary imaginary unit does not possess: they **define the plane** spanned by vectors (say, e_1 and e_2). Later, we will see how to implement this in practice, using the outer product.

1.1.6 Graphical Representation of Bivectors

Let us find out how to represent a bivector graphically. The obvious option is to try to do so with an oriented parallelogram (a square for $e_1 e_2$). However, **the shape of an oriented area that represents a bivector is not important**; we should keep the magnitude of the area and the orientation. For $e_1 e_2$, an oriented circle of radius $|e_1 e_2|/\sqrt{\pi} = 1/\sqrt{\pi}$ is often a practical choice. To justify our claims, look at $e_1 e_2 = e_1 \wedge e_2 \Rightarrow (e_1 + e_2) \wedge e_2 = e_1 e_2$, which can illustrate the fact that the shape is unimportant. Note that two vectors, apart from defining the plane, define the parallelogram as well. The outer product of such vectors (*bivector*) has magnitude equal to the parallelogram's area (see below), while the orientation we define as in Fig. 1.5. ◆ E04: For the leftmost parallelogram in Fig. 1.5, show that the formula $|a \wedge b| = |a||b||\sin\alpha|$ gives the area of the parallelogram. ◆

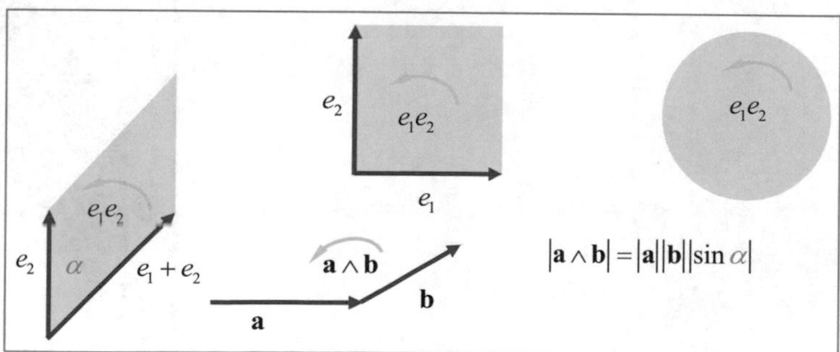

Fig. 1.5 Graphical representation of a bivector

As previously, for the antisymmetric part of the geometric product, we can write

$$ab - ba = ab_{\perp} - b_{\perp}a = 2ab_{\perp}$$

and see immediately that it anticommutes with a, $b_{\|}$, and $b_{\perp}$; therefore, it anticommutes with b and consequently with all vectors from the plane spanned by the vectors a and b. Obviously, it commutes with all vectors perpendicular to that plane. In addition, we have

$$(ab_{\perp})^2 = ab_{\perp}ab_{\perp} = -aab_{\perp}b_{\perp} = -a^2b_{\perp}^2,$$

meaning that this quantity is negative in Euclidean spaces. Consequently, the antisymmetric part of the geometric product is not a vector; it can square to a negative real number in a Euclidean vector space. And it is not a scalar: it anticommutes with some vectors. Note that from

$$ab + ba = 2ab_{\|}, \quad ab - ba = 2ab_{\perp},$$

we can derive many interesting properties of products, including their magnitudes, just using (see E04) $|b_{\|}| = |b||\cos\varphi|$, $|b_{\perp}| = |b||\sin\varphi|$.

Consider three vectors in $\Re^3$ that sum to zero (Fig. 1.6). From $\mathbf{a} + \mathbf{b} + \mathbf{c} = 0$, it follows that $\mathbf{a} \wedge \mathbf{b} = \mathbf{b} \wedge \mathbf{c} = \mathbf{c} \wedge \mathbf{a}$. To see this, it is enough to look at the expressions $(\mathbf{a} + \mathbf{b} + \mathbf{c}) \wedge \mathbf{a}$ and $(\mathbf{a} + \mathbf{b} + \mathbf{c}) \wedge \mathbf{b}$ (E05: Check this). However, we can deduce it easily, without calculation; it is enough to look at Fig. 1.6: each pair of vectors defines a parallelogram of area equal to double the area of the triangle, while all pairs give the same orientation of a bivector. This is important, since we often can draw conclusions easily just from geometric observations. In the formula for the area of a parallelogram, the sine function appears. Therefore, we see that the previous equalities are just the *law of sines*. Namely, if we recall that bivectors are shape-independent, we see that all of our three bivectors have the same factor $\hat{B}$ (the unit bivector). Now we have

$$\hat{B}ab\sin\gamma = \hat{B}bc\sin\alpha = \hat{B}ac\sin\beta \Rightarrow \frac{\sin\gamma}{c} = \frac{\sin\alpha}{a} = \frac{\sin\beta}{b}.$$

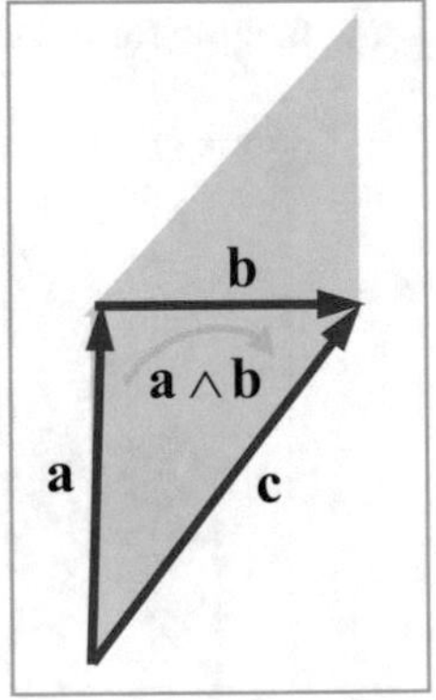

Fig. 1.6 Each pair of vectors defines the same bivector

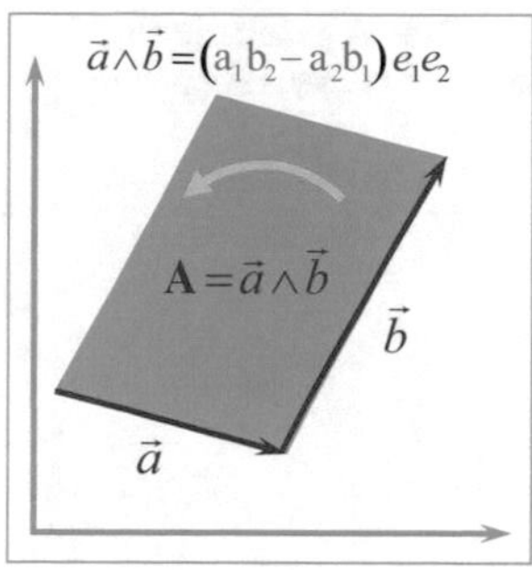

Fig. 1.7 An outer product in 2D

Consider an outer product in 2D. For vectors $\vec{a}$ and $\vec{b}$ we have ($\langle\rangle_2$ means *grade* 2) (Fig. 1.7)

$$\vec{a}\vec{b} = \vec{a}\cdot\vec{b} + \vec{a}\wedge\vec{b} = a_1a_2 + b_1b_2 + (a_1b_2 - a_2b_1)e_1e_2,$$
$$\cos\varphi = \frac{a_1a_2 + b_1b_2}{\sqrt{a_1^2 + a_2^2}\sqrt{b_1^2 + b_2^2}}, \quad \sin\varphi = \frac{a_1b_2 - a_2b_1}{\sqrt{a_1^2 + a_2^2}\sqrt{b_1^2 + b_2^2}},$$
$$\mathbf{A} = \left\langle\vec{a}\vec{b}\right\rangle_2 = \vec{a}\wedge\vec{b}, \quad S = |\vec{a}|\left|\vec{b}\right||\sin\varphi| = |a_1b_2 - a_2b_1|.$$

We can generalize these relations to 3D without effort.

1.1.7 Equations of Geometric Objects

Using the outer product, we can write down equations of different geometric objects. For example, the equation of a line passing through the endpoint of the vector $\vec{a}$ in the direction of the unit vector $\hat{u}$ (note that vectors $\vec{x}-\vec{a}$ and $\hat{u}$ are parallel) is

$$\left(\vec{x}-\vec{a}\right)\wedge\hat{u} = 0.$$

A possible solution of this equation is $\vec{x}-\vec{a} = \alpha\,\hat{u},\quad \alpha\in\mathbb{R}$. Note that the shape of a bivector is not important. Writing $\vec{x}\wedge\hat{u} = \vec{a}\wedge\hat{u}$, we see that this linear equation is just a statement about equality of bivectors (Fig. 1.8). Obviously, we can find the vector $\vec{d}$ orthogonal to the line, giving the distance to the line $\left|\vec{d}\right|$. From $\vec{d}\wedge\hat{u} = \vec{d}\hat{u}$ and $\vec{a}\wedge\hat{u} = \vec{a}\hat{u} - \vec{a}\cdot\hat{u}$, it follows that

$$\vec{d}\hat{u} = \vec{a}\wedge\hat{u} \Rightarrow \vec{d} = \left(\vec{a}\wedge\hat{u}\right)\hat{u} = \vec{a} - \left(\vec{a}\cdot\hat{u}\right)\hat{u}.$$

Bivectors of the form $a\wedge b$ (a *blade*; see below; a and b are vectors) can define a plane. Consider the bivector e_1e_2 and form the outer product in $Cl3$

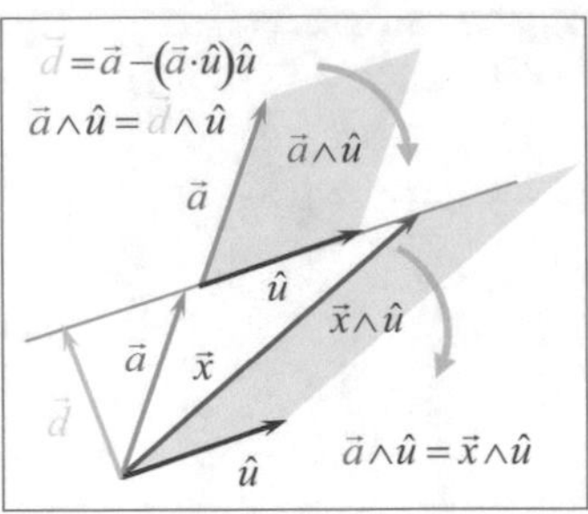

Fig. 1.8 A geometric object represented by an equation

$$(e_1 \wedge e_2) \wedge (a_1 e_1 + a_2 e_2 + a_3 e_3) = a_3 e_1 e_2 e_3.$$

We can infer that the outer product of a bivector with a vector gives the possibility to eliminate components of the vector that do not belong to the plane defined by the bivector. Therefore, we can define the plane of the bivector B (2D subspace) by the relation $B \wedge x = 0$. In our example, solutions are all vectors of the form $x = a_1 e_1 + a_2 e_2$. This is a general way to define subspaces. ◆ E06: Find the 2D subspace of the 3D vector space defined by the bivector $e_1 e_3 + e_2 e_3$. ◆

1.1.8 The Lorentz Force and Space Inversion

In physics, we have the familiar formula for the Lorentz force

$$\mathbf{F} = q(\mathbf{E} + \mathbf{v} \times \mathbf{B}).$$

Applying *space inversion* (all vectors change sign), we get

$$\mathbf{E} + \mathbf{v} \times \mathbf{B} \rightarrow -\mathbf{E} + (-\mathbf{v}) \times (-\mathbf{B}) = -\mathbf{E} + \mathbf{v} \times \mathbf{B},$$

which means that the force **F** fails to transform to −**F**. Using geometric algebra, we can define

$$\mathbf{F} = q(\mathbf{E} + j\mathbf{B} \wedge \mathbf{v})$$

($j \equiv e_1 e_2 e_3$; check that $j\mathbf{B}$ is a bivector), which transforms as

$$\mathbf{E} + j\mathbf{B} \wedge \mathbf{v} \rightarrow -\mathbf{E} + (-j)(-\mathbf{B}) \wedge (-\mathbf{v}) = -\mathbf{E} - j\mathbf{B} \wedge \mathbf{v},$$

giving $\mathbf{F} \rightarrow -\mathbf{F}$. The introduction of the bivector $j\mathbf{B}$ and the outer product solves the problem (as well as many others).

1.2 Complex Numbers

Imagine a unit bivector that defines a plane and has the properties of a (noncommutative) imaginary unit (in that plane). This is powerful: we can use the formalism of complex numbers in any plane, in any dimension. How? Let's return to our bivector e_1e_2 and the vector xe_1 + ye_2. If we multiply our vector by e_1 from the left, we get

$$e_1(xe_1 + ye_2) = x + ye_1e_2 = x + yI, \quad I = e_1e_2.$$

Consequently, we have a complex number. ◆ E07: What do we get if we multiply from the right?◆ We can define a complex conjugation $I \to -I = I^* = I^\dagger$, and then using $I^2 = -1$, it follows that

$$(x + yI)(x - yI) = x^2 + y^2.$$

◆ What if $e_2^2 = -1$ or $e_2^2 = 0$? See Sects. 2.2 and 2.3. ◆

The inverse of a complex number z is

$$z^{-1} = \frac{1}{x + yI} = \frac{x - yI}{(x + yI)(x - yI)} = \frac{x - yI}{x^2 + y^2} = \frac{z^*}{zz^*} = \frac{z^\dagger}{|z|^2}.$$

We will show that de Moivre's formula is also valid (Sect. 2.6), and we will show how to define functions of a complex argument (Sect. 1.13), etc. We have seen that bivectors like e_1e_2 can rotate vectors from their plane, anticommuting with them. For instance,

$$\begin{aligned} &e_1e_2e_1 = -e_2, \quad e_1e_1e_2 = e_2, \quad e_1e_1e_2 = -e_1e_2e_1, \\ &e_1e_2e_2 = e_1, \quad e_2e_1e_2 = -e_1, \quad e_2e_1e_2 = -e_1e_2e_2. \end{aligned}$$

Acting from the left, the bivector rotates vectors by $-\pi/2$, while acting from the right, it rotates by $\pi/2$. The imaginary unit $i = \sqrt{-1}$ also rotates a plane. For example, starting from the point $(1, 0)$ and multiplying by i, we get the point $(0, 1)$; multiplying by i again, we get $(-1, 0)$, etc. The imaginary unit rotates a plane by the angle $\pi/2$. The imaginary unit is commutative. Therefore, multiplication from the left or from the right makes no difference; the rotation angle is $\pi/2$. The point $(1, 0)$ in the e_1e_2 plane can be represented by the unit vector e_1, the point $(0, 1)$ by the vector e_2, etc. Consequently, the bivector $I = e_1e_2$ can rotate the plane e_1e_2 by angles $\pm\pi/2$, depending on the side on which the multiplication takes place.

The anticommutativity of bivectors thus becomes an advantage; it allows us to choose an orientation of rotations (see Fig. 1.9). Actually, noncommutativity of the geometric product opens a completely new world of mathematics, with endless possibilities. In geometric algebra, we have plenty of imaginary units; the unit pseudoscalar j in $Cl3$ is the commutative one.

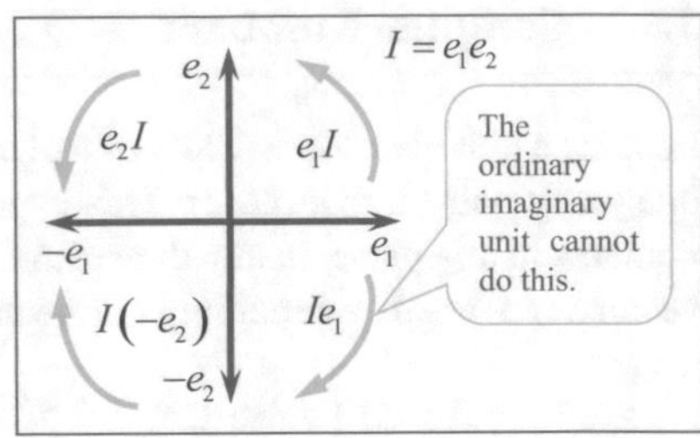

Fig. 1.9 The power of a bivector as a noncommutative imaginary unit

1.3 The Cross Product and Duality

◆ E08: The reader may show that every linear combination of unit bivectors in *Cl*3 can be expressed as an outer product of two vectors. ◆ This is not necessarily true in *Cl*4. Take, for example, $e_1e_2 + e_3e_4$ (a 2-vector). ◆ E09: Prove that there are no two vectors in *Cl*4 with the property $a \wedge b = e_1e_2 + e_3e_4$. ◆ In 3D, for each plane we have exactly one orthogonal unit vector (up to a sign), while this is not true in higher dimensions. For example, in 4D, the plane defined by the bivector e_1e_2 has orthogonal unit vectors e_3 and e_4 (as well as their linear combinations). Taking the bivector e_1e_2 in $\Re^3$ and multiplying it by $-e_1e_2e_3 \equiv -j$, we get $-e_1e_2j = e_3$. We can replace the cross product of arbitrary vectors via $\mathbf{a} \times \mathbf{b} \rightarrow -j\mathbf{a} \wedge \mathbf{b}$; however, we cannot put the equality sign here, $\mathbf{a} \times \mathbf{b}$ and $-j\mathbf{a} \wedge \mathbf{b}$ **behave differently under space inversion** ($j \rightarrow -j$). We can take $-j\mathbf{a} \wedge \mathbf{b}$ as a **new definition of** the cross product valid in *Cl*3; however, $\mathbf{a} \wedge \mathbf{b}I^{-1}$, where I is a unit pseudoscalar (see Sect. 1.4), is valid in any dimension. For example, in 2D, the element $\mathbf{a} \wedge \mathbf{b}I^{-1}$ is just a real number (show this, $I = e_1e_2$), while in 4D or higher, we can take advantage of the concept of *duality* (see below). In addition, the cross product is not associative, in contrast to $-j\mathbf{a} \wedge \mathbf{b}$. The cross product of vectors (*Gibbs*) requires the right-hand rule and the use of perpendiculars to surfaces. This is not necessary with bivectors. For example, we can completely omit objects such as "rotational axis." Find the GP of two vectors $\mathbf{a}$ and $\mathbf{b}$ in $\Re^3$ and show that it can be expressed formally as (Fig. 1.10)

$$\mathbf{ab} = \mathbf{a} \cdot \mathbf{b} + (\mathbf{a} \times \mathbf{b})e_1e_2e_3 = \mathbf{a} \cdot \mathbf{b} + (\mathbf{a} \times \mathbf{b})j, \quad \mathbf{a} = \sum_{i=1}^{3} \mathrm{a}_i e_i, \quad \mathbf{b} = \sum_{i=1}^{3} \mathrm{b}_i e_i.$$

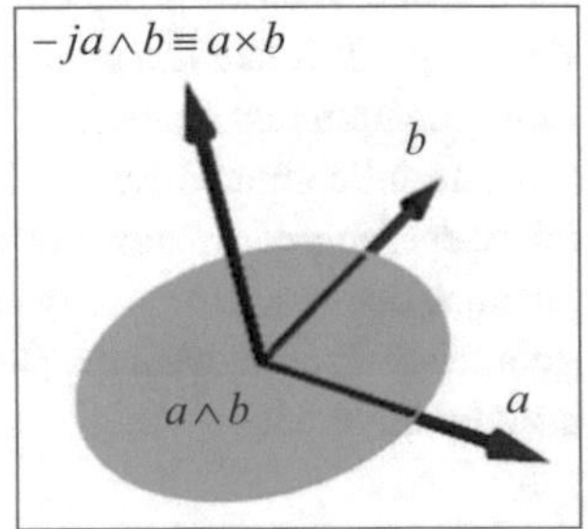

Fig. 1.10 The cross product (a new definition) and duality

1.3.1 Duality Operation

The *duality* operation (see Appendix 5.8) in *Cl*3 is $A^{\Delta} = -jA$, where j is the unit pseudoscalar and A is an element of the algebra (the duality is usually denoted by $*$, but that might be confused with complex conjugation). This important operation relates subspaces in a vector space (like planes and vectors in 3D).

◆ Show that $B = -j\mathbf{x}$ is a bivector in *Cl*3. What can you say about the relationship between the vector $\mathbf{x}$ and the bivector B? ◆ The cross product appears in many applications of mathematics and physics; consequently, it is important to have the possibility to translate formulas easily, which we can accomplish using the duality operation. In addition, our geometric intuition should be changed; in geometric algebra, we have oriented objects, like bivectors, that are not a part of standard mathematics. In fact, standard mathematics needs such objects; mathematicians define tensors, differential forms, etc., which are usually unintuitive. It is especially difficult to imagine a tensor. A bivector is easy to imagine and manipulate. Having a clear geometric picture, we can understand and solve problems by relying on our powerful geometric intuition. This property of geometric algebra might be seen as an ideal of many mathematicians throughout history.

We have seen that $\mathbf{a} \times \mathbf{b}$ is an axial vector in the vector algebra, having the strange property of invariance under space inversions. All vectors should change its orientation under space inversion. By introducing bivectors, this problem disappears; vectors change sign under space inversion; bivectors are invariant. In the presentation below, we will see how this strange property of the cross product introduces an asymmetry in otherwise symmetric theories (space inversion, mirror symmetry; see Sect. 2.7.8). Formulations with bivectors remove this problem.

1.4 Algebra

Consider 2D vector spaces again. All possible outer products of vectors expressed in an orthonormal basis can provide linear combinations of "numbers" 1, e_1, e_2, and $e_1 \wedge e_2 = e_1 e_2$ (we call any linear combination of these "numbers" a *multivector*). The outer product is anticommutative; consequently, all terms that have some unit vector repeated disappear. The "numbers" 1, e_1, e_2, and $I = e_1 e_2$ form a basis of a 2^2-dimensional *linear space* (the so-called *Clifford basis*), and we call such a linear space an *algebra* (*Clifford algebra*). When the geometric meaning is in the forefront, we use the name *geometric algebra* (due to Clifford himself). The element 1 is a real scalar. We also have two vectors and one bivector (in the terminology of geometric algebra, this bivector is called a *pseudoscalar* in the algebra, specifically, an element of the algebra with *maximum grade*). Note that the real scalars form a subspace (real numbers, of *grade* zero; see below), the vectors define 1D subspaces (of *grade* 1; do not confuse with a *subalgebra*), and the pseudoscalars define a 2D subspace (the space itself, of *grade* 2). ◆ Show that the pseudoscalar $I = e_1 e_2$ anticommutes with all vectors in 2D, while in *Cl*2 we have $I^2 = -1$. Find the dual elements of the elements from the Clifford basis of *Cl*2. ◆

In $\mathfrak{R}^3$, we have the basis of the algebra (*Cl*3)

$$(1, e_1, e_2, e_3, e_2e_3, e_3e_1, e_1e_2, e_1e_2e_3) = (1, e_1, e_2, e_3, je_1, je_2, je_3, j),$$

where $j \equiv e_1 \wedge e_2 \wedge e_3 = e_1e_2e_3$ is the unit *pseudoscalar*. ◆ E10: Show that j commutes with all elements of the Clifford basis in *Cl*3, and that $j^2 = -1$. Prove that the pseudoscalars in any dimension are all proportional to the unit pseudoscalar. ◆ Consequently, the pseudoscalar j is a perfect (commutative) imaginary unit in *Cl*3. Such a pseudoscalar will appear also in *Cl*7, *Cl*11, ... (see E10). This has far-reaching consequences. However, one should be careful here: the commutativity property of the pseudoscalar j generally excludes "walking through" products, except for the geometric product. Real numbers do not have this "problem": they can "walk through" all products. For example, we have $je_1e_3 = e_1je_3 = e_1e_3j$, that is, the geometric product allows "walking through." However, this is not generally valid with, say, the inner product (for the meaning of this calculation, see Sect. 1.5):

$$j(e_1 \cdot e_3) = (e_1 \cdot e_3)j = 0 \neq e_1 \cdot (je_3) = e_1 \cdot (e_1e_2) = e_2.$$

Here we have an example of a *mixed product*. However, if we consider the commutativity without "walking through," we have $jM = Mj$ for all multivectors M in *Cl*3.

In 3D, we have $\mathbf{a} \wedge \mathbf{b} \wedge \mathbf{c} \wedge \mathbf{d} = 0$ for four arbitrary vectors. Namely, the outer product is distributive and associative (see the literature or prove it itself; see problem 40). Consequently, if two vectors are parallel, the relation is true due to the anticommutativity of the outer product. Otherwise, we have, for example, $\mathbf{d} = \alpha\mathbf{a} + \beta\mathbf{b} + \gamma\mathbf{c}$, $\alpha, \beta, \gamma \in \mathbb{R}$, consequently, our statement is true due to distributivity and anticommutativity. We have shown that $e_1 \wedge e_2 = e_1e_2$. However, one should be careful when writing products with repeated indices. For example, $e_1 \wedge e_2 \wedge e_2 = 0$, but $e_1e_2e_2 = e_1$. ◆ E11: Show that

$$e_k \wedge e_l \wedge e_m = j\varepsilon_{klm}, \quad j = e_1e_2e_3,$$

where ε_{klm} is the Levi-Civita symbol in 3D, defined by

$$\varepsilon_{123} = \varepsilon_{231} = \varepsilon_{312} = -\varepsilon_{213} = -\varepsilon_{132} = -\varepsilon_{321} = 1,$$

while for repeated indices, ε_{klm} gives zero. E12: Show that the maximum grade of a multivector cannot be larger than the dimension of the vector space. Show that number of elements in the Clifford basis with grade k is equal to the binomial coefficient

$$\binom{n}{k},$$

where n is the dimension of the vector space. ◆ For real scalars, we have $k = 0$, giving one real scalar in the basis (e.g., 1). The same is true for $k = n$: there is just one element with grade n in the basis, which gave rise to the name "*pseudoscalar*." ◆ E13: Show that the number of elements in the Clifford basis of an n-dimensional vector space is equal to 2^n. ◆

1.4.1 Grades and Parity of a Multivector

An important concept is the *parity* of a multivector, which refers to the parity of its grades. The set of all elements of even grades defines a subalgebra. ◆ E14: Show that the geometric product of any two even elements is even. ◆ This is not true for the odd part of the algebra. For example, $e_1 j = e_2 e_3$, i.e., two odd-grade elements (grades 1 and 3) yielded an even one (grade 2). Products of bivectors from the *Cl*3 Clifford basis have grade 0 (when one bivector is squared) or 2 (when the bivectors are different). For example, $(e_1 e_2)^2 = e_1 e_2 e_1 e_2 = -e_1 e_2 e_2 e_1 = -1$, $(e_1 e_2)(e_2 e_3) = e_1 e_2 e_2 e_3 = e_1 e_3$.

We usually denote grades of a multivector M by $\langle M \rangle_r$, where r is the grade. For grade 0, we use $\langle M \rangle$; for example, $a \cdot b \equiv \langle ab \rangle$. An element of grade 0 is a real number, and it is independent of the order of multiplication. To see this, we can use the fact that every multivector is a linear combination of elements of the Clifford basis, whose product is a real number only when those elements have the same indices (in any order). Elements with the same indices commute, and therefore, we have $\langle AB \rangle = \langle BA \rangle$, which leads to the possibility of *cyclical changes*, such as $\langle ABC \rangle = \langle CAB \rangle$. This is a beneficial relationship. For example, consider the inner product $a \cdot b$ to see what happens if we apply the transformation $a \to -nan$ and $b \to -nbn$ (n is a unit vector). Note that the result of such a transformation is a vector. ◆ E15: Show this, decomposing the vector a into components parallel and perpendicular to n. ◆ The inner product of two vectors is just the zero grade of their geometric product, so we have, using cyclical changes,

$$(nan) \cdot (nbn) = \langle nannbn \rangle = \langle nabn \rangle = \langle abnn \rangle = \langle ab \rangle = a \cdot b.$$

Such a transformation does not change the inner product, and thus we have an example of an *orthogonal transformation* (this one is a *reflection*). A transformation $X \to nXn$ (n is a unit vector) generally does not change the grades of X. For instance, if we have $X = ab$, then it follows that

$$nabn = nannbn = (nan)(nbn);$$

that is, we have a geometric product of two vectors again. This is a very important conclusion. To see that this is generally valid, recall that each multivector is a linear combination of elements of the Clifford basis. Therefore, we have, for example, $e_1(e_1 e_3)e_1 = e_3 e_1 = -e_1 e_3$, and the grade is still 2. If the grade of an element is changed by some transformation, then the result of such a transformation is a new type of element; however, we do not want this generally. We usually want to transform vectors to vectors, bivectors to bivectors, etc., since they represent geometric objects.

As we have stated already, the oriented elements of the Clifford basis, when multiplied, give a real number only if they have the same indices of unit vectors. Therefore, in Euclidean spaces we have $\langle MM^\dagger \rangle \geq 0$, where equality holds only for $M = 0$ (see Sect. 1.10). For example, from $e_1 e_2 e_2 e_1 = 1$ it follows that

$$\langle (\alpha e_1 + \beta e_1 e_2)(\alpha e_1 + \beta e_2 e_1) \rangle = \langle \alpha^2 - 2\alpha\beta e_2 + \beta^2 \rangle = \alpha^2 + \beta^2 > 0.$$

1.4.2 Projection and Rejection

To conclude this chapter, let us find the *projection* and the *rejection* of a vector a (we announced this possibility earlier) with respect to a unit vector n. We have

$$a = n^2 a = n(n \cdot a + n \wedge a) = nn \cdot a + nn \wedge a = a_{\|} + a_{\perp},$$

where the **geometric product in mixed products is to be executed last**. For general formulas, see Sect. 1.11.3.

1.5 Mixed Products

In geometric algebra, we frequently have various combinations of products of elements (mixed products; see Sect. 3.4). For example, consider the product (a combination of the geometric product and the outer product)

$$a(b \wedge c) = a(bc - cb)/2 = (abc - acb)/2.$$

We can take the advantage of the obvious (and useful) relation $ab = 2a \cdot b - ba$ to show that (left to the reader)

$$a(b \wedge c) - (b \wedge c)a = 2(a \cdot b)c - 2(a \cdot c)b.$$

Here we have a situation in which the grade of an element (bivector) is decreased, and it is customary to write such an operation as the inner product, that is, a kind of **contraction** (grade lowering; the outer product generally raises grades). Thus, **we extend the definition of the inner product**. For $B = b \wedge c$ we have

$$a \cdot B \equiv (aB - Ba)/2 \Rightarrow a \cdot (b \wedge c) = (a \cdot b)c - (a \cdot c)b = a \cdot bc - a \cdot cb,$$

where we assume that the inner product has a higher precedence. This is a useful and important formula. If a is orthogonal to both b and c, we have $a \cdot (b \wedge c) = 0$; for example, $e_1 \cdot (e_2 \wedge e_3) = 0$. It is not difficult to show that

$$a \wedge B = (aB + Ba)/2,$$
$$aB = a \cdot B + a \wedge B.$$

◆ E16: Find $e_1 \cdot (e_1 e_2)$ and $e_1 \wedge (e_1 e_2)$. You can find an example of a geometric interpretation in the Sect. 6.1. ◆

1.5.1 A Useful Formula

Here is another useful relation (without a proof for now; note the grade lowering of a k-vector in brackets on the left side, i.e., a contraction):

$$\mathbf{x} \cdot (\mathbf{a}_1 \wedge \cdots \wedge \mathbf{a}_k) = \sum_{i=1}^{k} (-1)^{i+1} \mathbf{x} \cdot \mathbf{a}_i \big(\mathbf{a}_1 \wedge \cdots \wedge \breve{\mathbf{a}}_i \wedge \cdots \wedge \mathbf{a}_k\big),$$

where $\breve{\mathbf{a}}_i$ means that the vector $\mathbf{a}_i$ is missing in the outer product. ◆ E16: Find $e_1 \cdot (\mathbf{a} \wedge \mathbf{b})$. ◆

We will discuss some other mixed products and their properties below. It appears that products like $e_1 \cdot (e_1 e_2)$ have a clear geometric meaning (can you see it now?), which helps to explain the usefulness of geometric algebra.

1.6 Important Concepts

Before we dive into *Cl*3, let us introduce several general concepts.

1. *versor* $\rightarrow$ the geometric product of any number of vectors
2. *blade* $\rightarrow$ the outer product of vectors
3. *involution* $\rightarrow$ any function with the property $\mathrm{f}[\mathrm{f}(X)] = X$
4. *inverse* $\rightarrow$ (of X) the element Y such that $XY = 1$, $Y = X^{-1}$
5. *nilpotent* $\rightarrow N^2 = 0, N \neq 0$
6. *idempotent* $\rightarrow X^2 = X$
7. *zero divisors* $\rightarrow XY = 0$, $X, Y \neq 0$

(X, Y, N are multivectors). Now we will explain the listed terms in some details.

1.6.1 Versor

An example of a versor is *abc* if the factors are vectors. For the geometric product of two vectors, we generally have grades 0 and 2. For verification techniques that some multivector is a versor, see [12] and [23]. The geometric product of a versor and its *reverse* (*abc* $\rightarrow$ *cba*) is a real number.

1.6.2 Blade

An example of a blade is $a \wedge b \wedge c$ if the factors are vectors. For verification techniques that some multivector is a blade, see [12] and [23]. A blade is *simple* (in some basis) if it can be reduced to the outer product of the basis vectors (up to a real factor). For the basis (e_1, e_2, e_3), the bivector $e_1 e_2$ is simple, while $e_1 e_2 + e_1 e_3 = e_1(e_2 + e_3)$ is not. In 4D, we have 2-vectors that are not blades, for example $e_1 e_2 + e_3 e_4$. Some authors (*Baylis*) call k-vectors that square to a real number *simple* (such as all k-vectors from *Cl*3); otherwise, they are *compound*.

While versors like ab generally have grades 0 and 2, the blade $a \wedge b$ has grade 2 and defines a 2D subspace. ◆ E17: Show that every *homogeneous versor* (i.e., it has a single grade) is a blade. Show that every blade can be transformed to a versor with orthogonal vectors as factors. ◆ Every blade in *Cl*3 that is an outer product of three linearly independent vectors is proportional to the unit pseudoscalar (show this if you have not done so already).

◆ Consider an arbitrary set of indices of unit vectors of an orthonormal basis, some of which can be repeated. Find an algorithm for sorting indices, taking into account the skew-symmetry for different indices. The goal is to find the overall sign. After sorting, the unit vectors of the same index are multiplied, reducing thus to a single unit vector (up to sign) or a real number (± 1). For example, consider the orthonormal basis of the signature (2,1)

$$e_1^2 = e_2^2 = 1, \quad e_3^2 = -1 \Rightarrow$$
$$e_3 e_2 e_3 e_1 e_2 e_3 = -e_1 e_3 e_2 e_3 e_2 e_3 = e_1 e_2 e_3 e_3 e_2 e_3 = e_1 e_2^2 e_3 e_3 e_3 = e_1 e_3 e_3^2 = -e_1 e_3. \; ◆$$

The elements of the Clifford basis are simple blades. We have seen that any linear combination of unit bivectors in *Cl*3 defines a plane (that is, it can be represented as an outer product of two vectors that span the plane). However, this is a special property of 3D vector spaces and cannot be generalized. ◆ E18: Multiply every element of the *Cl*3 Clifford basis by the pseudoscalar j. What do you get? Figure 1.11 can help in thinking. You can use *GAViewer* and see what your products look like. ◆

1.6.3 Involutions

In geometric algebra, there are three *involutions* that are the most frequently utilized (however, they are not the only ones possible; see Sect. 4.2); these just change the signs of elements from the Clifford basis.

The grade involution $M \to \overleftarrow{M}$ changes the sign of each basis vector of the vector space (*space inversion*). In this way, all even elements remain unchanged, while odd ones change sign. Consider a general multivector M in *Cl*3 (x and n are vectors):

$$M = t + x + jn + bj, \quad t, b \in \mathbb{R}.$$

The grade involution gives

$$\overleftarrow{M} = t - x + jn - bj.$$

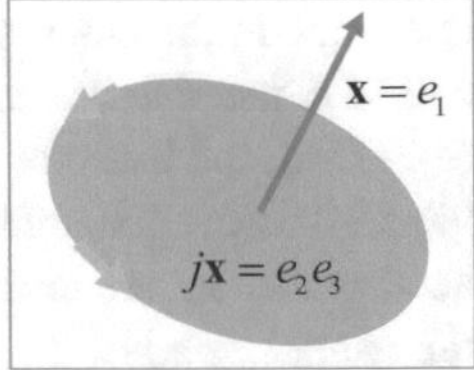

Fig. 1.11 The dual of an element from the Clifford basis

The grade involution is an *automorphism*, which means that $\overleftarrow{(MN)} = \overleftarrow{M}\overleftarrow{N}$ (E19: Show this). The elements $\left(M + \overleftarrow{M}\right)/2 \equiv \langle M \rangle_+$ and $\left(M - \overleftarrow{M}\right)/2 \equiv \langle M \rangle_-$ are the *even* and *odd* parts of the multivector M (find them for general M in *Cl*3).

The reverse involution $M \to M^\dagger$ is an anti*automorphism* ($(MN)^\dagger = N^\dagger M^\dagger$; E19: Show this) that reverses factors in a geometric product:

$$M^\dagger = t + x - jn - bj.$$

The elements $(M + M^\dagger)/2 \equiv \langle M \rangle_R$ and $(M - M^\dagger)/2 \equiv \langle M \rangle_I$ are called the *real* and *imaginary* parts of the multivector M (see below; find them for general M in *Cl*3).

The *Clifford conjugation* (*involution*) $M \to \bar{M}$ is an anti*automorphism* ($\overline{MN} = \bar{N}\bar{M}$; E19: Show this) that is a combination of the grade involution and the reverse involution:

$$\bar{M} = t - x - jn + bj = \overleftarrow{M}^\dagger.$$

We call the elements $(M + \bar{M})/2 \equiv \langle M \rangle_S$ and $(M - \bar{M})/2 \equiv \langle M \rangle_V$ the scalar and the vector parts of the multivector M (*complex scalars* and *complex vectors in Cl*3, due to j; see below. Find them). ◆ E20: What is the result of applying all three involutions to a multivector? What if we apply any two of them? Each involution changes the sign of some grades. If the overall sign of the grade is given in the form $(-1)^{\mathrm{f}(r)}$, where r is a grade, find the function f for each involution. Often, we need to check the properties of some product, sum, etc. What is the multivector if $M = \mathrm{inv}(M)$ or $\mathrm{inv}_1(M) = \pm\, \mathrm{inv}_2(M)$, where inv stands for any of the three involutions just defined? Show that for versors V, the relation

$$V = v_1 v_2 \cdots v_k \Rightarrow \overleftarrow{V} = (-v_1)(-v_2)\cdots(-v_k) = (-)^k V$$

is valid. Show that the multivector $\overleftarrow{V} x V^\dagger$ is a vector if x is a vector. ◆

1.6.3.1 The Klein Four-Group

The set of the identity involution, the grade involution, the reverse involution, and the Clifford conjugation is closed under convolution, which means that any two involutions from the set produce an involution from the set (*closure*), with the properties

1. there exists a unit element (the identity involution);
2. convolutions are associative;
3. inverses exist; involutions are inverses of themselves ($\mathrm{f}^2 = I$, where I is the identity involution);
4. closure.

With these properties, we have the structure of a *group*; in this case, it is the so called *Vierergruppe* (*Klein four-group, Klein group*). Please study this group as an abstract object and also as applied to involutions.

1.6.4 Inverses

An important consequence of the geometric multiplication of vectors is the existence of the inverse of a vector (and many other elements of geometric algebra); that is, we can divide by a vector. This is obvious from $e_i e_i = e_i^2 = 1 \Rightarrow e_i^{-1} = e_i$; however, with the geometric product, we can write $e_j e_i e_i e_j = e_j e_i^2 e_j = e_j^2 = 1 \Rightarrow \left(e_i e_j\right)^{-1} = e_j e_i$. Note that this is not possible with ordinary products (Gibbs). First, how should we define $e_j \cdot e_i \cdot e_i \cdot e_j$? Second, among other issues, the cross product is not even associative. For vectors in general (*null vectors* square to zero and do not have an inverse), we have

$$a^{-1} = a/a^2,$$

which means that the unit vector is the inverse of itself. The existence of the inverse has far-reaching consequences; it significantly distinguishes the geometric product from the ordinary scalar and cross products. Now we can solve equations such as $ab = c$:

$$ab = c \Rightarrow a = cb^{-1}.$$

We can define inverses for other multivectors as well. For example, it is easy to find the inverse of the versor ab:

$$(ab)^{-1} = ba/(abba) = ba/a^2b^2.$$

Here we take advantage of the fact that the geometric product of a versor and its reverse gives a real number. There exist multivectors without an inverse (we will see this in the text). The existence and definition of an inverse are not always simple and obvious. However, this task is relatively easy in *Cl*3. It is important to note that the existence of an inverse depends on the possibility to define the magnitude (norm, amplitude) of a multivector, which is sometimes ambiguous. For a general approach, see the references.

1.6.5 Nilpotents

The geometric product allows the existence of multivectors different from zero that square to zero. They are *nilpotents* in the algebra, having an important role in applications. For example, when formulated in *Cl*3, an electromagnetic wave in vacuum is just a nilpotent in the algebra. As an example of a nilpotent we have $N = e_1 + je_3$:

$$(e_1 + e_1 e_2)^2 = e_1(1 + e_2)e_1(1 + e_2) = e_1 e_1(1 - e_2)(1 + e_2) = 0.$$

◆ Display the nilpotent N graphically (you can use *GAViewer*). ◆

Nilpotents do not have an inverse. If $N \neq 0$ is a nilpotent and M is its inverse, then from $NM = 1$ it follows that $N^2M = N \Rightarrow 0 = N$. We will find the general form of nilpotents from *Cl*3 in Sect. 2.2.

1.6.6 Idempotents

Idempotents have the simple property $p^2 = p$. ◆ Show that the multivector $(1 + e_1)/2$ is an idempotent. ◆ In fact, every multivector of the form $(1 + \mathbf{f})/2$, $\mathbf{f}^2 = 1$, is an idempotent. We will find the general form of idempotents from *Cl*3 in Sect. 2.3. The *trivial idempotent* is 1. ◆ E21: Show that the trivial idempotent is the only one with an inverse.◆

1.6.7 Zero Divisors

◆ Multiply $(1 + e_1)(1 - e_1)$. ◆ There are multivectors different from zero whose geometric product is zero (*zero divisors*). Although this property differs from the properties of real numbers, it turns out to be very useful in many applications. Due to the existence of zero divisors, we have to be careful with cancellation of factors in geometric algebra (see Sect. 4.3).

We will frequently use the idempotents $u_\pm = (1 \pm e_3)/2$, which are zero divisors, since $u_+u_- = u_-u_+ = 0$. ◆ Find $(\alpha u_+ + \beta u_-)^n$, $\alpha, \beta \in \mathbb{R}$, $n \in \mathbb{N}$.◆

1.6.8 Addition of Different Grades

We should mention that addition of quantities like $\mathbf{x} + j\mathbf{n}$ (or other elements of different grades) is not a problem, as some people complain. We just add objects of different grades, and therefore, as with complex numbers, **such a sum preserves the separation of grades**. For example, if we add $2\mathbf{x}_1 + j\mathbf{n}_1$ to $\mathbf{x}_2 + 2j\mathbf{n}_2$, we get $(2\mathbf{x}_1 + \mathbf{x}_2) + j(\mathbf{n}_1 + 2\mathbf{n}_2)$, which means that we still have grades 1 and 2. Here the sum is to be understood as a **relation** between different subspaces. Let us clarify this a little bit for *Cl*3. Real numbers have grade zero and define the subspace of points. Vectors define oriented lines, bivectors define oriented plains, and pseudoscalars define oriented volumes. For example, the bivector B defines an oriented plane by the relation $B \wedge \mathbf{x} = 0$. In that plane, we can find a unit bivector $\hat{B}$ that possesses a number of interesting properties: it squares to -1, it is oriented, it rotates vectors in the plane, etc. As an example, for the bivector

$$B = e_1e_2 + e_2e_3 = e_2 \wedge (e_3 - e_1),$$

the vectors e_2 and $e_3 - e_1$ span the plane (the blue one in Fig. 1.12). The relation $B \wedge \mathbf{x} = 0$ gives vectors $\mathbf{x}$ as a linear combination of vectors e_2 and $e_3 - e_1$. ◆ E22: Find $BB^\dagger$.◆ We see that the unit bivector $\hat{B} = B/\sqrt{2}$ has a clear geometric interpretation. However, it is also an operator that rotates vectors in the plane it defines, as well as an imaginary unit for complex numbers defined in its plane. A multivector of the form $\alpha + B$, $\alpha \in \mathbb{R}$, is the sum of different grades, but there is no way to "blend" real scalars and bivectors in sums; they are always separated. However, together, as a sum, they are powerful, for example as *rotors* or *spinors* (see Sects. 1.9 and 1.14).

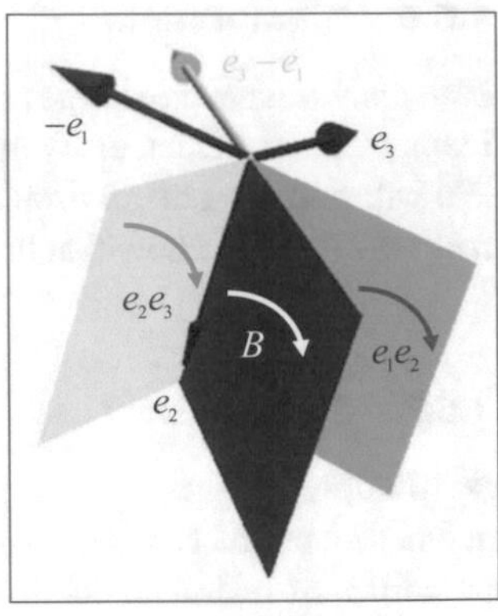

Fig. 1.12 The example of a bivector in 3D, it defines a plane

1.6.9 Lists of Coefficients

Every multivector can be expressed as a list of coefficients in the Clifford basis. As an example, we can use the multivector $3 - e_2 + e_1e_2$ in $Cl2$, whose list of coefficients is $(3, 0, -1, 1)$. It is clear that we can add and subtract such lists, find a rule to multiply them, etc. Summing elements of different grades is equivalent to making such a list, as is familiar with complex numbers; namely, we can represent a complex number $\alpha + i\beta$ as an ordered pair of real numbers (α, β). Thus, the separation of grades is associated with the separation of list positions.

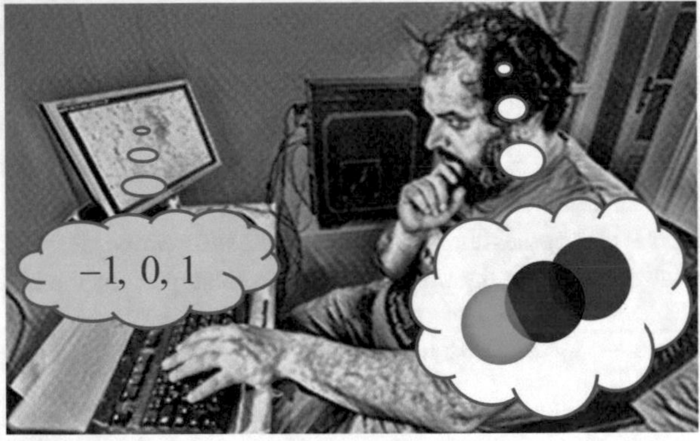

Lists of coefficients can be useful in practical implementations on computers. However, the spirit of geometric algebra is in calculations without coefficients. In expressing vectors as lists of coefficients, we introduce operations with matrices (such as the Pauli matrices), which are suitable for modern computers. However, our brains are not like computers: we have powerful geometric intuition, which means that we should calculate using geometric objects, which is possible due to the development of geometric algebra.

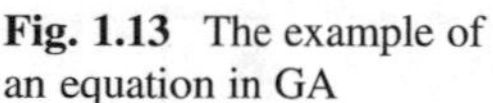

Fig. 1.13 The example of an equation in GA

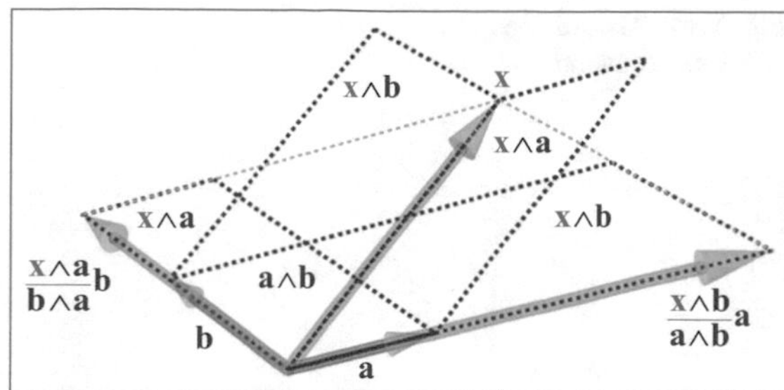

1.7 Examples of Solving Equations

To find the real numbers α and β such that $\mathbf{x} = \alpha\mathbf{a} + \beta\mathbf{b}$ in $\Re^3$, we use the outer product

$$\mathbf{x} \wedge \mathbf{a} = \alpha\mathbf{a} \wedge \mathbf{a} + \beta\mathbf{b} \wedge \mathbf{a} = \beta\mathbf{b} \wedge \mathbf{a},$$
$$\mathbf{x} \wedge \mathbf{b} = \alpha\mathbf{a} \wedge \mathbf{b} + \beta\mathbf{b} \wedge \mathbf{b} = \alpha\mathbf{a} \wedge \mathbf{b}.$$

Note that the bivectors $\mathbf{x} \wedge \mathbf{a}$ and $\mathbf{b} \wedge \mathbf{a}$ define the same plane and that both are proportional to the unit bivector in that plane; that is, their ratio is a real number (a unit bivector divided by itself gives 1). Therefore, we have

$$\mathbf{x} = \frac{\mathbf{x} \wedge \mathbf{b}}{\mathbf{a} \wedge \mathbf{b}}\mathbf{a} + \frac{\mathbf{x} \wedge \mathbf{a}}{\mathbf{b} \wedge \mathbf{a}}\mathbf{b}.$$

We can display such expressions graphically (Fig. 1.13).

1.7.1 Quadratic Equations

Consider now the quadratic equation $X^2 + X + 1 = 0$.

◆ E23: Show that $X = -\exp(\pm I\pi/3)$, $I = e_1e_2$, is the solution. Can you find a solution for an arbitrary quadratic equation? ◆ Pay attention to the fact that we can interpret the expression $X^2 + X + 1$, with the above solution, as the operator that acting on a vector v gives zero. This means that we have the sum of the vector (v), the rotated vector (Xv), and a twice rotated vector (X^2v). We can arrange these three vectors into a triangle (Fig. 1.14).

Regarding rotations and an exponential form, see Sect. 1.9. Here you can feel free to treat expressions as complex numbers, with the imaginary unit $i \leftrightarrow I = e_1e_2$, that is, you can use the trigonometric form of a complex number. You will find an explanation of such an approach in the next chapters.

Fig. 1.14 The solution of a quadratic equation

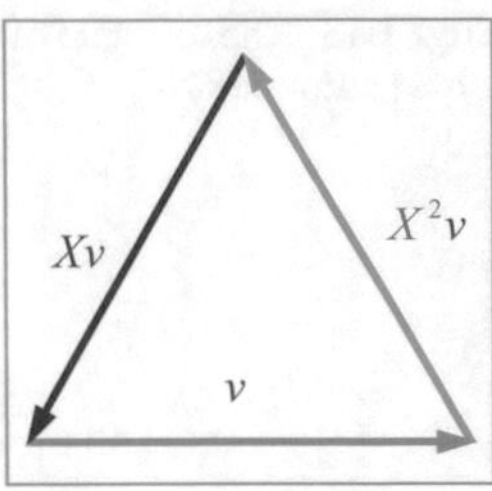

1.8 Geometric Product of Vectors in Trigonometric Form

Consider the square of a bivector in $\mathfrak{R}^n$ (for other signatures, see the literature; the main ideas are the same):

$$(a \wedge b)(a \wedge b) = (ab - a \cdot b)(a \cdot b - ba) =$$
$$-ab^2 a - (a \cdot b)^2 + a \cdot b(ab + ba) =$$
$$(a \cdot b)^2 - a^2 b^2 = -a^2 b^2 \sin^2\theta,$$

where we have made use of $(a \cdot b)^2 = a^2 b^2 \cos^2\theta$. Another way to see this is to start from the form $a \wedge b = ab_\perp$:

$$(ab_\perp)^2 = ab_\perp ab_\perp = -a^2 b_\perp^2 = -a^2 b^2 \sin^2\theta.$$

We see that the square of a bivector in $\mathfrak{R}^n$ is a negative real number. Now we can define the magnitude of a bivector as

$$|a \wedge b| = |a||b||\sin\theta|.$$

The geometric product of two vectors can be written as

$$ab = |a||b|\hat{a}\hat{b} = |a||b|\left(\hat{a} \cdot \hat{b} + \hat{a} \wedge \hat{b}\right) = |a||b|\left(\cos\theta + \hat{B}\sin\theta\right), \ \hat{B} = \frac{\hat{a} \wedge \hat{b}}{\sin\theta},$$

$$\hat{B}^2 = -1,$$

$$ab = |a||b|\exp\left(\hat{B}\theta\right).$$

Note that we have a similar formula for complex numbers (see Sect. 3.5 and Appendix 5.1). However, the situation is quite different here, since the unit bivector $\hat{B}$ is not just an “imaginary unit”: it defines the plane spanned by the vectors $\hat{a}$ and $\hat{b}$. This is a great advantage compared to the ordinary complex numbers; it brings a clear geometric meaning to expressions. For example, in the formulation of quantum mechanics in geometric algebra (see Sect. 2.10), we can use the real numbers only; there is no need for $\sqrt{-1}$. In addition, we can see the geometric meaning of expressions directly, which makes the new formulation powerful. This also provides new insights, which otherwise would be hidden or difficult to reach.

1.8.1 Merging Multiplication Tables

Here we have the opportunity to answer the question about the multiplication tables from Sect. 1.1.1. We have seen that the multiplication tables for the scalar and cross products are almost complementary. As we know, the geometric product of two vectors can be decomposed into symmetric and antisymmetric parts. Then we can find their magnitudes, which have the functions sine and cosine as factors, and all this gives us the "unified" multiplication table

$$\begin{array}{cccc} \cdot & e_1 & e_2 & e_3 \\ e_1 & 1 & 0 & 0 \\ e_2 & 0 & 1 & 0 \\ e_3 & 0 & 0 & 1 \end{array} \quad \oplus \quad j\otimes \quad \begin{array}{cccc} \times & e_1 & e_2 & e_3 \\ e_1 & 0 & e_3 & -e_2 \\ e_2 & -e_3 & 0 & e_1 \\ e_3 & e_2 & -e_1 & 0 \end{array} \quad \rightarrow \quad \begin{array}{cccc} \mathrm{GP} & e_1 & e_2 & e_3 \\ e_1 & 1 & je_3 & -je_2 \\ e_2 & -je_3 & 1 & je_1 \\ e_3 & je_2 & -je_1 & 1 \end{array}$$

We see that the new multiplication table has bivectors as nondiagonal elements ($\oplus$ and $\otimes$ are just for fun). In fact, looking at these tables, one can get nice insights about our 3D space and geometric algebra in general. It is important to appreciate the fact that in the cross product table we have axial vectors, while in the geometric product table we have new objects: bivectors. The idea that the product of two vectors should be a vector gave rise to the definition of the strange cross product. The geometric product of two vectors generally gives a sum of a real number and a bivector, possessing all-important properties of both the scalar and cross products. However, it brings new, far-reaching possibilities, and the story continues...

1.9 Reflections, Rotations, Spinors, Quaternions...

The reader is now, perhaps, convinced that the geometric product is the natural, and indeed the inevitable, way to multiply vectors. One way or another, the magic is still to come.

1.9.1 Bivectors as Rotors

Before we proceed with general formalisms, let us play a little with the bivector $I = e_1e_2$ in $Cl3$. Note that I commutes with vectors perpendicular to its plane ($Ie_3 = e_3I$) and anticommutes with vectors coplanar to its plane ($Ie_1 = -e_1I$). We have seen that the ordinary imaginary unit i can rotate a complex plane (x, y). Using the bivector I, we can rotate vectors as well. For example,

$$e_1I = e_1e_1e_2 = e_2,$$

which rotates just as the imaginary unit does. However, due to anticommutativity, we can multiply from the left,

$$Ie_1 = e_1e_2e_1 = -e_2,$$

to get the rotation by $-\pi/2$. Note that $Ie_3 = j$, and we have just a new element, not a rotation. ◆ What will happen if we multiply from both sides? ◆ We have

$$Ie_1I = e_1e_2e_1e_1e_2 = e_1,$$

or

$$Ie_1I = -I^2e_1 = e_1.$$

We see that the vectors from the bivector plane are not changed. However,

$$Ie_3I = I^2e_3 = -e_3,$$

which means that the vector $\alpha_1e_1 + \alpha_2e_2 + \alpha_3e_3$ will change to $\alpha_1e_1 + \alpha_2e_2 - \alpha_3e_3$, a *vector reflected in the bivector plane* (Fig. 1.15). We can try all this again, but using the reverse of the bivector this time. Defining the unit bivector $\hat{B} = -I = e_2e_1$, we have

$$\hat{B}e_1\hat{B}^\dagger = e_2e_1e_1e_1e_2 = -e_1,$$

which is rotation by π, the angle **doubled** compared to the angle between e_1 and e_2. We also have

$$\hat{B}e_3\hat{B}^\dagger = e_2e_1e_3e_1e_2 = e_3,$$

and this is exactly what we expect from the rotation in the I plane. Now we are getting somewhere.

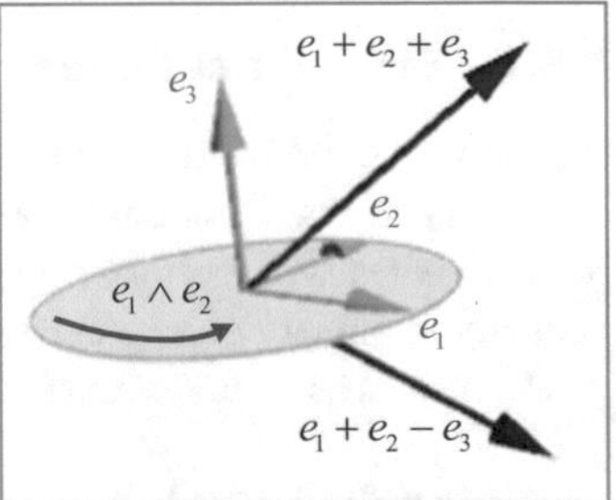

Fig. 1.15 A vector reflected across the plane defined by a bivector

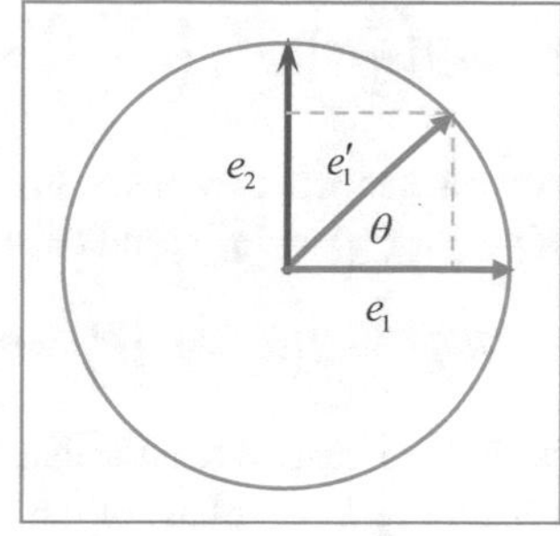

Fig. 1.16 The components of a rotated vector

How to rotate by an angle θ? From Fig. 1.16, we have

$$e_1' = e_1 \cos\theta + e_2 \sin\theta,$$

and then, after rearranging, it follows that

$$e_1' = \left(\cos\theta + \hat{B}\sin\theta\right)e_1 = \exp\left(\theta\hat{B}\right)e_1,$$

where we made use of the relation for the geometric product in trigonometric form. Note that multiplication from the right will give rotation by $-\theta$:

$$e_1\left(\cos\theta + \hat{B}\sin\theta\right) = \left(\cos\theta - \hat{B}\sin\theta\right)e_1 = \exp\left(-\theta\hat{B}\right)e_1.$$

For vectors perpendicular to $\hat{B}$, we have

$$\left(\cos\theta + \hat{B}\sin\theta\right)e_3 = e_3\cos\theta - j\sin\theta,$$

and again we have no rotation! However, we can try the "sandwich" form, as before. We have

$$\exp\left(\theta\hat{B}\right)e_1\exp\left(-\theta\hat{B}\right) = \exp\left(2\theta\hat{B}\right)e_1,$$

which is rotation by the angle 2θ. If we want rotation by the angle θ, we should write

$$\exp\left(\frac{\theta}{2}\hat{B}\right)e_1\exp\left(-\frac{\theta}{2}\hat{B}\right) = \exp\left(\theta\hat{B}\right)e_1,$$

and this is how the famous half-angles appear! We call the element $R = \exp\left(\theta\hat{B}/2\right)$ a *rotor*, with the properties $R^\dagger = \exp\left(-\theta\hat{B}/2\right)$ and $RR^\dagger = 1$. By checking the perpendicular vectors

$$\exp\left(\frac{\theta}{2}\hat{B}\right)e_3\exp\left(-\frac{\theta}{2}\hat{B}\right) = e_3\exp\left(\frac{\theta}{2}\hat{B}\right)\exp\left(-\frac{\theta}{2}\hat{B}\right) = e_3,$$

we get just what we need. For the vector v, we can find components coplanar with the bivector ($v_\parallel$) and perpendicular to it ($v_\perp = v - v_\parallel$). Then we have

$$RvR^\dagger = R(v_\parallel + v_\perp)R^\dagger = Rv_\parallel R^\dagger + v_\perp,$$

which is a general relation that we can generalize to higher dimensions easily. We can use reflections in the plane of the bivector I to find components of a vector:

$$IvI = I(\alpha_1 e_1 + \alpha_2 e_2 + \alpha_3 e_3)I = \alpha_1 e_1 + \alpha_2 e_2 - \alpha_3 e_3 = v_\parallel - v_\perp,$$
$$v_\parallel = (v + IvI)/2, \quad v_\perp = (v - IvI)/2.$$

1.9.2 Invariants of Rotations

The vector $e_3 \perp I$ is invariant under the rotation by the rotor $R = \exp(-\theta I/2), I = e_1 e_2$. In addition, $RjR^\dagger = jRR^\dagger = j$; the pseudoscalar is also invariant (real numbers are invariant, too). For the bivector I, we have $RIR^\dagger = IRR^\dagger = I$. If we denote the rotation of an element e by $\mathrm{R}(e)$, we can look for all elements of the algebra such that $\mathrm{R}(e) = \lambda e$, $\lambda \in \mathbb{R}$. We already found such elements, and we call them *eigenblades*, all with the *eigenvalue* $\lambda = 1$. What is the geometric meaning of this? For the real numbers, it is obvious: the number 2 rotated is still the number 2. For vectors, we have already seen that components perpendicular to the rotation plane (defined by the bivector) do not change. If we imagine the bivector I as an oriented unit circle, it is clear that it will not change under the rotation. For the pseudoscalar j, we can imagine the cylindrical oriented unit volume, with the unit bivector as the base. The base is not affected by rotation, the perpendicular unit vector is an invariant as well, and we can write

$$RjR^\dagger = RIe_3R^\dagger = RIR^\dagger Re_3R^\dagger = Ie_3 = j$$

or (this seems to be trivial, but such relations are important in general, Fig. 1.17)

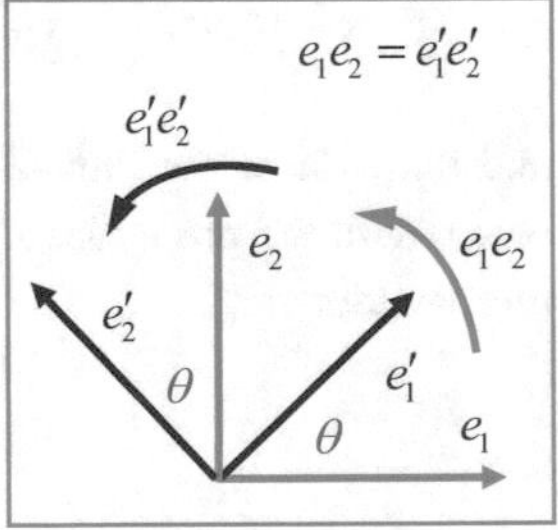

Fig. 1.17 The bivector as an invariant of a rotation

$$RjR^\dagger = Re_1e_2e_3R^\dagger = Re_1R^\dagger Re_2R^\dagger Re_3R^\dagger = e_1'e_2'e_3 = Ie_3.$$

We see that it is really easy to find invariants for rotors; we just have to find elements of the algebra that commute with the bivector I, and that is really easy.

Note that the vectors e_1', e_2', e_3 are orthogonal, so we can write

$$e_1 \wedge e_2 \wedge e_3 = e_1' \wedge e_2' \wedge e_3 = Re_1 \wedge e_2 \wedge e_3R^\dagger = \left(Re_1R^\dagger\right) \wedge \left(Re_2R^\dagger\right) \wedge \left(Re_3R^\dagger\right),$$

or

$$\mathrm{R}(e_1 \wedge e_2 \wedge e_3) = \mathrm{R}(e_1) \wedge \mathrm{R}(e_2) \wedge \mathrm{R}(e_3).$$

Such a transformation (*outermorphism*) preserves the outer product. This is an example of a *linear transformation* (see Sect. 1.17).

1.9.3 The Formalism of Reflections and Rotations

Consider now the powerful formalism of geometric algebra applied to reflections and rotations more generally (we are still in $\Re^n$; details for other signatures can be found in the literature). For the vector a and the unit vector n in $\Re^3$ (just to make things easier; generalizations are straightforward), we can find the *projection* (parallel to n) and the *rejection* (orthogonal to n) of the vector a. We have $a = a_\parallel + a_\perp$, $a_\parallel = (a \cdot n)n = a \cdot nn$; therefore, we can write

$$a' = -nan = -n\left(a_\parallel + a_\perp\right)n = -\left(a_\parallel - a_\perp\right)nn = a_\perp - a_\parallel.$$

This means that the vector a is reflected in the plane **B** orthogonal to n (generally a hyperplane; Figure 1.18). We can omit the minus sign, and then we have the reflection across the vector n. Namely, the reflection of $-a$ is $-n(-a)n$, and we have a ray reflected off the mirror **B** (Fig. 1.19). Recall that reflections do not change the grade of the reflected object. We should mention that in physics we are often interested in reflections at 3D surfaces, so we can slightly adjust the picture Fig. 1.19 to get Fig. 1.20. We use the fact that $j^2 = -1$, whence for the reflection of the incident ray **a**, it follows that

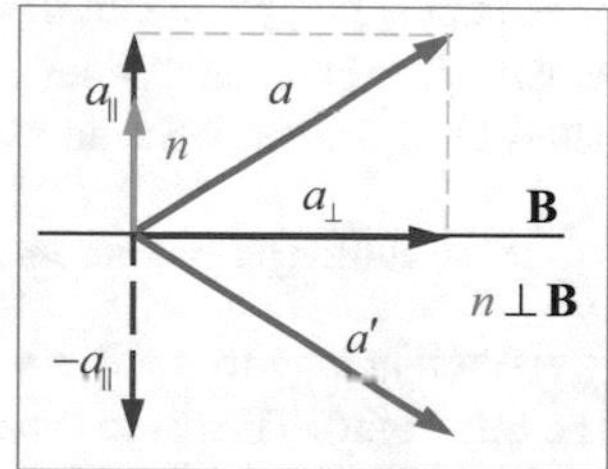

Fig. 1.18 The reflection of a vector using the sandwich form

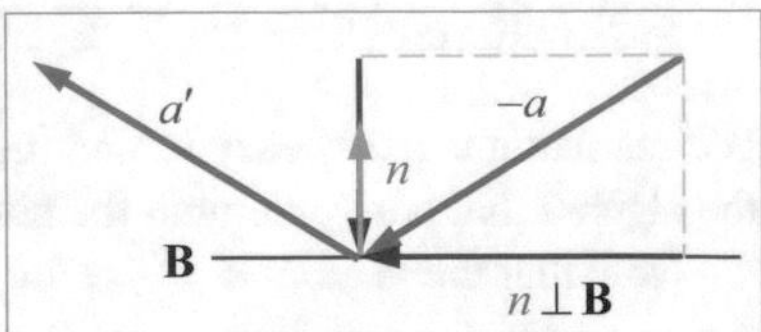

Fig. 1.19 The reflection across a vector using the sandwich form

Fig. 1.20 The reflection of a vector on the plane defined by a bivector

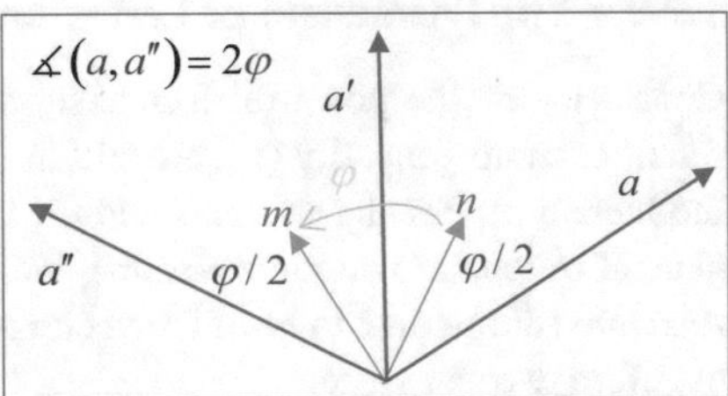

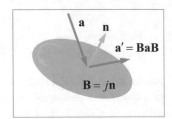

Fig. 1.21 Two reflections give a double angle

$$\mathbf{a}' = -\mathbf{nan} = j^2\mathbf{nan} = j\mathbf{na}j\mathbf{n} = \mathbf{BaB},$$

where the unit bivector **B** defines the reflection plane.

What if we apply two consecutive reflections, using two unit vectors m and n? There is a well-known theorem that states that two consecutive reflections give a rotation. In Figure 1.21, we see that after the reflection across n, we have $a \to a'$. Then by reflection across m we have $a' \to a''$. If the angle between the unit vectors m and n is φ, then the rotation angle of the vector a is 2φ (prove this). Analogously, if we want to rotate a vector by an angle φ, we need to use unit vectors that make the angle $\varphi/2$. We see again how half-angles appear, which are characteristic in the description of spin in quantum mechanics. It appears that there is nothing "quantum" about the half-angle; it is simply a part of the geometry of our 3D space (with the geometric product). We will discuss this later. Now we can write an expression for a rotation as

$$a'' = m(nan)m = mnanm.$$

Another way to rotate a vector is to construct a rotation operator that operates from the left (right). Thanks to the existence of an *inverse* of a vector, this is easy to achieve:

$$a'' = \left(a''a^{-1}\right)a \equiv Oa, \quad O = a''a^{-1}, \quad \measuredangle(a, a'') = 2\varphi.$$

However, the method that uses reflections is very general (**rotates any element of any algebra**), elegant, and has a "sandwich" form that is actually common and preferable in geometric algebra, as we demonstrated above. In particular, it facilitates generalizations to higher dimensions.

1.9.4 A Special Rotor Construction

We will construct the rotor that rotates a unit vector a into the unit vector b in the "sandwich" form (see [21]). We define the unit vector $n = (a + b)/|a + b|$. Then from the obvious relation

$$b = nan = bnanb,$$

we see that we can reflect the vector a,

$$-nan = -\frac{a+b}{|a+b|}a\frac{a+b}{|a+b|} = -\frac{a+b}{|a+b|}a\frac{ab+1}{|a+b|}b = -\frac{a+b}{|a+b|}\frac{b+a}{|a+b|}b = -b,$$

and then reflect the vector $-b$ using the unit vector b:

$$-b(-b)b = b.$$

Our rotor is now (Fig. 1.22)

$$R = bn = b\frac{a+b}{|a+b|} = \frac{1+ba}{|a+b|},$$

which we can check:

$$Ra R^{\dagger} = \frac{1+ba}{|a+b|}a\frac{1+ab}{|a+b|} = \frac{a+b}{|a+b|}\frac{a+b}{|a+b|}b = b.$$

Using $a \cdot b = \cos\theta$, we get

$$ba = \cos\theta - \frac{a \wedge b}{\sin\theta}\sin\theta = \cos\theta - \hat{B}\sin\theta,$$

$$|a+b| = \sqrt{2(1+\cos\theta)}, \quad 1 + \cos\theta = 2\cos^2(\theta/2),$$

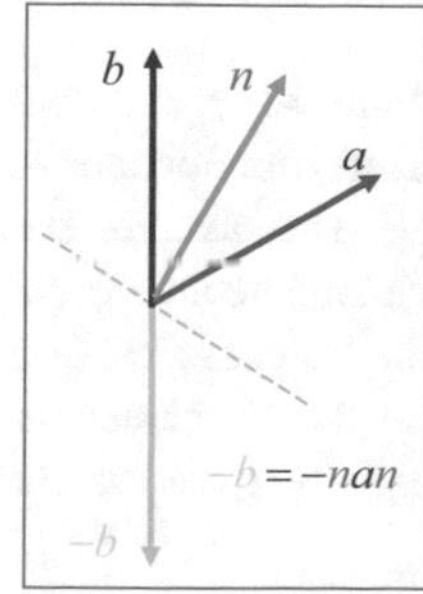

Fig. 1.22 A special rotor construction

$$R = \frac{1 + ba}{|a + b|} = \frac{1 + \cos\theta}{\sqrt{2(1 + \cos\theta)}} - \frac{\hat{B}\sin\theta}{\sqrt{2(1 + \cos\theta)}} = \cos(\theta/2) - \hat{B}\sin(\theta/2)$$
$$= \exp\left(-\hat{B}\theta/2\right).$$

1.9.5 Rotors as Geometric Products of Two Unit Vectors

Let's analyze the term *mnanm*. Generally, geometric products of two unit vectors consist of grades 0 and 2. Therefore, it belongs to the even part of the algebra (subalgebra), which means that the product of any two of these elements will result in an element of the even part of the algebra. We denote this product by $R = mn$ (*rotor* in the text). Now we have

$$a'' = RaR^{\dagger}, RR^{\dagger} = mnnm = 1 = R^{\dagger}R, \quad R^{\dagger} = R^{-1},$$

where $R^{\dagger}$ means *reverse* ($mn \to nm$). For the rotation angle φ, we need the unit vectors with the angle $\varphi/2$ between them. We have $m\,n = m \cdot n + m \wedge n$, where $|m \wedge n| = |\sin(\varphi/2)|$. Using the unit bivector $\hat{B} \equiv n \wedge m/\sin(\varphi/2)$ (note the order of vectors), we have

$$mn = m \cdot n + m \wedge n = \cos(\varphi/2) - \hat{B}\sin(\varphi/2) = \exp\left(-\hat{B}\varphi/2\right).$$

The minus sign is just due to the convention (a positive rotation is counterclockwise). In *Cl*3, we can write a unit bivector $\hat{B}$ as $j\mathbf{w}$, where $\mathbf{w}$ is the unit vector parallel to the rotation axis; however, we cannot generalize such formulas immediately. The rotor reverse is

$$R^{\dagger} = nm = \exp\left(\hat{B}\varphi/2\right).$$

Therefore, the rotation is finally

$$a'' = RaR^{\dagger} = \exp\left(-\frac{\varphi}{2}\hat{B}\right)a\exp\left(\frac{\varphi}{2}\hat{B}\right).$$

This is the general formula; you can write it in any dimension. If a commutes with $\hat{B}$, the rotation transformation has no effect on a. If a anticommutes with $\hat{B}$, we have the operator form

$$a'' = \exp\left(-\varphi\hat{B}\right)a = a''a^{-1}a.$$

Note that $a''a^{-1}$ contains the bivector defined by a'' and a, which anticommutes with a. Thus, the operator form is not general, but we can use it in some selected plane.

The bivector $\hat{B}$ defines the rotation plane, and it is clear that the rotor R does not change vectors perpendicular to this plane. Note that we do not need rotation matrices, the Euler angles, or any other known formalism. Once you define the unit bivector, it will do the entire necessary job. You can imagine it as a small spinning wheel that does exactly what we need. Note that two different consecutive rotations in *Cl*3 make a rotation again (E24: Show this). This produces a *group structure*, but here we will not talk about it.

Example Rotate the vector $e_1 + e_2 + e_3$ in the plane e_1e_2 by the angle φ. We have

$$\exp\left(-\frac{\varphi}{2}e_1e_2\right)(e_1+e_2+e_3)\exp\left(\frac{\varphi}{2}e_1e_2\right).$$

Taking advantage of the fact that the vector e_3 commutes with the bivector e_1e_2, while e_1 and e_2 anticommute, we can write (Fig. 1.23)

$$\begin{aligned}\exp\left(-\frac{\varphi}{2}e_1e_2\right)(e_1+e_2+e_3)\exp\left(\frac{\varphi}{2}e_1e_2\right) &= e_3+\exp(-\varphi e_1e_2)(e_1+e_2)\\ &= e_3+(\cos\varphi-e_1e_2\sin\varphi)(e_1+e_2)\\ &= e_3+(e_1\cos\varphi+e_2\sin\varphi)+(-e_1\sin\varphi+e_2\cos\varphi).\end{aligned}$$

For vectors in the plane e_1e_2, we recognize the rotation matrix

$$\begin{pmatrix}\cos\varphi & -\sin\varphi\\ \sin\varphi & \cos\varphi\end{pmatrix},$$

where the columns represent images of the unit vectors. To get a rotation by the angle $-\varphi$, we use the bivector $e_2e_1 = -e_1e_2$.

Consider the rotation

$$\exp\left(-\frac{0.7\pi}{2}e_1e_2\right)a\exp\left(\frac{0.7\pi}{2}e_1e_2\right)$$

and the corresponding rotation matrix

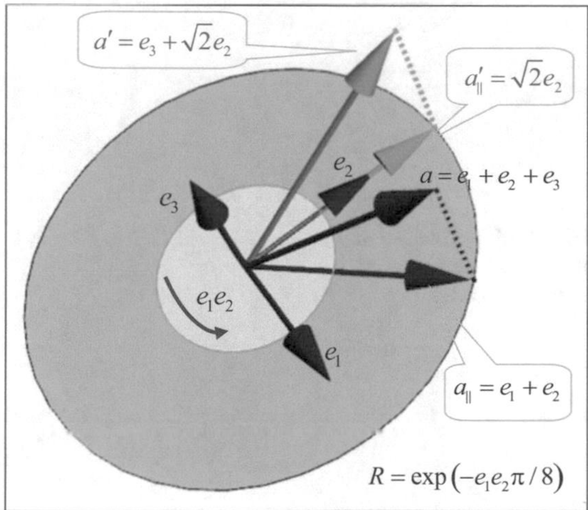

Fig. 1.23 An example of rotation of a vector in GA

$$\begin{pmatrix} -0.588 & -0.809 \\ 0.809 & -0.588 \end{pmatrix}.$$

◆ What can you say about the geometric interpretation, that is, what can you deduce by just looking at the matrix? Try now to construct a rotation matrix for an arbitrary plane. Try to repeat the process in 4D. ◆ The ease in performing rotations in geometric algebra is something previously unseen. There are no special cases, no vague matrices: just follow the simple application of rotors to any multivector. Many prefer quaternions, but they do not possess geometric clarity. In addition, they are limited to 3D (and they are included in this formalism; see Sect. 1.9.10). If the benefits of geometric algebra were elegance and power of rotations only, it would be worth of effort. However, it gives us much, much more.

1.9.6 Small Rotations

We can factor any rotation into small rotations

$$R = e^{I\varphi/2} = \underbrace{e^{I\varphi/2n}\cdots e^{I\varphi/2n}}_{n},$$

which we can use in practice, for example in interpolations. Consider the rotation of the vector e_2 by a small angle in the plane $e_1 e_2$ (Figs. 1.24 and 1.25). Recall the definition

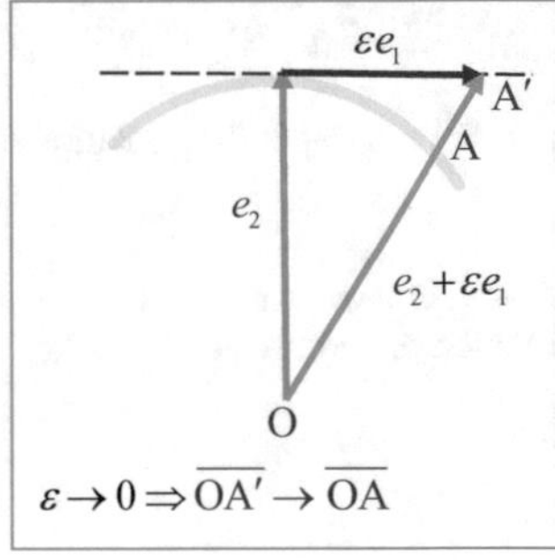

Fig. 1.24 A small rotation of a vector

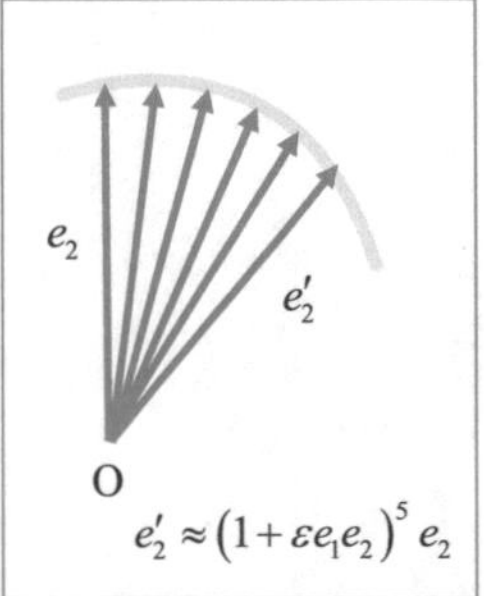

Fig. 1.25 A rotation of a vector as a sequence of small rotations

$$e^x = \lim_{n\to\infty}\left(1+\frac{x}{n}\right)^n$$

and let $1 + \varepsilon e_1 e_2$ be an operator, where ε is a small real number. Acting from the left, we have

$$(1+\varepsilon e_1 e_2)e_2 = e_2 + \varepsilon e_1,$$

and thus we obtain an approximate small rotation of the vector e_2. Note the sign of the number ε: for $\varepsilon < 0$ we will have a counterclockwise rotation. The operator $1 + \varepsilon e_1 e_2$ rotates all vectors in the plane through the same angle. Consequently, by successive applications to e_2, we rotate e_2 first, then we rotate the rotated vector, and so on. (We can neglect small changes in the magnitude). This justifies the definition of the exponential form of the rotor: each rotation is the composition of a large number of small successive rotations.

Of course, all this is well defined for infinitely small rotations, so for the bivector B, we have

$$e^B = \lim_{n\to\infty}\left(1+\frac{B}{n}\right)^n.$$

1.9.7 Rotors and Eigenblades

Recall that the rotor $\exp\left(-\hat{B}\varphi/2\right)$ will not change the bivector $\hat{B}$, which is an invariant of the rotation. Rotations are linear orthogonal transformations, described in linear algebra by matrices. To find invariants of such transformations, we traditionally study results of the action of matrices on **vectors only**. For a matrix A (representing a linear transformation), we seek vectors x such that $Ax = \lambda x$, which provides solutions for eigenvalues $\lambda \in \mathbb{C}$. In fact, the complex λ in the matrix formulation may indicate that we have a rotor in GA. We have seen that in geometric algebra, there exist invariants of rotations (bivectors, or any blade). Instead of the concept of eigenvectors, we can introduce the concept of *eigenblades* (which includes eigenvectors). This allows the reduction of the set of eigenvalues of a transformation to the set of real numbers, and it gives a geometric meaning to the concept of eigenblades. We will discuss linear transformations in Sect. 1.17.

1.9.8 Rotors Are Strange

The rotor $-R$ has the same effect as the rotor R, but the direction of the rotation is not the same. For example, the vector e_1 can be rotated to $-e_2$ clockwise by $\pi/2$ or counterclockwise by $3\pi/2$; therefore, we see that a rotor clearly shows the direction of rotation (try this with matrices), for example

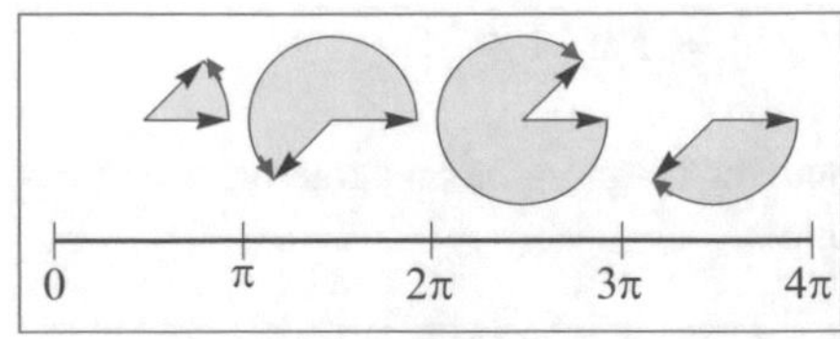

Fig. 1.26 Rotors are unit spinors

$$-e^{I\varphi/2} = e^{-I\pi}e^{I\varphi/2} = e^{-I(2\pi-\varphi)/2}.$$

The minus sign disappears due to the "sandwich" form. We have two possible rotors for each rotation (find out what is meant by a *double cover* of a group) (Fig. 1.26).

Note that due to the half-angle, the rotor

$$\exp\left(-\varphi\hat{B}/2\right) = \cos\left(\varphi/2\right) - \hat{B}\sin\left(\varphi/2\right)$$

has periodicity of 4π, instead of 2π. For such objects, we often use the name *unit spinor* (see Sect. 4.4). Geometric algebra is an ideal framework in which to study all unusual properties of rotations, but to do so would take a lot of space.

Example Let's rotate (see Sect. 3.2, [21]) some object in 3D around e_1 by $\pi/2$, then around e_2 by $\pi/2$. What do we get? ◆ Do this also using matrices. ◆ We have (note the order of rotors)

$$e^{-je_2\pi/4}e^{-je_1\pi/4} = \frac{1}{\sqrt{2}}(1-je_2)\frac{1}{\sqrt{2}}(1-je_1) = \cdots = \frac{1}{2} - \frac{1}{2}\sqrt{3}j\frac{e_1+e_2-e_3}{\sqrt{3}}$$
$$= e^{-j\mathbf{v}\pi/3}, \quad \mathbf{v} = \frac{e_1+e_2-e_3}{\sqrt{3}},$$

so we have a rotation by $2\pi/3$ around the vector $\mathbf{v}$. ◆ Try to rotate the basis vectors in your head. ◆

Question What is the meaning of $e^{I\pi} = -1$? In *Cl*2, for $I = e_1e_2$ ($\mathbf{v}$ is a vector in the e_1e_2 plane; you can choose $\mathbf{v} = e_1$, if you like), we have, due to anticommutativity,

$$e^{I\pi/2}\mathbf{v}e^{-I\pi/2} = e^{I\pi}\mathbf{v} = -\mathbf{v},$$

whence on multiplying on the right by $\mathbf{v}^{-1}$, we get a clear interpretation. The rotor $\exp(I\pi/2)$ transforms the vector $\mathbf{v}$ to the vector $-\mathbf{v}$, i.e., it rotates it by $-\pi$ (the sign is unimportant here). Of course, we also recognize the rotational properties of the ordinary imaginary unit in the complex plane (selected in advance). However, a bivector **defines** the rotation plane, and we could write identical relations, without change, in any dimension, in any plane. In fact, a bivector in the exponent of the rotor could depend on time; the formulas would be valid, and the rotation plane would change with the bivector. When rotations are expressed by elements of geometric algebra, the results always have a geometric meaning. Thus, generally, we do not need to spend time to interpret results; they are already in a form that can employ our geometric intuition. Try to do this with the "square root of minus one" and coordinates.

1.9.9 How to Rotate a Basis

Now we want to find the rotor in 3D that will transform the orthonormal coordinate basis e_i to the orthonormal coordinate basis f_i (see [21]). We need a rotor with the property $f_i = Re_iR^\dagger$. Note that for geometric products of basis vectors, we have $f_1f_2f_3 = e_1e_2e_3$, which means that the handedness must be the same (prove this, using the fact that rotations are orthogonal transformations with determinant 1). Defining $R = \alpha - \beta\hat{B}$, α, $\beta \in \mathbb{R}$, with $R^\dagger = \alpha + \beta\hat{B}$, where $\hat{B}$ is a unit bivector, and using the simple and useful relations in $Cl3$

$$\sum_i e_i^2 = 3, \quad \sum_i e_i\hat{B}e_i = -\hat{B}$$

(E25: Prove them), we have

$$\sum_i e_iR^\dagger e_i = 3\alpha - \beta\hat{B} = 4\alpha - R^\dagger.$$

The reader is left to find $\sum_i f_ie_i$ and to show that it is valid (note the elegance):

$$R = \frac{X}{\sqrt{XX^\dagger}}, \quad X = 1 + \sum_i f_ie_i.$$

What about rotation by π? E26: Show that every rotor can be expressed using the Euler angles

$$\exp(-je_3\phi/2)\exp(-je_1\theta/2)\exp(-je_3\psi/2).$$

1.9.10 Rotors and Quaternions

Let us mention the historical role of William Rowan Hamilton, who in the nineteenth century found an effective mechanism for rotations in 3D: *quaternions*. There is a connection between quaternions and the formalism described here. To be precise, we can easily relate the unit quaternions to the unit bivectors in $Cl3$. However, quaternions are like extended complex numbers: they do not have a clear geometric interpretation. Moreover, they exist only in 3D, while the formalism of geometric algebra is valid in any dimension. Every calculation in which we use quaternions can be easily translated into the language of geometric algebra, while the reverse is not true. However, quaternions are still used successfully to calculate rotations, for example in the computers of military and space vehicles, as well as in robotics. If you implement geometric algebra on your computer, quaternions are unnecessary. Moreover, according to NASA, rotors are about 20% faster.

Unit quaternions have the properties $\mathbf{IJK} = -1$, $\mathbf{IJ} = -\mathbf{JI}$, and so on, and the square of each of them is -1. It was enough to come up with objects that square to -1 and anticommute to describe rotations in 3D successfully. ◆ The reader can check that the substitutions

$$\mathbf{I} \rightarrow e_{23} = je_1, \quad \mathbf{J} \rightarrow e_{13} = -je_2, \quad \mathbf{K} \rightarrow e_{12} = je_3,$$

generate the multiplication table of the unit quaternions ◆ (the minus sign in front of je_2 is due to the fact that the Hamilton unit quaternions define a left coordinate system). Certainly, it is good to understand that the bivector $e_{12} = e_1e_2$ has a clear and direct geometric interpretation, while the unit quaternion **K** (like the imaginary unit or a matrix) has not. Unfortunately, the concept of geometric objects such as bivectors is often strange to those who are traditionally oriented.

Now we can write a quaternion in GA (and get a spinor; see Sect. 1.14):

$$q_0 + q_1\mathbf{I} + q_2\mathbf{J} + q_3\mathbf{K} \rightarrow q_0 + q_1je_1 - q_2je_2 + q_3je_3.$$

1.9.11 Rotors Are Universal

Once we know how to rotate vectors, we can rotate any element of a geometric algebra. Note an especially nice feature of geometric algebras: objects that perform transformations ("operators") are also elements of the algebra. Let us look at the rotation of a versor,

$$RabcR^{\dagger} = RaR^{\dagger}RbR^{\dagger}RcR^{\dagger} = \left(RaR^{\dagger}\right)\left(RbR^{\dagger}\right)\left(RcR^{\dagger}\right),$$

which clearly shows how the rotation of a versor can be reduced to rotations of individual vectors and vice versa. Every multivector is a linear combination of elements of the Clifford basis, whose elements are simple blades; therefore, they are versors. We see that our last statement is always true, due to linearity. The reader should practice rotations of different objects in *Cl*3. Find the term *gimbal lock* on the Internet (it is fun).

The unit pseudoscalar in *Cl*3 commutes with all elements of the algebra. Therefore, we can write

$$R = \exp\left(-\hat{B}\theta/2\right) \rightarrow \exp\left(-\hat{B}\theta/2\right)\exp(\varphi j),$$

where $\exp(\varphi j)$ is a complex phase. Due to commutativity, we have

$$\exp(\varphi j)a\exp(-\varphi j) = \exp(\varphi j)\exp(-\varphi j)a = a,$$

from which it follows that (see Sect. 1.14)

$$RaR^{\dagger} = \exp\left(-\hat{B}\theta/2\right)\exp(\varphi j)a\exp(-\varphi j)\exp\left(\hat{B}\theta/2\right) = \exp\left(-\hat{B}\theta/2\right)a\exp\left(\hat{B}\theta/2\right).$$

1.9.12 Rotors as Arcs

It is interesting to look at the unit sphere in 3D and unit vectors originating at the center of the sphere (Fig. 1.27). Each rotation of the unit vector defines an arc on some great circle. Such arcs, if we take into account their orientation, can become a kind of vectors on the sphere, and composition of two rotations can be reduced to an addition

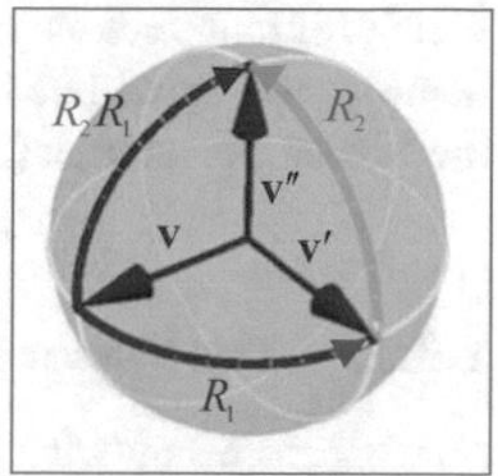

Fig. 1.27 Rotors as arcs

(noncommutative) of such vectors (see [4]). This is a nice tool for seeing the effect of two successive rotations. Just take a small sphere and play with it.

1.9.13 Rotors Are Unit Spinors

If we take an arbitrary element of the even part of *Cl*3, not only the rotors, in addition to a rotation, we get an additional effect, a dilatation, which is exactly the property of spinors. Spinors are closely associated with the even part of the algebra. Geometric algebra hides an unusual amount of mathematics (traditionally branched out in different disciplines) within itself. It is amazing how the redefinition of vector multiplication integrates different branches of mathematics into a single formalism. Spinors, tensors, Lie groups and algebras, various theorems of integral and differential calculus, are united, along with the theory of relativity (special and general), quantum mechanics, and the theory of quantum information, to an almost unbelievable extent. Many complex results of physical theories become simple here and acquire a new meaning. Maxwell's equations are reduced to three letters, with the possibility of inverting the derivative operator over the *Green functions*. Hard problems in electromagnetism become solvable elegantly (see [2]). The Kepler problem is nicely reduced to the problem of the harmonic oscillator (see Sect. 2.9.1). The Dirac theory formulates in *Cl*3 (see [5]) the minimal standard model in *Cl*7 ([44]). And so on. Geometric algebra has a good chance to become the mathematics of future. Unfortunately, it is difficult to break through traditional university (and especially high-school) programs.

1.9.14 Angles as Areas

Consider the rotor that rotates a vector by the angle θ, $R = \exp\left(-\theta\hat{B}/2\right)$. We see that a vector of length r sweeps the area

$$\frac{\theta}{2\pi}r^2\pi = r^2\frac{\theta}{2},$$

which for unit vectors gives just $\theta/2$. Consequently, the bivector $\theta\hat{B}/2$ suggests that the quantity $\theta/2$ should be interpreted as an area instead of an angle, since the magnitude of the bivector is an area (this idea is due to *Hestenes*).

1.9.15 The Versor Product

Starting from sandwich products (x and v are vectors)

$$x \to -vxv^{-1},$$

we can use the vectors $v_1, v_2, \ldots, v_k$ to define the versor $V = v_k \cdots v_2 v_1$ and the *versor product*

$$x \to (-1)^k VxV^{-1} = v_k \cdots v_2 v_1 x v_1^{-1} v_2^{-1} \cdots v_k^{-1} = \overleftarrow{V} x V^{-1},$$

where we introduced the grade involution $\overleftarrow{V}$. Generally, we can define (for any single-graded multivector M)

$$\mathrm{V}(M) \equiv (-1)^{mk} VMV^{-1},$$

where $m = \text{grade}(M)$. From the properties of the grade involution and the inverse, using associativity, we have

$$\overleftarrow{V}_2\left(\overleftarrow{V}_1 x V_1^{-1}\right)V_2^{-1} = \overleftarrow{V}_2\overleftarrow{V}_1 x V_1^{-1} V_2^{-1} = \overleftarrow{(V_2 V_1)}\, x (V_2 V_1)^{-1},$$

which means that the composition of versor products is a versor product. In Sect. 1.4.1, it is shown that the sandwich product with vectors does not change the grade of a transformed object; therefore, the versor product preserves the grade. It also preserves the form of the geometric product, as well as the outer product, the left (right) contraction, the scalar product, and some other products (such as *meet* and *join*). If $\circ$ stands for any of these products, we can write

$$\overleftarrow{V}(x \circ y)V^{-1} = \left(\overleftarrow{V} x V^{-1}\right) \circ \left(\overleftarrow{V} y V^{-1}\right).$$

The versor product plays an important role in transformations of geometric objects, including rotations, translations, etc., especially in the *conformal model* (see Sect. 2.12). Note that the versor product is a composition of reflections, and it is a well-known fact that *all Euclidean transformations can be made by multiple reflections in well-chosen planes*. For example, we know that two reflections give a rotation.

1.9.16 Special Versors

Note that in vxv^{-1} it is not important that v is a unit vector, since the magnitudes of v and v^{-1} cancel. Nevertheless, we can define the special versors of the form

$$V = (-1)^k n_1 n_2 \cdots n_k, \quad VV^\dagger = 1, \quad n_i^2 = 1.$$

Such versors, like the general ones, preserve the length of a vector,

$$a'^2 = VaV^\dagger VaV^\dagger = Va^2V^\dagger = a^2VV^\dagger = a^2,$$

as well as the inner product (*orthogonal transformations*),

$$2a' \cdot b' = 2\left(VaV^\dagger\right) \cdot \left(VbV^\dagger\right) = \left(VaV^\dagger VbV^\dagger + VbV^\dagger VaV^\dagger\right) = V(ab + ba)V^\dagger$$
$$= (ab + ba)VV^\dagger = 2a \cdot b.$$

In Fig. 1.28, we see successive reflections at mirrors. For three mirrors perpendicular to the unit vectors $n_i = e_i$, we have

$$k_3 = (-1)^3 e_3 e_2 e_1 k e_1 e_2 e_3 = -(-j)kj = -k,$$

which means that the light beam will propagate in the opposite direction after the three reflections.

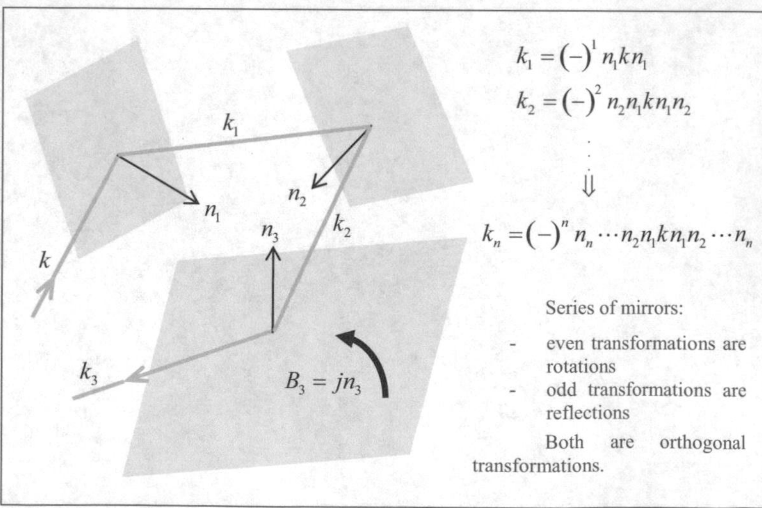
$k_1 = (-)^1 n_1 k n_1$
$k_2 = (-)^2 n_2 n_1 k n_1 n_2$
$\Downarrow$
$k_n = (-)^n n_n \cdots n_2 n_1 k n_1 n_2 \cdots n_n$
k_1
n_2
n_1
n_3
k_2
k
k_3
$B_3 = jn_3$
Series of mirrors:
- even transformations are rotations
- odd transformations are reflections
Both are orthogonal transformations.

Fig. 1.28 Reflections on a system of mirrors

Reflections have many applications in science, technology, media, and daily life. Geometric algebra provides a powerful technique for handling multiple reflections. As was shown, once you know the geometry of a mirror system (that is, the unit vectors perpendicular to the mirrors), you can predict the behavior of any incident light beam just by multiplying unit vectors. With the help of computers, this job becomes like child's play. You can even follow each step of the calculations to see the light beam's entire trajectory. Interactive applications can help to vary the parameters of a mirror system. There is no need to worry about coordinates, matrices, special cases, and all that. The geometric product takes care of everything. (See also *ray tracing*.) (Fig. 1.29)

The reader interested in various representations of rotations in 3D can search on Wikipedia or Wolfram MathWorld for the terms *Euler parameters*, *Euler–Rodrigues formula*, *rotation formalisms in three dimensions*, *quaternions and spatial rotations*, and *Cayley–Klein parameters*.

Rotors in geometric algebra are exceedingly useful in physics, from classical mechanics to advanced modern physics. They appear in astronomy, relativity, quantum mechanics (*spinors*), etc. One can study Fig. 1.30 and Fig. 1.31 to understand rotations better.

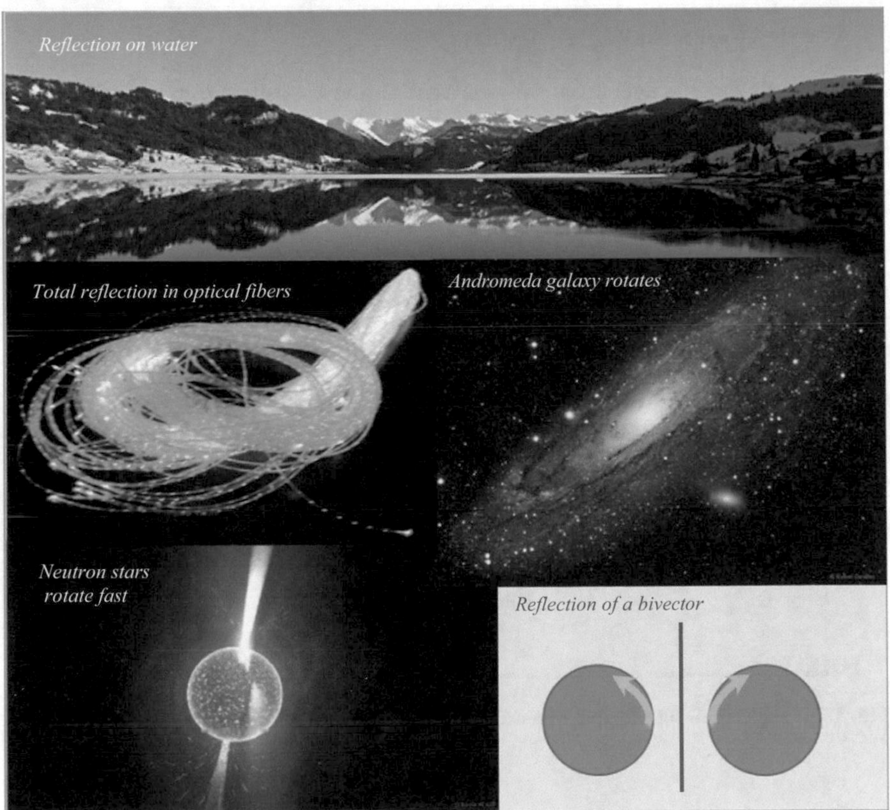

Fig. 1.29 Reflections and rotations

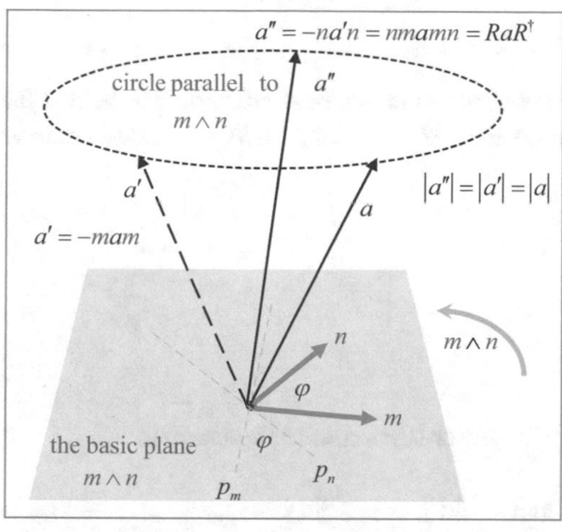

- $\mathbf{m} \wedge \mathbf{n}$ defines the basic plane, orientation, and the rotation angle
- $\mathbf{a}_\perp$ is invariant under rotations; only $\mathbf{a}_\parallel$ is rotated by 2φ
- The same formalism is valid in any dimension (in dimensions higher than 3 there is a subspace invariant under rotations).
- It is easy to obtain any composition of rotations in the same manner.
- The geometric product of vectors gives us the possibility to maintain rotations easily.

Fig. 1.30 Rotations of vectors using a bivector

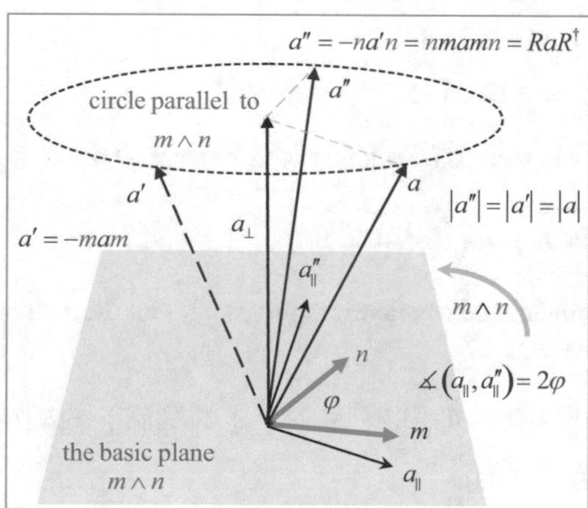

Fig. 1.31 Rotations of vectors as two reflections

1.10 The Scalar Product and the Magnitude of Multivectors

We define the *scalar product* of multivectors A and B as

$$A * B = \langle AB \rangle,$$

and we see that for vectors, it gives the inner product. If we recall that every multivector is a linear combination of elements from the Clifford basis, it is clear that only the parts of multivectors with the same indices give a nonzero contribution to the scalar product. In Euclidean vector spaces we have, for example, $e_1e_2(e_1e_2)^\dagger = e_1e_2e_2e_1 = 1$, which leads to the positive quantity

$$A * A^\dagger = \langle AA^\dagger \rangle \geq 0, \quad (1.1)$$

where equality means that all coefficients of the multivector A are zero, that is, that the multivector A is zero. In $Cl3$, for the multivector $M = t + \mathbf{x} + j\mathbf{n} + jb, j = e_1e_2e_3$, we have

$$M * M^\dagger = t^2 + |\mathbf{x}|^2 + |\mathbf{n}|^2 + b^2, \quad t, b \in \mathbb{R}$$

(check this).

1.10.1 The Multivector Norm

We can define a special function of a multivector (the *scalar magnitude*) as

$$\|A\| = \|A^\dagger\| = \sqrt{A * A^\dagger}, \quad A = \|A\|\hat{A}, \quad \|\hat{A}\| = 1, \quad \|A\| \in \mathbb{R},$$

and we see that this function possesses some interesting properties (here we use double brackets to avoid possible confusion with the MA). ◆ Show that $\langle AB^\dagger \rangle = \langle BA^\dagger \rangle$. ◆ First, we have

$$\|\alpha A\| = |\alpha|\|A\|, \quad \alpha \in \mathbb{R}. \quad (1.2)$$

Second, we have in general (check this using examples) $A * B \leq \|A\|\|B\|$, as well as

$$\|AB\| = \sqrt{\langle ABB^\dagger A^\dagger \rangle} = \sqrt{\langle A^\dagger ABB^\dagger \rangle}.$$

◆ Find possible examples and counterexamples (if any) for the relation $\|AB\| = \|A\|\|B\|$. ◆ We can also write

$$\begin{aligned} \|A + B\|^2 &= \|A\|^2 + 2A{*}B^\dagger + \|B\|^2 \leq \|A\|^2 + 2\|A\|\|B\| + \|B\|^2 \\ &= (\|A\| + \|B\|)^2. \end{aligned} \quad (1.3)$$

The relations (1.1), (1.2), and (1.3) define a *norm*, here the *multivector norm*. We can compare the multivector norm to a vector norm in Euclidean vector spaces. Namely, we can represent a vector as a list of coefficients and define a vector norm in $\mathfrak{R}^n$ as

$$v = (v_1, \ldots, v_n) \Rightarrow |v| = \sqrt{\sum_{i=1}^{n} v_i^2},$$

whence we see that the multivector norm has the same mathematical form, because we can represent multivectors as 2^n-dimensional lists.

We can try to find a bilinear transformation that preserves the multivector norm, such as $M \to XMY$. The square of the new multivector norm is

$$\langle XMYY^\dagger M^\dagger X^\dagger \rangle = \langle X^\dagger XMYY^\dagger M^\dagger \rangle,$$

and we immediately see one simple solution $XX^\dagger = YY^\dagger = 1$, which means that X and Y could be rotors. We will see (in Appendix 5.4) that the bilinear transformations that preserve the multivector amplitude $\sqrt{M\bar{M}}$ are associated with Lorentz transformations.

We already know that blades are versors (they can be expressed as the geometric product of orthogonal vectors), and for versors in nondegenerate vector spaces (there are no vectors that square to zero), we can write $V^\dagger V = \langle V^\dagger V \rangle$, from which it follows that

$$V^{-1} = V^\dagger \|V\|^{-2},$$

which means that all versors and blades in nondegenerate vector spaces are invertible. In *Cl*3, we will use the Clifford involution to define inverses based on the geometric product.

1.11 Contractions

We defined the inner product, which for vectors coincides with the usual scalar multiplication of vectors. In general, in geometric algebra, we can define various products that lower the grades of elements (the outer product raises them). It appears that the best choice is the *left contraction* (LC). For vectors, this is just the inner product, but generally, LC makes it possible to avoid various special cases, such as, for example, the inner product of a vector with a real number. Here we will mention just a few properties of the left contraction; see [23] for more details. ◆ Prove that for two single-graded (*homogeneous*) elements (including real numbers) A_k and B_l (of grades k and l) we have the outer product $A_k \wedge B_l = \langle A_k B_l \rangle_{l+k}$, $0 \le l+k \le n$ (a grade cannot be negative or greater than the dimension of the vector space). ◆ Now, the idea is that for any two single-graded elements, we define a multiplication $A_k \rfloor B_l = \langle A_k B_l \rangle_{l-k}$, $0 \le l-k \le n$, which gives a result of the grade grade(B_l) – grade(A_k), from which it immediately follows that the left contraction is zero for grade(B_l) < grade(A_k).

We will mostly discuss the left contraction on blades, including real numbers. For $\alpha \in \mathbb{R}$, vectors a, b, and blades A, B, C, we have

$$\alpha \rfloor B = \alpha B, \quad \text{(LC 1)}$$
$$B \rfloor \alpha = 0, \quad \text{grade}(B) > 0. \quad \text{(LC 2)}$$

For vectors, we have (grade 1 – 1 = 0)

$$a \rfloor b = \langle ab \rangle = a \cdot b,$$

that is, the left contraction (LC) is the same as the inner product for vectors. We will use two more relations (and see that they are correct and very useful):

$$a \rfloor (B \wedge C) = (a \rfloor B) \wedge C + \overleftarrow{B} \wedge (a \rfloor C), \quad \text{(LC 3)}$$
$$A \rfloor (B \rfloor C) = (A \wedge B) \rfloor C \quad \text{(LC 4)}$$

(for $(A \rfloor B) \rfloor C$, see [23]). From (LC 3), we see that

$$e_1 \rfloor (e_1 \wedge e_2) = (e_1 \rfloor e_1) \wedge e_2 + (-e_1) \wedge (e_1 \rfloor e_2) = e_2$$

(see Sect. 1.5), since $e_1 \rfloor e_1 = e_1 \cdot e_1 = 1$, $e_1 \rfloor e_2 = e_1 \cdot e_2 = 0$. From this simple example, we can infer that the left contraction $A \rfloor B$ gives a blade contained in the blade B and orthogonal to the blade A. We will see that this conclusion is justified and that it can help us to understand the previous relations. For general multivectors, we can use the linearity property to get

$$M \rfloor N = \sum_{r,s} \langle \langle M \rangle_r \langle N \rangle_s \rangle_{s-r}. \quad \text{(LC 5)}$$

We can also define the right contraction

$$M \lfloor N = \sum_{r,s} \langle \langle M \rangle_r \langle N \rangle_s \rangle_{r-s}.$$

However, for blades we have

$$B_r \lfloor A_s = \left(A_s{}^\dagger \rfloor B_r{}^\dagger \right)^\dagger = (-1)^{s(r+1)} A_s \rfloor B_r$$

(see [23]), and the right contraction is not essential. We can apply (LC 5) to the case in which the first factor is a vector, for example $e_1 \rfloor M = \sum_{1,s} \langle e_1 \langle M \rangle_s \rangle_{s-1}$, where for real numbers (grade zero) we have a negative grade, $0 - 1$, which means that real numbers do not contribute anything. Note that for real numbers, we can write formally

$$e_1 \rfloor \alpha = \left(e_1 \alpha - \overleftarrow{\alpha} e_1 \right) / 2 = 0,$$

which we can try for the other grades. For grade 1 (say orthonormal unit vectors), we have

$$e_1 \rfloor e_i = \left(e_1 e_i - \overleftarrow{e_i} e_1 \right) / 2 = (e_1 e_i + e_i e_1)/2 = \delta_{1i}.$$

Of course, this result is the same as $\langle e_1 e_i \rangle$. For grade 2 (bivectors), we have

$$e_1 \rfloor B = \left(e_1 B - \overleftarrow{B} e_1 \right) / 2 = (e_1 B - B e_1)/2.$$

For $B = e_1 e_i$ we have $e_1 e_1 e_i - e_1 e_i e_1 = 2 e_i$, while for B perpendicular to e_1 we have $e_1 B - e_1 B = 0$. Again, this result is the same as $\langle e_1 B \rangle_1$. Finally, for the unit pseudoscalar, we have

$$e_1 \rfloor j = \left(e_1 j - \overleftarrow{j} e_1\right)/2 = (e_1 j + j e_1)/2 = e_1 j = e_2 e_3 = \langle e_1 j \rangle_2.$$

Due to the linearity, for the vector x and the multivector M we can write generally

$$x \rfloor M = \left(xM - \overleftarrow{M} x\right)/2,$$

where $\overleftarrow{M}$ stands for the grade involution of the multivector M. The reader can check that we can write the similar relation

$$x \wedge M = \left(xM + \overleftarrow{M} x\right)/2,$$

from which it follows that

$$xM = x \rfloor M + x \wedge M.$$

Note that for the unit pseudoscalar in $Cl3$ we have $xj = x \rfloor j + x \wedge j = x \rfloor j$. In the relation $(A \wedge B) \rfloor C = A \rfloor (B \rfloor C)$, we have the subspace $A \wedge B$ left contracted with C. The relation means that we can find $B \rfloor C$ first, which is the subspace in C orthogonal to the subspace B. Then from that subspace we "contract" the subspace A. Simply stated, we can contract subspaces by parts. For example, $(e_1 \wedge e_2) \rfloor j = e_1 \rfloor (e_2 \rfloor j) = e_1 \rfloor (e_2 j) = e_1 \rfloor (-e_1 e_3) = -e_3$, which means that the left contraction gives the subspace in j orthogonal to $e_1 \wedge e_2$, i.e., $-e_3$.

1.11.1 Left Contractions and Permutations

For the left contraction we have (a_i are vectors)

$$a_1 \rfloor (a_2 \wedge a_3) = (a_1 \cdot a_2) a_3 - (a_1 \cdot a_3) a_2,$$
$$a_1 \rfloor (a_2 \wedge a_3 \wedge a_4) = (a_1 \cdot a_2)(a_3 \wedge a_4) - (a_1 \cdot a_3)(a_2 \wedge a_4) + (a_1 \cdot a_4)(a_2 \wedge a_3),$$

which means that there is a *pattern in indices*. Specifically, note the permutations $(1, 2, 3)$ and $(1, 3, 2)$ in the first relation, with signatures (signs of *permutations*) 1 and -1. From the second relation, we see the permutations $(1, 2, 3, 4)$, $(1, 3, 2, 4)$, and $(1, 4, 2, 3)$ with signatures 1, -1, and 1. This is not a coincidence; playing with indices opens a door to the wonderful land of geometric algebra, seen in different colors. Note that the indices here are not from coordinates.

1.11.2 Left Contraction and *k*-Vectors

Consider now the pseudoscalar I from the Clifford basis in Cln. For a unit vector we have

$$Ie_i = (-1)^{n-1} e_i I,$$

where the exponent is due to the fact that e_i anticommutes with $n-1$ basis vectors from the pseudoscalar I. Consequently, for vectors we can write

$$Ia = (-1)^{n-1} aI,$$

then similarly for any single graded element,

$$IA_r = (-1)^{(n-1)r} A_r I.$$

For example, in 4D we have

$$Ie_3e_4 = (-1)^{(4-1)2} e_3e_4I = e_3e_4I.$$

An element A_kI has grade $n-k$. Therefore, we can write

$$2a\rfloor(A_kI) = aA_kI - (-1)^{n-k}A_kIa = aA_kI - (-1)^{n-k}(-1)^{n-1}A_kaI = \left(aA_k + (-1)^k A_k a\right)I \Rightarrow$$
$$a\rfloor(A_kI) = (a \wedge A_k)I = a \wedge A_kI.$$

Due to linearity, we can apply this relation to any multivector. For example, in $Cl3$, we have

$$e_1\rfloor(e_2 j) = (e_1 \wedge e_2)j = -e_3.$$

For the outer product we have

$$2a \wedge (A_kI) = aA_kI + (-1)^{n-k}A_kIa = aA_kI + (-1)^{n-k}(-1)^{n-1}A_kaI = \left(aA_k - (-1)^k A_k a\right)I \Rightarrow$$
$$a \wedge (A_kI) = (a\rfloor A_k)I = a\rfloor A_kI.$$

We can use the inner product instead of the left contraction, as long as we do not multiply real numbers. From these relations, it follows that (for any multivector)

$$A\rfloor B = \left(A \wedge \left(BI^{-1}\right)\right)I,$$

which in $Cl3$ takes the form $A\rfloor B = -(A \wedge (Bj))j$. For example, (Fig. 1.32)

$$e_1\rfloor e_1e_3 = (e_1 \wedge (-e_1e_3j))j = (e_1 \wedge (-e_2))j = e_3,$$

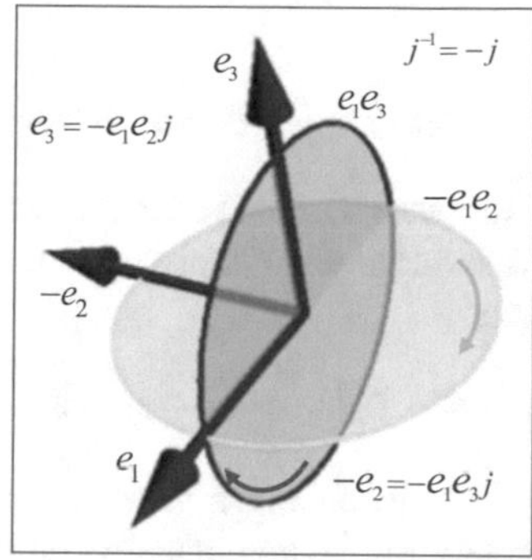

Fig. 1.32 An example of the left contraction

which we can check:

$$e_1\rfloor(e_1e_3)=\frac{e_1e_1e_3-e_1e_3e_1}{2}=\frac{e_3+e_3}{2}=e_3.$$

1.11.3 Projectors

We often need a projection of one subspace onto another. For example, we could find the projection of the vector x onto the unit vector n:

$$\mathrm{P}_n(x)=(x\cdot n)n=x\cdot nn.$$

Using the left contraction, we can write this as

$$\mathrm{P}_n(x)=\left(x\rfloor n^{-1}\right)n,$$

where the inverse is for the case of the general vector instead of the unit vector (the magnitudes just cancel). If we use the left contraction in nondegenerate algebras (without basis unit vectors that square to zero), we can define a projector as

$$\mathrm{P}_B(A)\equiv\left(A\rfloor B^{-1}\right)B.$$

For example, for $B=e_1e_2$ and $A=e_1+e_2+e_3$, we have $B^{-1}=e_2e_1$, giving

$$\begin{aligned}
e_1\rfloor B^{-1}&=(e_1e_2e_1-e_2e_1e_1)/2=-e_2,\\
e_2\rfloor B^{-1}&=(e_2e_2e_1-e_2e_1e_2)/2=e_1,\\
e_3\rfloor B^{-1}&=(e_3e_2e_1-e_2e_1e_3)/2=0,
\end{aligned}$$

and finally

$$\mathrm{P}_B(A) = (e_1 - e_2)e_1e_2 = e_1 + e_2.$$

Note that we could calculate directly, using (LC 5), $e_1 \rfloor B^{-1} = \langle e_1e_2e_1\rangle_1 = -e_2$, etc. We found the vector coplanar with the e_1e_2 plane. To get the orthogonal component, we just use

$$A - \mathrm{P}_B(A) = e_1 + e_2 + e_3 - (e_1 + e_2) = e_3.$$

Projections are very important; for example, we have seen their importance for rotations; the commutation properties of $A_{\|} = \mathrm{P}_B(A)$ and $A_{\perp} = A - \mathrm{P}_B(A)$ make transformations easier. Note that we can easily generalize all these relations to any dimension. (Fig. 1.33)

For vectors and bivectors, we can write (the geometric product executes last)

$$x = xB^{-1}B = x \cdot B^{-1}B + x \wedge B^{-1}B,$$

which gives the vector component contained in the bivector plane and the component orthogonal to the bivector plane. This could be useful for applying rotations to vectors (recall the commutation rules). ◆ Show that for vectors x, y and a blade B we have (see [23])

$$\mathrm{P}_B(y) \wedge \mathrm{P}_B(y) = \mathrm{P}_B(x \wedge y). \quad ◆$$

Projectors are exceptionally important in geometric algebra, since they represent relations between subspaces. With the commutation properties, calculations in geometric algebra turn out to be fast and elegant.

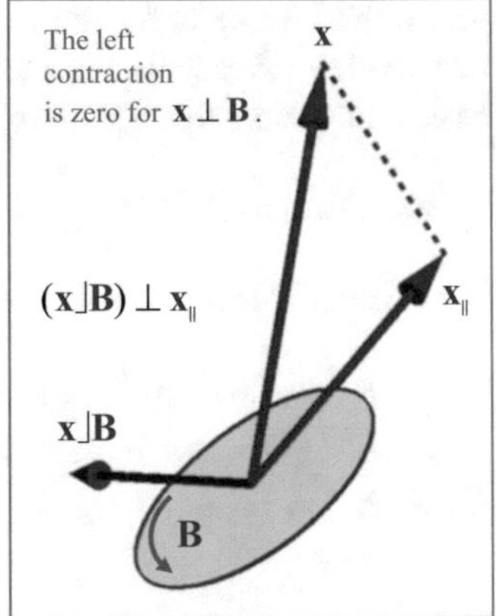

Fig. 1.33 A visualization of the left contraction

1.11.4 A Proof of a Special Formula

We can use (LC 3) to prove the relation for a vector (x) and a k-blade

$$x\rfloor(a_1 \wedge a_2 \wedge \cdots \wedge a_k) = \sum_{i=1}^{k} (-1)^{i-1}(x\rfloor a_i)a_1 \wedge a_2 \wedge \cdots \wedge \breve{a}_i \wedge \cdots \wedge a_k \quad \text{(LC 6)}$$

where $\breve{a}_i$ means that the vector a_i is missing from the outer product. ◆ Prove (LC 6), starting with $a_1 \wedge a_2 \wedge \cdots \wedge a_k = a_1 \wedge A$ (see [23]). ◆ We see that if we have $x\rfloor a_i = 0$ for all i, then $x\rfloor(a_1 \wedge a_2 \wedge \cdots \wedge a_k) = 0$. The special case for blades of grade 2 is

$$x\rfloor B = x\rfloor(a_1 \wedge a_2) = (x\rfloor a_1)a_2 - (x\rfloor a_2)a_1 = (x \cdot a_1)a_2 - (x \cdot a_2)a_1.$$

◆ Consider two bivectors from the Clifford basis in *Cln*. Show that
(a) $(e_i \wedge e_j)\rfloor(e_i \wedge e_j) = (e_i \wedge e_j)^2$,
(b) $(e_i \wedge e_j)\rfloor(e_j \wedge e_k) = 0, \quad k \neq i, j$,
(c) $(e_i \wedge e_j)\rfloor(e_k \wedge e_l) = 0$, for all different indices. (Hint: Relation (LC 5) is the best choice; however, other relations could be instructive.) ◆

From (LC 4), we have

$$(e_i \wedge e_j) \rfloor (e_k \wedge e_l) = e_i \rfloor [e_j \rfloor (e_k \wedge e_l)] = e_i \rfloor [e_j \cdot e_k e_l - e_k e_j \cdot e_l] = \delta_{jk}\delta_{il} - \delta_{jl}\delta_{ik}.$$

Let us show that the outcome of $A\rfloor B$ is perpendicular to the subspace A. Here, when we say "subspace A," we mean the blade A that defines the subspace by the relation $x \wedge A = 0$. Take the basis vector a from A; then $a \wedge A = 0$. Now using (LC 4), we have

$$a\rfloor(A\rfloor B) = (a \wedge A)\rfloor B = 0\rfloor B = 0$$

so a is orthogonal to $A\rfloor B$.

The left contraction can help us to define angles between subspaces (see [30]). Due to generality, the clear geometric interpretation, and benefits for use on computers (there are no exceptions, and therefore we do not need *if* loops), we should use the left contraction instead of the inner product (see [24]).

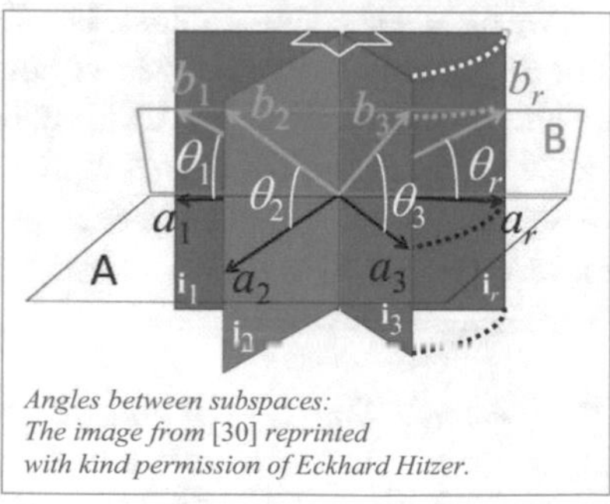

Angles between subspaces: The image from [30] *reprinted with kind permission of Eckhard Hitzer.*

1.12 Commutators and Orthogonal Transformations

We define the *commutator* as a new kind of product of multivectors (instead of the common $\times$, we use the character $\otimes$, to avoid possible confusion with the cross product):

$$[A,B]=A\otimes B\equiv(AB-BA)/2.$$

This product is not associative, i.e., $(A\otimes B)\otimes C\neq A\otimes(B\otimes C)$, but we have the *Jacobi identity*

$$(A\otimes B)\otimes C+(C\otimes A)\otimes B+(B\otimes C)\otimes A=0.$$

1.12.1 Commutators with Bivectors

We are especially interested in commutators with bivectors, since they appear in physics frequently. There is a general formula (B is a bivector, not necessarily a blade, X is a multivector)

$$BX=B\rfloor X+B\wedge X+B\otimes X$$

For a general proof, see [25]. Here we will just illustrate this, taking advantage of linearity and using the elements from the Clifford basis. If X is a real number, $X=\alpha$, we have $B\rfloor\alpha=0$, $B\wedge\alpha=B\alpha$, and $B\otimes\alpha=B\alpha-\alpha B=0$. If X is a blade, we can write generally

$$BX-B\otimes X=\frac{BX+XB}{2}=\{B,X\}=B\rfloor X+B\wedge X=\langle BX\rangle_{s-2}+\langle BX\rangle_{s+2},\quad s=\text{grade}(X).$$

For $X=x$ (a vector), we have $x=x_{\|}+x_{\perp}$ (with the respect to B), $Bx_{\|}=-x_{\|}B$, $Bx_{\perp}=x_{\perp}B$, $B\rfloor x=0$, and consequently,

$$\frac{Bx+xB}{2}=Bx_{\perp}=B\wedge x_{\perp},$$

which is true. For bivectors, we have several possibilities. For $X=B$, we have $B\wedge B=0$, and therefore $B^2=\langle B^2\rangle$. For $B=e_ie_j$ and $X=e_je_k$, $i\neq k$, we have $BX=-XB$, $B\wedge X=0$, and $\langle BX\rangle=\langle e_ie_k\rangle=0$. Other bivectors from the Clifford basis commute; their left contraction is zero, giving the valid relation $BX=B\wedge X$. For $X=X_k$, where X_k is the element of the Clifford basis of grade k, we can proceed as earlier. If $B=e_ie_j$ and $X_k=e_jA_{k-1}$ (up to a sign), we have $B\wedge X_k=0$, $BX_k=-X_kB$, and $\langle BX_k\rangle_{k-2}=0$, since grade$(BX_k)=k$. For $X_k=e_ie_jA_{k-2}$, we have

$$B\wedge X_k=0,\ BX_k=X_kB=B^2A_{k-2},\ \langle BX_k\rangle_{k-2}=\left\langle B^2A_{k-2}\right\rangle_{k-2}=B^2\langle A_{k-2}\rangle_{k-2}=B^2A_{k-2}.$$

If the grade of BX_k is $k+2$, we have $BX_k=X_kB$, $\langle BX_k\rangle_{k-2}=0$, and $BX_k=B\wedge X_k$. Note that in all our calculations, the commutation properties play a central role.

Nonzero commutators with a bivector maintain the grade of a multivector:

$$\mathrm{grade}(B) = 2 \Rightarrow \mathrm{grade}(X \otimes B) = \mathrm{grade}(X).$$

Again, we will just analyze the elements of the Clifford basis. If $BX_k = X_kB$, the commutator is zero. In *Cl*3, for example, we have $(e_1e_2) \otimes e_2 = e_1$ (due to anti-commutativity). We can use the previous results to prove our statement. Let's take, for example, $B = e_ie_j$, $X_k = e_jA_{k-1}$ (why?), $BX_k = -X_kB$ (why?), and $B \otimes X_k = BX_k = e_ie_je_jA_{k-1} = e_j^2 e_iA_{k-1}$, the element of grade *k*. ◆ Apply the commutator with $B = e_ie_j$ to each element of the Clifford basis in *Cl*3. ◆ We often use transformations of the type

$$e^{-B/2}Xe^{B/2}.$$

Consequently, if we take a small bivector of the form $B = \varepsilon\hat{B}, \hat{B}^2 = -1$, we can write

$$\exp(\pm B/2) = \exp\left(\pm\hat{B}\varepsilon/2\right) = \cos\left(\varepsilon/2\right) \pm \hat{B}\sin\left(\varepsilon/2\right) \approx 1 \pm \hat{B}\varepsilon/2;$$

then, ignoring the terms with ε^2, we get (check)

$$\exp\left(-\varepsilon\hat{B}/2\right)X\exp\left(\varepsilon\hat{B}/2\right) \approx X + \varepsilon X \otimes \hat{B}.$$

All small transformations reduce to the addition of a small commutator made by the bivector. Preservation of grades is important here, for we want to have a geometric object of the same type after the transformation. The last transformation is an orthogonal transformation that slightly changes the initial multivector. Essentially, we are trying to find the orthogonal transformation connected to the identity transformation, which means that we can implement it in small steps, starting with the identity transformation. Reflections do not meet this requirement: we cannot perform "a little reflection." We call such small transformations *perturbations*. In fact, we would like to implement perturbations of elements of a geometric algebra by rotors; they are powerful, connected to the identity transformation, and easy to generalize to any dimension. Note that orthogonal transformations do not permit one to just add a small vector δx to the vector x, since orthogonal transformations must preserve a vector's length. Therefore, we must have $x \cdot \delta x = 0$. Generally, such an element (δx) of a geometric algebra has the form $\delta x = x \rfloor \delta B$, where δB is a small bivector. We can prove this using (LC 4):

$$x \cdot (x \rfloor \delta B) = x \rfloor (x \rfloor \delta B) = (x \wedge x) \rfloor \delta B = 0.$$

Now it follows that

$$\delta x = x \rfloor \delta B = (x\delta B - \delta Bx)/2 = x \otimes \delta B,$$

and we have the desired element in the form of the commutator.

It may seem that the restriction on rotations is too strict, since it looks as though we cannot perform a simple translation of a vector. However, this just means that we need to find a way **to describe translations by rotations**, which is possible to achieve in geometric algebra (see [23]).

1.13 Functions of a Complex Argument

For the vector $\mathbf{r} = xe_1 + ye_2$ in $\mathfrak{R}^2$, we can write

$$\mathbf{r} = e_1(x + ye_1e_2) = e_1(x + yI), \quad I = e_1e_2,$$

meaning that we get a complex number $x + yI$, however, with the noncommutative "imaginary unit." The first thing people complain about is, *Yes, but your imaginary unit is not commutative, and quantum mechanics cannot be formulated without the imaginary unit...* We see immediately that such a "critic" has given an opinion on something he knows almost nothing about. First, quantum mechanics works nicely (and even better) with real numbers, without the imaginary unit. However, one should learn geometric algebra, and then learn the formulation of quantum mechanics in the language of geometric algebra. We can do without the ordinary imaginary unit; in fact, many relations obtain a clear geometric meaning in the language of geometric algebra, providing new insights into the theory. Second, noncommutativity of our bivector $I = e_1e_2$ actually becomes an advantage; it enriches the theory of complex numbers and, as we continue to repeat until even songbirds get bored, gives it a clear geometric meaning. For our complex number $z = e_1\mathbf{r}$, we have (due to anticommutativity) $z^* = \mathbf{r}e_1$, and consequently,

$$\begin{aligned} zz^* &= e_1\mathbf{r}\mathbf{r}e_1 = r^2e_1e_1 = r^2 = x^2 + y^2, \\ z + z^* &= e_1\mathbf{r} + \mathbf{r}e_1 = 2e_1 \cdot \mathbf{r} = 2x, \\ z - z^* &= e_1\mathbf{r} - \mathbf{r}e_1 = 2e_1 \wedge \mathbf{r} = 2yI, \end{aligned}$$

etc. We see that operations on complex numbers are, without any problem, confined to the operations in geometric algebra.

1.13.1 The Cauchy–Riemann Equations

We can define the derivative operator in $\mathfrak{R}^2$ as

$$\nabla \mathrm{f} \equiv e_1 \frac{\partial \mathrm{f}}{\partial x} + e_2 \frac{\partial \mathrm{f}}{\partial y}$$

and we introduce a complex field $\psi(x, y) = u(x, y) + Iv(x, y)$. A simple calculation shows (check) that the derivative of the field is

$$\nabla\psi = \nabla u + \nabla(Iv) = \nabla u - I\nabla v = e_1\left(\frac{\partial u}{\partial x} - \frac{\partial v}{\partial y}\right) + e_2\left(\frac{\partial v}{\partial x} + \frac{\partial u}{\partial y}\right).$$

Therefore, if we want the derivative to be zero identically (*analyticity*), the *Cauchy–Riemann* equations immediately follow. Note how the anticommutativity of unit vectors gives the correct signs. The analyticity condition in geometric algebra has a simple form, $\nabla\psi = 0$; we can immediately generalize it to higher dimensions. Moreover, this is just the right moment to stop and think. Let advocates of the traditional approach do all that we have done just using the commutative imaginary unit. Actually, it is amazing how this good old imaginary unit has accomplished so much, given the modest possibilities! Nevertheless, it is time to rest a little, let bivectors, pseudoscalars, etc., do the job. We should note, to dispel any possibility of confusion, that the choice of the plane e_1e_2 was insignificant here. We can take a bivector like $(e_1 + e_2)(e_1 - e_3)$, normalize it, and get a new "imaginary unit," but in the new plane. We can do that in $\mathfrak{R}^4$ as well, and take, for example, $I = e_3e_4$, and all formulas will be valid. The plane e_1e_2 is just one of infinitely many of them; however, the geometric relationships in each of them are the same. We can solve the problem in the plane e_1e_2 and then rotate all results to the plane we want; we have powerful rotors in geometric algebra. This literally means that we do not have to be experts in matrix calculations; an advanced high-school student can perform calculations. We can rotate any object, not just vectors. Linear algebra is the mathematics of vectors and operators, while geometric algebra is the mathematics of subspaces and operations on them. Anyone who deals with mathematics should understand how important this could be.

1.13.2 Taylor Expansion and Analytic Functions

We will show now how to obtain solutions of the equation $\nabla\psi = 0$ using series in z (see [21]). First, note an obvious relation for vectors,

$$abc + bac = (ab + ba)c = 2a \cdot bc,$$

where the inner product has priority. The operator ∇ is acting as a vector (expressions like $\mathbf{r}\nabla$ are possible, but then we usually write $\dot{\mathbf{r}}\dot{\nabla}$, which does not mean the time derivative, but indicates the element on which the derivative operator acts, giving the desired order in products of the unit vectors). Therefore, taking advantage of the previous relation (a very useful calculation)

$$\nabla z = \nabla(e_1\mathbf{r}) = 2e_1 \cdot \nabla\mathbf{r} - e_1\nabla\mathbf{r},$$

and then using (x and y are independent coordinates)

$$\frac{\partial(xe_1)}{\partial x} = e_1\frac{\partial x}{\partial x} = e_1, \quad \frac{\partial(ye_2)}{\partial x} = e_2\frac{\partial y}{\partial x} = 0,$$

it follows that

$$\nabla\mathbf{r} = e_1\frac{\partial(xe_1 + ye_2)}{\partial x} + e_2\frac{\partial(xe_1 + ye_2)}{\partial y} = 2,$$

$$2e_1 \cdot \left(e_1\frac{\partial}{\partial x} + e_2\frac{\partial}{\partial y}\right) = 2\frac{\partial}{\partial x},$$

$$2e_1 \cdot \nabla\mathbf{r} = 2\frac{\partial}{\partial x}(xe_1 + ye_2) = 2e_1,$$

and finally

$$\nabla z = 2e_1 - 2e_1 = 0.$$

Now we have

$$\nabla (z - z_0)^n = n(z - z_0)^{n-1} \nabla (e_1 \mathbf{r} - z_0) = 0,$$

and as a result, the Taylor expansion about z_0 automatically gives an analytic function—again, in any plane, in any dimension. It is not only that geometric algebra contains all the theory of functions of complex variables (including integral theorems, as a special case of the *fundamental theorem of integral calculus in geometric algebra*; see [25]), but it also extends and generalizes it to any dimension. Is that not a miracle? Recall that we were just wondering how to multiply vectors. If somebody still has a desire to pronounce the sentence *Yes, but...*, she (or he) could go back to the beginning of the text and see how all this began. The time of geometric algebra is yet to come, hopefully. The Dark Ages of matrices and coordinates will disappear and will be substituted by the time of synergy of algebra and intuitively clear geometry. Students will learn much faster and with greater pleasure. Moreover, when we succeed in teaching a completely new type of computer to "think" in this magical language (imagine a computer that knows how to perform operations on subspaces, without coordinates), children will be able to play with oriented geometric objects as they now play car racing games or other computer games. We will learn properties of triangles, circles, spheres, and other shapes through play, on computers, interactively. The language of geometric algebra is so powerful that it can automate even the process of proving theorems or the writing of code (there is still a good deal of work to do; however, the possibilities are there). We have reason to think that geometric algebra is not just "another formalism"; in fact, it offers the possibility to reexamine the very concept of a number.

1.14 Spinors

Let us find even elements of geometric algebra that in the "sandwich" form do not change the grade of vectors (i.e., a vector transforms to a vector; see also [13] and [25]) and that rotate and dilate them. In geometric algebra, such elements are usually called *spinors*. Such a definition of spinors is in accordance with the definitions common in physics. We are looking for multivectors ψ with the property ($\mathbf{v}$ is a vector) $\psi \mathbf{v} \psi^\dagger = \rho R \mathbf{v} R^\dagger, \quad \rho \in \mathbb{R}$, which is precisely a rotation of the vector $\mathbf{v}$ with a dilatation. If we define $S \equiv R^\dagger \psi$, using $S^\dagger = \psi^\dagger R$ and $RR^\dagger = 1$, then the previous relation becomes

$$S \mathbf{v} S^\dagger = \rho \mathbf{v},$$

and we will find the general form of the even element S. We see that the element S induces a pure dilation of the vector $\mathbf{v}$. Therefore, it must commute or anticommute with $\mathbf{v}$. Real numbers and pseudoscalars of odd dimensions commute, while pseudoscalars of even dimensions anticommute with all vectors (see E10). Other grades do not possess such a general property. Consequently, the element S is, generally, either a real scalar or a pseudoscalar or combination of both: $S = \alpha + \beta I$, $\alpha, \beta \in \mathbb{R}$, which is an interesting and important result. Using the definition of S, it follows that

$$S\mathbf{v}S^{\dagger} = \alpha^2\mathbf{v} + \alpha\beta\left(I\mathbf{v} + \mathbf{v}I^{\dagger}\right) + \beta^2 I\mathbf{v}I^{\dagger} = \rho\mathbf{v}.$$

Note that the pseudoscalar part must vanish for odd dimensions if we want S to be even.

In $Cl3$ (the signature $p = 3$, $q = 0$), the pseudoscalar $I = j$ commutes with all elements of the algebra, and its reverse is $I^{\dagger} = -j$. Consequently, the middle term disappears, and we have

$$\alpha^2 + \beta^2 = \rho \Rightarrow \psi = (\alpha + j\beta)R.$$

Now it is easy to check that

$$\psi\mathbf{v}\psi^{\dagger} = (\alpha + j\beta)R\mathbf{v}(\alpha - j\beta)R^{\dagger} = (\alpha + j\beta)(\alpha - j\beta)R\mathbf{v}R^{\dagger} = \left(\alpha^2 + \beta^2\right)R\mathbf{v}R^{\dagger}$$
$$= \rho R\mathbf{v}R^{\dagger}.$$

Note that the element $(\alpha + j\beta)R$ has all grades. However, we usually choose $\beta = 0$ in order to stay in the even part of the algebra. We can interpret this result as

$$(\alpha + j\beta)R = R\rho\ \exp(\varphi j),$$

which means that the element $R\rho$ is multiplied by a commutative complex phase. For $\varphi = 0$, we have an element from the even part of the algebra. The complex phase disappears in the "sandwich" form, due to $\exp(\varphi j)^{\dagger} = \exp(-\varphi j)$. In general, note that

$$\mathbf{v}I^{\dagger} = (-1)^{n-1}I^{\dagger}\mathbf{v}, \quad \mathbf{v}I^{\dagger} = (-1)^{(n-1)(n+2)/2}I\mathbf{v}, \quad II^{\dagger} = (-1)^{q}$$

(prove it, at least for signatures (3, 0) and (1, 3)), and we can find solutions (find them) dependent on the parity of the number $(n-1)(n+2)/2$.

Spinors in geometric algebra, as elsewhere, can be defined by (left) *ideals of the algebra* (see Appendix 5.3 and [7]), usually without restriction to the even part of an algebra.

1.15 A Bit of "Ordinary" Physics

In geometric algebra, we can solve kinematic problems quite generally, using simple and intuitively clear calculations. Consider the problem of uniformly accelerated motion in classical mechanics (constant acceleration; see [27]). The problem is easy to reduce to the relations (Fig. 1.34)

$$\mathbf{v} = \mathbf{v}_0 + \mathbf{a}t, \quad \mathbf{v} + \mathbf{v}_0 = 2\mathbf{r}/t,$$

where the second relation defines the average speed vector $\bar{\mathbf{v}} = \mathbf{r}/t$. Thus, we have

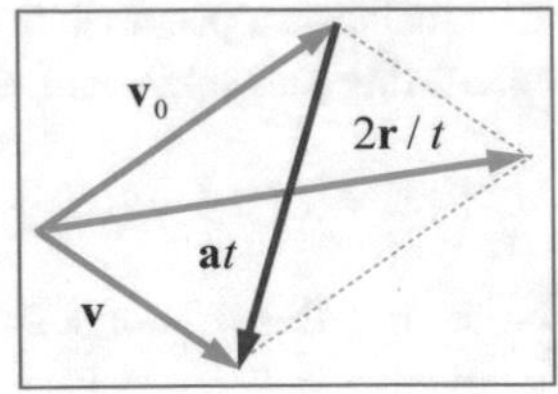

Fig. 1.34 A uniformly accelerated motion

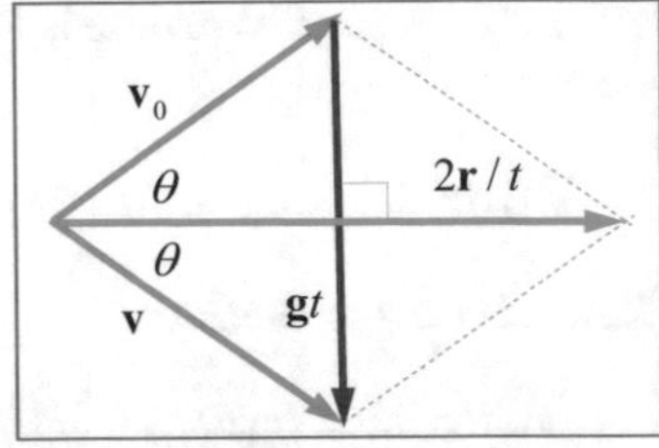

Fig. 1.35 The projectile motion problem

$$(\mathbf{v}+\mathbf{v}_0)(\mathbf{v}-\mathbf{v}_0) = 2\mathbf{ra} \Rightarrow v^2 - v_0^2 + \mathbf{v}_0\mathbf{v} - \mathbf{v}\mathbf{v}_0 = v^2 - v_0^2 + 2\mathbf{v}_0 \wedge \mathbf{v}$$
$$= 2(\mathbf{r}\cdot\mathbf{a} + \mathbf{r}\wedge\mathbf{a}).$$

By comparison of the scalar and the bivector parts, we get

$$v^2 - v_0^2 = 2\mathbf{r}\cdot\mathbf{a}, \quad \mathbf{v}_0 \wedge \mathbf{v} = \mathbf{r}\wedge\mathbf{a},$$

that is, the law of conservation of energy and the surface of the parallelogram theorem. For the projectile motion problem ($\mathbf{a}=\mathbf{g}$), we have (Fig. 1.35)

$$\mathbf{r}\cdot\mathbf{g} = 0 \Rightarrow v^2 = v_0^2 \Rightarrow |\mathbf{v}_0 \wedge \mathbf{v}| = v_0^2 \sin(2\theta) = |\mathbf{r}\wedge\mathbf{g}| = rg \Rightarrow r = \frac{v_0^2}{g}\sin(2\theta).$$

This is the well-known relation for the range. Note how the properties of the geometric product lead to simple and general calculations, without coordinates.

1.15.1 The Kepler Problem

As another example, we will consider the *Kepler problem*. If we observe *binaries* in an inertial reference frame (the *two-body problem*, Fig. 1.36), we can use their radius vectors and masses to define (G is the universal gravitational constant)

$$\mathbf{r} \equiv \mathbf{r}_2 - \mathbf{r}_1, \quad \mu \equiv G(m_1+m_2).$$

Newton's second law gives

$$\ddot{\mathbf{r}} = -\mu|\mathbf{r}|^{-3}\mathbf{r},$$

whence defining the bivector $L = \mathbf{r}\wedge\dot{\mathbf{r}}$, it follows that $\dot{L} = \dot{\mathbf{r}}\wedge\dot{\mathbf{r}} + \mathbf{r}\wedge\ddot{\mathbf{r}} = 0$, which means that $L = \text{const}$. We will not develop this further; there is a more effective way to

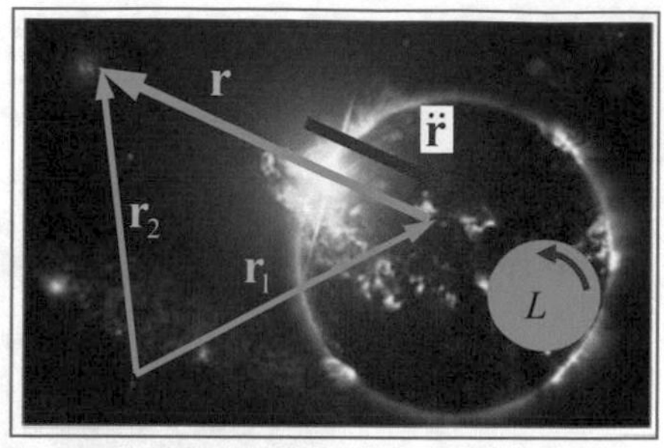

Fig. 1.36 The Kepler problem

treat this problem using *eigenspinors* (see Sect. 2.8). Let us just touch upon the fact that here we need just a few more steps to get the *Laplace–Runge–Lenz vector*. As a rule, immediately after posing the problem, after a few lines, we obtain nontrivial conclusions that textbooks usually place as a difficult part at the end. The examples here are to show how to obtain solutions without writing coordinates. Unfortunately, research shows (see [26]) that many physics students see vectors mainly as series of numbers (coordinates), which blurs the connection between linear algebra and geometry, which is a sad reflection of current educational systems, regardless of their location on the planet. With the geometric product, algebra and geometry go hand in hand. Instead of treating vectors as key elements of the algebra, we have a whole range of objects that are not vectors, which have a clear geometric meaning. We calculate with subspaces, in any dimension. Something like that is impossible to achieve by just manipulating coordinates. We emphasize this: it is impossible! The Russian physicist Lev Landau, famous for his mathematical skills, ended up in Stalin's Lubyanka prison. On his release, he remarked that his captivity had been welcome, since he had learned to do tensor calculus in his head. Physicists of the future will be even more skilled than Landau; they will use linear transformations in geometric algebra instead of tensor calculus. They will calculate faster, regardless of the dimension of the space, without arid coordinates, and with a clear geometric interpretation at every step. Landau was also notorious for his method of accepting new students. He would say to a young candidate, "Here, solve this integral." Many failed. In geometric algebra, there is a theorem (*the fundamental theorem of calculus*) that combines all known integral theorems we have in physics, including complex area. Landau would be surprised! He was a typical representative of the mathematics of the twentieth century, although the new mathematics already existed in his time. It existed, but was sadly almost completely neglected and forgotten. Part of the price paid (and we still pay it) is recurrent rediscovery of what was neglected and forgotten. Pauli discovered its matrices, and we have continued to use matrices. There is a frequent complaint that geometric algebra is noncommutative and that this discourages people. What about matrices? Not only are they noncommutative, they are unintuitive! Then Dirac discovered his matrices, perfect for geometric algebra. Again, we continued with matrices. Moreover, many authors on various occasions have rediscovered spinors, while giving them different names. Then we decided to construct fast spacecraft equipped with computers and found that we have problems with matrices. Then we started to use quaternions and improved things to some extent. We can find a number of other indications; after all, it is obvious that numerous problems simply disappear when we introduce the geometric product instead of the products of Gibbs. Despite everything, one of the distinguished authors in the field of geometric algebra, Garret Sobczyk, wrote in an e-mail:

> I am surprised that after 45 years working in this area, it is still not generally recognized in the scientific community. However, I think that ... it deserves general recognition ... Too bad Clifford died so young, or maybe things would be different now.

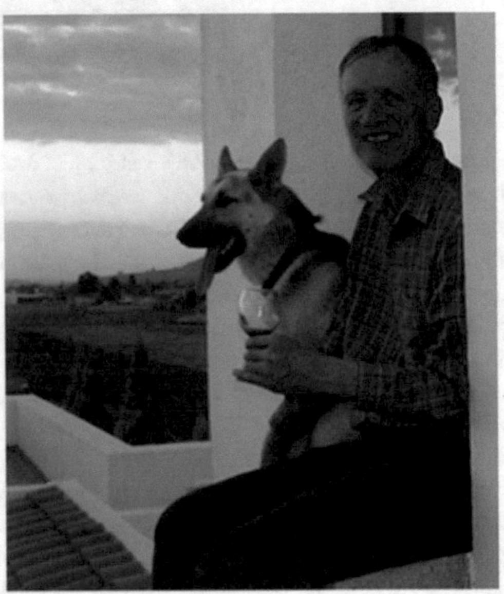

Garret Sobczyk

1.16 Words and Sentences

Let us look, just for fun, at how "words" in geometric algebra can have a geometric content, for example, the "word" **abba** in $Cl3$ (a real scalar). From $\mathbf{ab} = S + A$ (the symmetric and antisymmetric parts), we obtain

$$\begin{aligned}\mathbf{abba} &= a^2b^2 = (S+A)(S-A) = S^2 - A^2 = S^2 + |A|^2 \\ &= (\mathbf{a}\cdot\mathbf{b})^2 + |\mathbf{a}\wedge\mathbf{b}|^2 = a^2b^2\left(\cos^2\theta + \sin^2\theta\right),\end{aligned}$$

and we have a well-known trigonometric identity. This is, of course, just a game; however, it is important to develop intuition about the geometric content written in expressions in geometric algebra. Due to the properties of the geometric product, the structure of expressions is quickly revealed, for example relations between subspaces, being an element of a subspace, orthogonality, being parallel. We have seen that we can represent vectors by the Pauli matrices in 3D (see also Sect. 4.1.1). Somebody could give us the word **abba** expressed in the language of the Pauli matrices, and we could resolve it into the symmetric and antisymmetric parts (as is customary). But then try to derive the sine and cosine of the angle and the basic trigonometric identity. If you succeed (it is possible), how would you interpret that angle? Moreover, how would you even come up with the idea to look for an angle, just by looking at matrices? Again, it is possible, but one has to work hard (or choose to learn GA).

1.16.1 Geometric Content of Expressions

With geometric algebra, interpretation is natural and straightforward. This is the main idea: **the language of coordinates hides an important geometric content**. True, physicists know that the Pauli matrices have something to do with the orientation of the spin, but generally, the problem of the geometric interpretation remains. As another example, take the unit vectors $m = (e_1 + e_2)/\sqrt{2}$ and $n = (e_2 + e_3)/\sqrt{2}$ in $Cl3$. It is not difficult to imagine or draw them (Fig. 1.37); there is the plane spanned by the vectors and the bivector $B = m \wedge n$ in it (the bivector defines the plane). Imagine again that we are using the Pauli matrices, however, as before, without the awareness that they represent vectors in 3D (we cannot even know this if we do not accept the geometric product of vectors). Someone could investigate linear combinations of the Pauli matrices, come up with the idea to look at the antisymmetric part of products of matrices, something like $(\hat{\sigma}_m\hat{\sigma}_n - \hat{\sigma}_n\hat{\sigma}_m)/2$, where $\hat{\sigma}_m = (\hat{\sigma}_1 + \hat{\sigma}_2)/\sqrt{2}$ and $\hat{\sigma}_n = (\hat{\sigma}_2 + \hat{\sigma}_3)/\sqrt{2}$. We should now calculate this, in order to compare the needed calculation with matrices and simple calculation of the outer product (in fact, there is no need to calculate the outer product; we have the geometric picture without effort). In any case, the bivector is

$$\begin{aligned} m \wedge n &= (e_1 + e_2) \wedge (e_2 + e_3)/2 \\ &= (e_1e_2 + e_1e_3 + e_2e_3)/2. \end{aligned}$$

Fortunately, computers can help here with matrices (you see the problem?); the antisymmetric part of the matrix product is

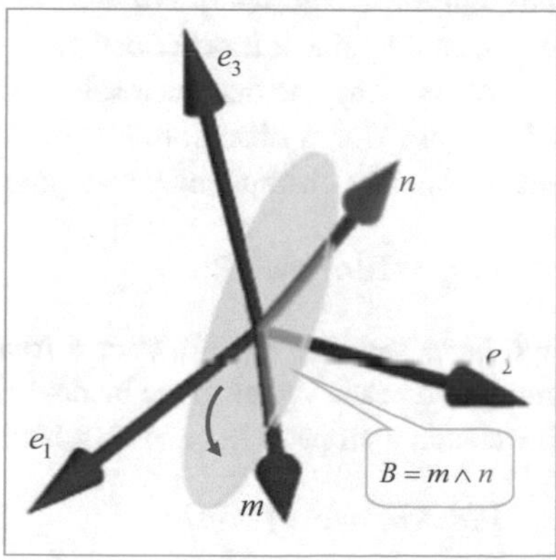

Fig. 1.37 In geometric algebra, we always have a clear geometric interpretation

$$\begin{pmatrix} i & -1+i \\ 1+i & -i \end{pmatrix}/2.$$

Now, how should we interpret this matrix as a plane, without relating to vectors in 3D or finding the angle between—what? It is easy to express multivectors from *Cl*3 via the Pauli matrices, but it takes time to express 2 × 2 complex matrices in the form of multivectors. **The language of matrices blurs the geometric content**! In quantum mechanics, with the Pauli matrices, we need the imaginary unit, and then people say that the imaginary unit is necessary to formulate theories of the subatomic world. This often leads to a philosophical debate and questions about the "real nature" of the world we live in. In the language of geometric algebra, the **imaginary unit is unnecessary:** quantum mechanics can be formulated beautifully and elegantly using real numbers only, with the apparent geometric interpretation. All philosophy of the "imaginary unit in quantum mechanics" follows from a clumsy mathematics. Together with real numbers, complex numbers and quaternions could be of interest in traditionally formulated quantum mechanics; however, they all are natural parts of *Cl*3 (see the text).

In the article [1], Baez comments: "instead of being distinct alternatives, real, complex and quaternionic quantum mechanics are three aspects of a single unified structure." Spot on! Geometric algebra was not mentioned in the cited article; nonetheless, it is clear that geometric algebra elegantly unites all three aspects.

1.17 Linear Transformations

We are often interested in transformations of elements of an algebra (e.g., vectors, bivectors...) to other elements in the same space. Among them, certainly the most interesting are linear transformations. Consider a linear transformation F (note the text format) that transforms vectors into vectors, with the property

$$\mathrm{F}(\alpha a + \beta b) = \alpha\mathrm{F}(a) + \beta\mathrm{F}(b), \quad \alpha, \beta \in \mathbb{R}.$$

Note that $\mathrm{F}(\alpha a) = \alpha\mathrm{F}(a), \quad \alpha \in \mathbb{R}$, means that a line through the origin remains a line through the origin, while $\mathrm{F}(a + b) = \mathrm{F}(a) + \mathrm{F}(b)$ means that the parallelogram rule is unaffected by linear transformations.

We can imagine that the result of a linear transformation is, for example, the rotation of a vector with a dilatation. For such a simple picture, we do not need vector components. Another example may be a pure rotation:

$$\mathrm{F}(a) = \mathrm{R}(a) \equiv RaR^{\dagger}.$$

We have seen that the effect of a rotation on a blade is the same as the action of the rotation on each vector in the blade; therefore, we require that our linear transformations have such a property in general, which means that

$$\mathrm{F}(a \wedge b) = \mathrm{F}(a) \wedge \mathrm{F}(b).$$

Acting on a vector, the linear transformation F gives back a vector, and we see that the form of the outer product is preserved. Such a transformation has a special name: *outermorphism*.

We can also define linear transformations on other blades (see [23]). Here we postulate that

$$\mathrm{F}(\alpha) = \alpha, \quad \alpha \in \mathbb{R},$$

which is justified by many reasons. For example, we want the origin to be unaffected by a linear transformation. This means that $\mathrm{F}(x \cdot y) = x \cdot y$, so it might be thought that linear transformations preserve the inner product in general, that is, that all linear transformations are orthogonal. Of course, this is not true, since the orthogonality of transformations is defined as $x \cdot y = \mathrm{F}(x) \cdot \mathrm{F}(y)$. What about the geometric product? Is it possible to write $\mathrm{F}(xy) = \mathrm{F}(x)\mathrm{F}(y)$, where F is an outermorphism? Due to $x \cdot y \in \mathbb{R}$, we have

$$\begin{aligned}&\mathrm{F}(xy) = \mathrm{F}(x \cdot y + x \wedge y) = x \cdot y + \mathrm{F}(x \wedge y) = x \cdot y + \mathrm{F}(x) \wedge \mathrm{F}(y),\\&\mathrm{F}(x)\mathrm{F}(y) = \mathrm{F}(x) \cdot \mathrm{F}(y) + \mathrm{F}(x) \wedge \mathrm{F}(y),\end{aligned}$$

which means that such a relation holds for orthogonal transformations (*outermorphisms*). The orthogonal transformations preserve the length of a vector, as well as angles between them, which is easy to show. From $x \cdot y = \mathrm{F}(x) \cdot \mathrm{F}(y)$, we have $x \cdot x = \mathrm{F}(x) \cdot \mathrm{F}(x) \Rightarrow |x| = |\mathrm{F}(x)|$, whence follows the preservation of angles (show this).

However, note that it is still possible that for $x \cdot y \neq \mathrm{F}(x) \cdot \mathrm{F}(y)$ angles (or lengths, but not both) are preserved. For example, consider a linear transformation in *Cl*2 that rotates vectors by $\pi/2$, doubling their lengths. Then we have

$$\mathrm{F}(e_1) \cdot \mathrm{F}(e_2) = 2e_2 \cdot 2(-e_1) = 0,$$

which means that the transformation preserves angles. However,

$$\mathrm{F}(e_1) \cdot \mathrm{F}(e_1) = (2e_2) \cdot (2e_2) = 4 \neq e_1 \cdot e_1 = 1,$$

which means that the transformation changes vector lengths. We also have

$$\mathrm{F}(e_1 \wedge e_2) = \mathrm{F}(e_1) \wedge \mathrm{F}(e_2) = 2e_2 \wedge 2(-e_1) = 4e_1 \wedge e_2,$$

and we can try the geometric product

$$\begin{aligned}&\mathrm{F}(e_1 e_2) = \mathrm{F}(e_1)\mathrm{F}(e_2) = 4e_2(-e_1) = 4e_1 e_2 = 4e_1 \wedge e_2,\\&\mathrm{F}(e_1 e_1) = \mathrm{F}(1) = 1 \neq \mathrm{F}(e_1)\mathrm{F}(e_1) = 4e_2 e_2 = 4.\end{aligned}$$

Our linear transformation is not an orthogonal transformation; it changes lengths of vectors. In fact, we could conclude this from $\mathrm{F}(e_1 \wedge e_2) = 4e_1 \wedge e_2$, which means that the *determinant* (see below) of the transformation is 4; orthogonal transformations have determinant 1 or -1, which means that a unit pseudoscalar ($e_1 e_2$) is not changed, up to a sign. If we imagine a unit pseudoscalar as an oriented volume defined by the unit vectors of an orthonormal basis, it is obvious that a change of lengths alone, or change of angles alone, could change the magnitude of the pseudoscalar. We have good reasons to demand the preservation of a pseudoscalar magnitude as a criterion for orthogonal transformations. Note that without dilatation, our transformation becomes orthogonal (a pure rotation).

The action of two successive transformations can be written as $\mathrm{F}(\mathrm{G}(a)) \equiv \mathrm{FG}a$, which is handy for manipulations of expressions. If for a linear transformation $\mathrm{F}: V \to W$ (V and W are vector spaces) there exists a linear transformation $\bar{\mathrm{F}}: W \to V$, we shall call it the *transposed transformation* (*adjoint*). Here we will consider transformations $\mathrm{F}: V \to V$ only. We say that a linear transformation $\bar{\mathrm{F}}$ is transposed, since we can always find its matrix representation in some basis and see that the matrix of $\bar{\mathrm{F}}$ is just the transposed matrix of F (see [27]). Having two elements A and B of a geometric algebra, we define the *scalar product* $A * B \equiv \langle AB \rangle = \langle BA \rangle$ (see Sect. 1.10), which for vectors gives $a * b \equiv \langle ab \rangle = \langle ba \rangle = a \cdot b$. Now we can give an **implicit definition of the adjoint:**

$$\langle a\bar{\mathrm{F}}(b)\rangle = \langle \mathrm{F}(a)b\rangle,$$

for any two vectors a and b. We can also write this as $a \cdot \bar{\mathrm{F}}(b) = \mathrm{F}(a) \cdot b$. To see what this means, we can imagine a simple example. Let us specify $a = e_1$, $b = \alpha_1 e_1 + \alpha_2 e_2$, and imagine that F rotates vectors by $\pi/2$ in the plane $e_1 e_2$. Then we have, for example, $\mathrm{F}(a) = a e_1 e_2 = e_1 e_1 e_2 = e_2$ (by the way, note that multiplication by the bivector $e_1 e_2$ from the right looks as if vector is rotated by the right palm counterclockwise, and opposite for the left side). Now we have

$$\mathrm{F}(e_1) \cdot (\alpha_1 e_1 + \alpha_2 e_2) = e_2 \cdot (\alpha_1 e_1 + \alpha_2 e_2) = \alpha_2.$$

According to Fig. 1.38, we see that our linear transformation transforms the vector $a = e_1$ to $\mathrm{F}(e_1) = e_2$, and the inner product gives $\alpha_2 = \cos\varphi$. But it is clear that we can take the

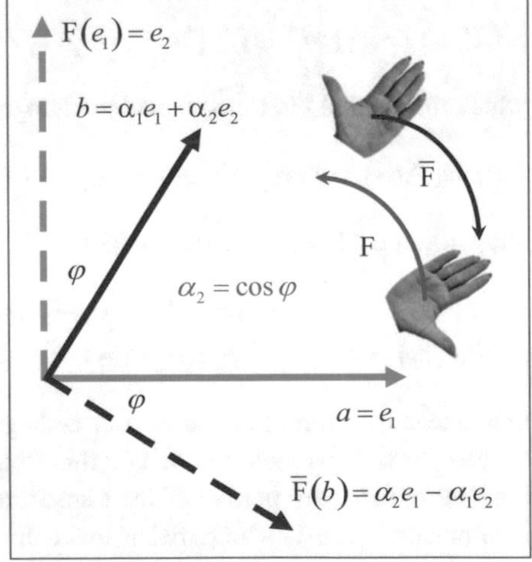

Fig. 1.38 A rotation as a simple linear transformation

vector b, rotate it clockwise, and the inner product with the vector $a = e_1$ will give the same result. Therefore, the adjoint transformation is just rotation by $-\pi/2$ in the plane e_1e_2 (for a general result for rotations, see below). Then we have

$$\bar{\mathrm{F}}(b) = e_1e_2b = e_1e_2(\alpha_1 e_1 + \alpha_2 e_2) = \alpha_2 e_1 - \alpha_1 e_2.$$

Let e^i be the *reciprocal basis vectors* (see Sect. 3.3) with the property

$$e^i \cdot e_j = \delta_{ij}.$$

Here we are using an orthonormal basis of positive signature, and accordingly,

$$e^i = e_i \Rightarrow e^i \cdot e_j = e_i \cdot e_j = \delta_{ij}.$$

The previous definition is motivated by two facts; first, we want to use the *Einstein summation convention* (we will indicate a summation by indices using Greek letters)

$$e^\mu e_\mu \equiv \sum_{i=1}^{n} e^i e_i,$$

and second, we want the ability to generalize easily to general frames. The explicit form of the adjoint transformation can be found using an orthonormal basis

$$e_i \cdot \bar{\mathrm{F}}(a) = \mathrm{F}(e_i) \cdot a,$$

where $\mathrm{F}(e_i) \cdot a$ is the ith component of $\bar{\mathrm{F}}(a)$. Thus we have (recall that the inner product has priority)

$$\bar{\mathrm{F}}(a) = e^\mu a \cdot \mathrm{F}(e_\mu),$$

where summation over μ is understood. The designation $\bar{\mathrm{F}}$ is not common, while F^{T} or $\mathrm{F}^\dagger$ is; however, sometimes we use $\underline{\mathrm{F}}$ for linear transformations; then a nice symmetry in expressions could occur if we use $\bar{\mathrm{F}}$. Furthermore, $\bar{\mathrm{F}}(a)$ is not a matrix or a tensor, and this designation highlights the difference. There cannot be confusion with the Clifford conjugation in the text, since we are consistently using the *italic* format for multivectors. For the adjoint transformation of the product of transformations (*function composition*), we have

$$\overline{\mathrm{FG}}(a) = \bar{\mathrm{G}}\bar{\mathrm{F}}(a).$$

Instead of a proof, we will take an illustrative example. Consider the two rotations

$$\mathrm{F}(a) = \exp(-e_2e_3\pi/4)a\exp(e_2e_3\pi/4), \quad \mathrm{G}(a) = \exp(-e_1e_2\pi/4)a\exp(e_1e_2\pi/4),$$

and note that

$$\mathrm{G}(e_3) = e_3,$$
$$\mathrm{e}_1 \cdot \bar{\mathrm{G}}(e_2) = \mathrm{e}_2 \cdot \mathrm{G}(e_1) = \mathrm{e}_2 \cdot \mathrm{e}_2 = 1 \Rightarrow \bar{\mathrm{G}}(e_2) = \mathrm{e}_1,$$
$$\mathrm{e}_2 \cdot \bar{\mathrm{G}}(e_1) = \mathrm{e}_1 \cdot \mathrm{G}(e_2) = \mathrm{e}_1 \cdot (-\mathrm{e}_1) = -1 \Rightarrow \bar{\mathrm{G}}(e_1) = -\mathrm{e}_2;$$

hence we see that $\bar{\mathrm{G}}(a) = \exp(e_1e_2\pi/4)a\exp(-e_1e_2\pi/4)$, which means that *the rotation adjoint is just the rotation inverse*. Now we have

$$\begin{aligned}\exp(-e_2e_3\pi/4)\exp(-e_1e_2\pi/4) &= \frac{1}{2}(1-e_2e_3)(1-e_1e_2)\\ &= \frac{1}{2}(1-e_1e_2+e_3e_1-e_2e_3)\end{aligned}$$
$$= \frac{1-je_1+je_2-je_3}{2} = \frac{1}{2} - \frac{\sqrt{3}}{2}j\frac{e_1-e_2+e_3}{\sqrt{3}}$$
$$= \cos(\pi/3) - j\frac{e_1-e_2+e_3}{\sqrt{3}}\sin(\pi/3) = \exp\left(-\hat{B}\pi/3\right)$$

meaning that $\overline{\mathrm{FG}} = \exp\left(\hat{B}\pi/3\right)$. Now we can check that

$$\bar{\mathrm{G}}\bar{\mathrm{F}} = \exp(e_1e_2\pi/4)\exp(e_2e_3\pi/4) = \frac{1}{2}(1+e_1e_2)(1+e_2e_3) = \exp\left(\hat{B}\pi/3\right) = \overline{\mathrm{FG}}$$

Note that $\bar{\mathrm{F}}\bar{\mathrm{G}}$ would produce an incorrect sign for e_2.

Transformations with the property $\bar{\mathrm{F}} = \mathrm{F}$ are *symmetric*. The important symmetric transformations are $\bar{\mathrm{F}}\mathrm{F}$ and $\mathrm{F}\bar{\mathrm{F}}$.

1.17.1 Determinants

For a unit pseudoscalar I, the *determinant of a linear transformation* is defined as

$$\mathrm{F}(I) \equiv I\det\mathrm{F},\quad \det\mathrm{F} \in \mathbb{R}$$

This definition is in a full agreement with the standard definition. Note that this relation looks like an eigenvalue relation. In fact, this is true: the unit pseudoscalar is invariant (eigenblade), and the determinant is a real eigenvalue. An example is a rotation in *Cl*3:

$$\mathrm{R}(j) = RjR^\dagger = jRR^\dagger \Rightarrow \det\mathrm{R} = RR^\dagger = 1,\quad j = e_{123},$$

which we expect for rotors (for rotation matrices, too). Again, note the power of the formalism: without components, without matrices, by a simple manipulation, we get an important result. The unit pseudoscalar represents an oriented volume; therefore, a linear transformation of the unit pseudoscalar is simply reduced to its multiplication by a real number (the shape is unimportant; Fig. 1.39). To find the determinant of the transposed transformation, we will take an example again, which may be more instructive than a formal proof. Consider the linear transformation

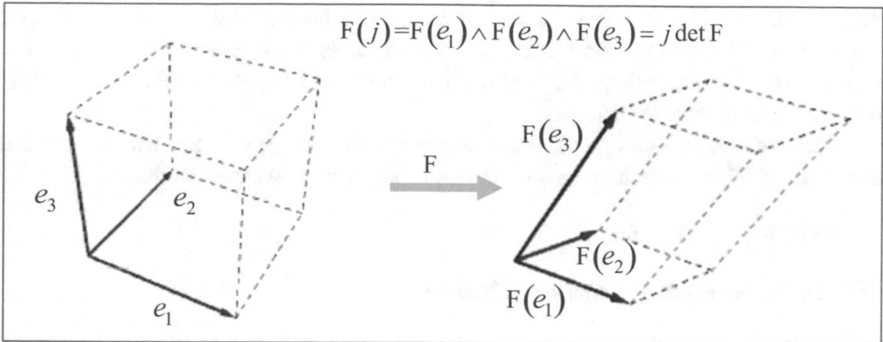

Fig. 1.39 A linear transformation of the pseudoscalar gives a determinant

$$\mathrm{F}(e_1) = \alpha_{11}e_1 + \alpha_{21}e_2, \quad \mathrm{F}(e_2) = \alpha_{12}e_1 + \alpha_{22}e_2, \quad \mathrm{F}(e_3) = e_3.$$

Note that the matrix of the linear transformation ($\mathrm{F}(e_k)$ is in column k) is

$$\begin{pmatrix} \alpha_{11} & \alpha_{12} & 0 \\ \alpha_{21} & \alpha_{22} & 0 \\ 0 & 0 & 1 \end{pmatrix} \xrightarrow{T} \begin{pmatrix} \alpha_{11} & \alpha_{21} & 0 \\ \alpha_{12} & \alpha_{22} & 0 \\ 0 & 0 & 1 \end{pmatrix}.$$

To find the adjoint transformation, we can use the transposed matrix or the formula

$$\bar{\mathrm{F}}(e_j) = e^{\mu} e_j \cdot \mathrm{F}(e_\mu),$$

from which it follows that

$$\bar{\mathrm{F}}(e_1) = \alpha_{11}e_1 + \alpha_{12}e_2, \quad \bar{\mathrm{F}}(e_2) = \alpha_{21}e_1 + \alpha_{22}e_2, \quad \bar{\mathrm{F}}(e_3) = e_3.$$

Now we can demonstrate how the outer product takes care of everything

$$\begin{aligned} &\mathrm{F}(j) = \mathrm{F}(e_1) \wedge \mathrm{F}(e_2) \wedge \mathrm{F}(e_3) = (\alpha_{11}e_1 + \alpha_{21}e_2) \wedge (\alpha_{12}e_1 + \alpha_{22}e_2) \wedge e_3 \\ &= (\alpha_{11}\alpha_{22}e_1 \wedge e_2 + \alpha_{12}\alpha_{21}e_2 \wedge e_1) \wedge e_3 = (\alpha_{11}\alpha_{22} - \alpha_{12}\alpha_{21})e_1 \wedge e_2 \wedge e_3 = (\alpha_{11}\alpha_{22} - \alpha_{12}\alpha_{21})j. \end{aligned}$$

We have the same result for the adjoint, and consequently, $\det \mathrm{F} = \det \bar{\mathrm{F}} = \alpha_{11}\alpha_{22} - \alpha_{12}\alpha_{21}$.

For the composition of transformations, we have

$$(\mathrm{FG})(I) = \mathrm{FG}(I) = \mathrm{F}(I \det \mathrm{G}) = (\det \mathrm{G})\mathrm{F}(I) = I \det \mathrm{F} \det \mathrm{G}$$

which is a well-known theorem for determinants; however, recall how much effort we need to prove this using matrix theory. Here, the proof is almost trivial. A beginner needs a great deal of time to become skilled with matrices. Finally, she (he) has a tool that cannot elegantly cope even with rotations. That time could have been used to learn the basics of geometric algebra and thereby obtain a powerful tool for many branches of mathematics. Geometric algebra today, thanks to many talented and hardworking people, has become a well-developed theory, with applications in many areas of mathematics, physics, engineering, biology, the study of brain function, computer graphics, ray tracing, robotics, among many others.

Here we give, without proof, a useful relation (the reader can prove it; however, it could be useful to check it on various examples). For two bivectors we have

$$\langle B_1 \bar{\mathrm{F}}(B_2)\rangle = \langle \mathrm{F}(B_1)B_2\rangle.$$

This can be extended to arbitrary multivectors as

$$\langle A\bar{\mathrm{F}}(B)\rangle = \langle \mathrm{F}(A)B\rangle.$$

1.17.2 Inverse of a Linear Transformation

Now we will define the inverse of a linear transformation. For a vector x we have

$$\mathrm{F}x = y \Rightarrow x = \mathrm{F}^{-1}y$$

if the inverse of the linear transformation exists. For the multivector M, we have the general relation $xM = x\rfloor \mathrm{M} + x \wedge M$, giving $xj = x\rfloor j$. Now we can write

$$\begin{aligned}
&x\bar{\mathrm{F}}(j) = x\rfloor\bar{\mathrm{F}}(j) = x\cdot\bar{\mathrm{F}}(e_1)\bar{\mathrm{F}}(e_2\wedge e_3) + x\cdot\bar{\mathrm{F}}(e_2)\bar{\mathrm{F}}(e_3\wedge e_1) + x\cdot\bar{\mathrm{F}}(e_3)\bar{\mathrm{F}}(e_1\wedge e_2)\\
&= e_1\cdot\mathrm{F}(x)\bar{\mathrm{F}}(je_1) + e_2\cdot\mathrm{F}(x)\bar{\mathrm{F}}(je_2) + e_3\cdot\mathrm{F}(x)\bar{\mathrm{F}}(je_3)\\
&= \bar{\mathrm{F}}(e_1\cdot\mathrm{F}(x)je_1 + e_2\cdot\mathrm{F}(x)je_2 + e_3\cdot\mathrm{F}(x)je_3) = \bar{\mathrm{F}}(j\mathrm{F}(x)) \Rightarrow\\
&x = \mathrm{F}^{-1}y = \frac{\bar{\mathrm{F}}(j\mathrm{F}(x))}{\bar{\mathrm{F}}(j)} = \frac{\bar{\mathrm{F}}(jy)}{j\det\bar{\mathrm{F}}} = \frac{\bar{\mathrm{F}}(jy)j^{-1}}{\det\mathrm{F}} = \frac{\bar{\mathrm{F}}(j^{-1}y)j}{\det\mathrm{F}}
\end{aligned}$$

where $j^{-1} = -j$. It generally follows that

$$\mathrm{F}^{-1}(A) = I\bar{\mathrm{F}}(I^{-1}A)(\det\mathrm{F})^{-1}, \quad \bar{\mathrm{F}}^{-1}(A) = I\mathrm{F}(I^{-1}A)(\det\mathrm{F})^{-1}.$$

1.17.3 Examples of Linear Transformations

For rotors in $Cl3$, we have $\mathrm{R}(a) = RaR^{\dagger}$, which applied to any multivector gives $\mathrm{R}(M) = RMR^{\dagger}$ and $\bar{\mathrm{R}}(M) = R^{\dagger}MR$; consequently, using $\det\mathrm{R} = 1$, we have

$$\mathrm{R}^{-1}(M) = jR^{\dagger}j^{-1}MR = R^{\dagger}MR = \bar{\mathrm{R}}(M);$$

that is, the inverse of a rotation is equal to the adjoint rotation. This is actually a possible definition of orthogonal transformations (transformations with determinant ± 1).

Consider the linear transformation $\mathrm{F}e_i = \alpha e_i$, $\alpha \in \mathbb{R}$. We have $\mathrm{F}I = \alpha^n I$, which means the scaling of the space. For $\alpha = -1$, we have the space inversion and $\det\mathrm{F} = (-1)^n$, which is negative for odd dimensions. Because of $\det\mathrm{R} = 1$, we see that for odd dimensions we cannot compensate for the space inversion using rotations. Consider now a projector in 2D subspace $I = e_1e_2$, with the properties

$$\mathrm{P}e_1 = e_1, \quad \mathrm{P}e_2 = 0.$$

We have

$$\mathrm{P}(I) = \mathrm{P}(e_1 \wedge e_2) = \mathrm{P}(e_1) \wedge \mathrm{P}(e_2) = 0,$$

and thus $\det\mathrm{P} = 0$. We see that P, acting on bivectors, produces zero bivectors, while acting on vectors generally gives a vector: $\mathrm{P}(e_1 + e_2) = e_1$.

We can treat reflections as linear transformations as well, so let's examine the reflection $\varpi(a) = e_1e_2ae_1e_2$. We have

$$\varpi(e_1) = e_1e_2e_1e_1e_2 = e_1, \quad \varpi(e_2) = e_1e_2e_2e_1e_2 = e_2, \quad \varpi(e_3) = e_1e_2e_3e_1e_2 = -e_3,$$

which defines the reflection in the e_1e_2 plane. Acting on the pseudoscalar, we get

$$\varpi(j) = \varpi(e_1) \wedge \varpi(e_2) \wedge \varpi(e_3) = -j, \quad \det\varpi = -1.$$

An even number of reflections is a rotation. For nice examples, see [21].

1.17.4 Eigenvectors and Eigenblades

The reader should be familiar with the concept of eigenvalues and eigenvectors. Briefly, for an operator (matrix) m, we can define eigenvalues λ_i and eigenvectors v_i as

$$mv_i = \lambda_i v_i, \quad \lambda_i \in \mathbb{C}.$$

In geometric algebra, we would say that a linear transformation has the *eigenvector* e and the *eigenvalue* λ if

$$\mathrm{F}(e) = \lambda e,$$

which entails

$$\det(\mathrm{F} - \lambda I) = 0.$$

Therefore, we have a polynomial equation (the *secular equation*). Generally, the secular equation has roots over the complex field; however, we (usually) have geometric algebra over the field of real numbers, and it is not desirable to introduce complex numbers. For example, how should we interpret the product $\sqrt{-1}e_1$, which is not an element of the algebra? Fortunately, this is not necessary in geometric algebra; we can give a completely new meaning to complex solutions. For this purpose, we introduce the concept of an *eigenblade*. To be precise, vectors are just elements of the algebra with grade 1; however, in geometric algebra, we also have grades 2, 3, ..., which are not defined in the ordinary theory of vector spaces. It is therefore natural to extend the definition of eigenvalues to other elements of the algebra. For a blade B_r with grade r, we define

$$F(B_r) = \lambda B_r, \quad \lambda \in \mathbb{R}.$$

In fact, we have already seen such relations; that is, for $B_r = I$, we have the eigenvalue detF, as a consequence of $F(I) = I \det F$. Accordingly, the unit pseudoscalars are eigenblades of linear transformations. To explain the concept of an eigenblade, we will take an excellent example from [21]. If we specify the linear transformation with the property

$$F(e_1) = e_2, \quad F(e_2) = -e_1$$

(do you recognize rotation?), it is not difficult to find a solution of the eigenvalue problem using matrices. The matrix of the transformation F is

$$\begin{pmatrix} 0 & -1 \\ 1 & 0 \end{pmatrix},$$

with eigenvalues $\pm i$, $i = \sqrt{-1}$, and eigenvectors $e_1 \pm ie_2$ (E27: Use the secular equation and prove this). In geometric algebra, for the blade $e_1 \wedge e_2$, we have (note the elegance)

$$F(e_1 \wedge e_2) = F(e_1) \wedge F(e_2) = e_2 \wedge (-e_1) = e_1 \wedge e_2.$$

Therefore, the blade $e_1 \wedge e_2$ is an eigenblade with (real) eigenvalue 1. Our blade is an invariant of the transformation, but we know this already from the rotor formalism! There is no need for the imaginary unit i: we have our blade. In fact, if you see the imaginary unit in the eigenvalue problem with matrices, there is an eigenblade behind it (see Appendix 5.9). Note that vectors in the plane defined by $e_1 \wedge e_2$ are changed by the transformation; however, the unit bivector is not. You see simple mathematics and an important result. In standard methods, using matrices, there are no blades at all. Why? Simply put, there is no geometric product there. Therefore, try to find such a result using matrices and compare. All those who like to comment on geometric algebra by sentences such as "Yes, but the imaginary unit in quantum mechanics..." should think twice about this simple example, and when they come to the conclusion that "it does not make sense...," well, what can I say? Just think again. This is the question of how we understand the very concept of a number. Probably Grassmann and Clifford directed us well; the time of their ideas is yet to come.

If the basis vectors e_i and e_j are eigenvectors of a linear transformation F, then

$$e_i \cdot F(e_j) = e_i \cdot (\lambda_j e_j) = \lambda_j e_i \cdot e_j.$$

E28: Apply the previous relations to symmetric linear transformations and show that their eigenvectors with different eigenvalues must be orthogonal.

Euclidean 3D Geometric Algebra (*Cl*3)

Miroslav Josipović

M. Josipović, *Geometric Multiplication of Vectors*,
Compact Textbooks in Mathematics,
https://doi.org/10.1007/978-3-030-01756-9_2

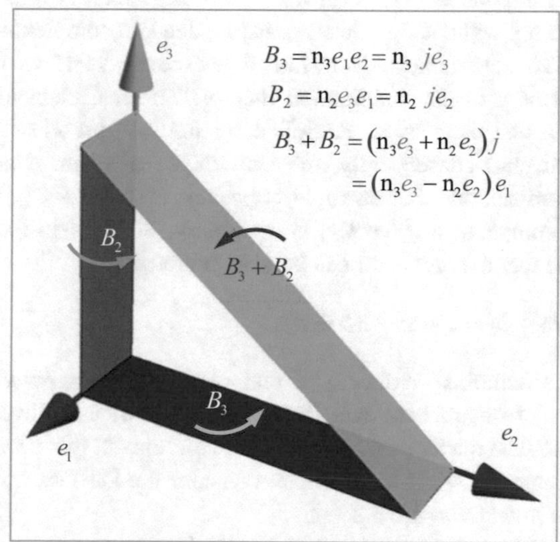

Addition of bivectors in *Cl*3

2.1 The Structure of Multivectors in *Cl*3

Generally, a multivector in *Cl*3 can be rewritten as

$$M = t + \mathbf{x} + j\mathbf{n} + jb, \quad t, b \in \mathbb{R}, \quad j = e_1 e_2 e_3,$$

The original version of this chapter was revised. A correction to this chapter can be found at https://doi.org/10.1007/978-3-030-01756-9_7.

where for vectors and complex vectors we are using the **bold** format here. We have already seen that the unit pseudoscalar j commutes with all elements of the algebra and squares to -1, and therefore, it is an ideal replacement for the ordinary imaginary unit (there are many "imaginary units" in GA). A pseudoscalar with such properties will appear again in *Cl*7, *Cl*11, ... (see E10). Here we often use a very useful form of a multivector:

$$M = Z + \mathbf{F}, \quad Z = t + bj, \quad \mathbf{F} = \mathbf{x} + j\mathbf{n}.$$

The element Z obviously commutes with all elements of the algebra (it belongs to the *center of the algebra*). This feature makes it a *complex scalar*. A complex scalar is really acting as a complex number, as we shall see below. This is the reason that we write $Z \in \mathbb{C}$, although obviously, we have to change the meaning of the symbol $\mathbb{C}$, that is, we replace the ordinary imaginary unit by the pseudoscalar j. An element $\mathbf{F}$ is a *complex vector*, with (*real*) *vectors* as components. The choice of the designation ($\mathbf{F}$), as well as that for complex scalars, is not without significance, that is, due to a complex mixture of electric and magnetic fields in electromagnetism. Here, when we say "real," we mean a real scalar or vector, as well as a linear combination thereof. The real elements are defined as $M = M^{\dagger}$. When a real element is multiplied by the unit pseudoscalar j, we get an *imaginary* element, and consequently, the sum of a real element and an imaginary element gives a complex one. For example, $\mathbf{x}$ (*vector*) is real, $t + \mathbf{x}$ (*paravector*) is real, $t + j\mathbf{n}$ (*spinor*) is complex, $j\mathbf{n}$ (*bivector*) is imaginary, $\mathbf{F} = \mathbf{x} + j\mathbf{n}$ (*complex vector*) is complex, etc. Note that a multivector can be also expressed as

$$M = t + \mathbf{x} + j\mathbf{n} + bj = t + \mathbf{x} + j(b + \mathbf{n});$$

that is, it is just a complex number with real components (*paravectors*). ◆ Use an involution (which?) to extract the real (imaginary) part of a multivector. How about Z and $\mathbf{F}$? Or $t + j\mathbf{n}$? The reader should write all involutions in this new form; you could use complex conjugation. ◆ As an example, consider the Clifford involution (*Clifford conjugation*, *main involution*) $\bar{M} = Z - \mathbf{F}$:

$$(M + \bar{M})/2 = Z \equiv \langle M \rangle_S \quad \text{(the } \textit{scalar part}\text{)},$$
$$(M - \bar{M})/2 = \mathbf{F} \equiv \langle M \rangle_V \quad \text{(the } \textit{vector part}\text{)}$$

(see E19). Due to the commutativity of the complex scalar Z, we have

$$M\bar{M} = (Z + \mathbf{F})(Z - \mathbf{F}) = Z^2 - \mathbf{F}^2 = (Z - \mathbf{F})(Z + \mathbf{F}) = \bar{M}M,$$

where (here we use the designations for vector lengths in *italics*, like $|\mathbf{x}| = x$)

$$Z^2 = t^2 - b^2 + 2tbj, \quad \mathbf{F}^2 = x^2 - n^2 + j(\mathbf{xn} + \mathbf{nx}) = x^2 - n^2 + 2j\mathbf{x} \cdot \mathbf{n}.$$

Here is the result to remember: **the square of a complex vector is a complex scalar**. This means that the element $M\bar{M}$ is a complex scalar. For the moment, we will denote **any** involution of M by $\underline{M}$, while we call $M\underline{M}$ the square of the *amplitude*.

Theorem 1 *The product $M\bar{M}$ is the **only** (square of the) amplitude of the form $M\underline{M}$ that satisfies $M\underline{M} = \underline{M}M \in \mathbb{C}$ for all elements M from Cl3.*

Proof: We have

$$(Z+\mathbf{F})(\underline{Z}+\underline{\mathbf{F}}) = Z\underline{Z}+Z\underline{\mathbf{F}}+\underline{Z}\mathbf{F}+\mathbf{F}\underline{\mathbf{F}},$$
$$(\underline{Z}+\underline{\mathbf{F}})(Z+\mathbf{F}) = Z\underline{Z}+Z\underline{\mathbf{F}}+\underline{Z}\mathbf{F}+\underline{\mathbf{F}}\mathbf{F}.$$

Therefore, we have two possibilities:

$$\underline{Z}=Z,\quad \underline{\mathbf{F}}=-\mathbf{F},\quad \text{and}\quad \underline{Z}=-Z,\quad \underline{\mathbf{F}}=\mathbf{F},$$

which differ only in the overall sign. Every involution that changes a complex vector in a different way (up to overall sign) changes the bivector or the vector part only. Consequently,

$$\mathbf{F}\underline{\mathbf{F}} = (\mathbf{x}+j\mathbf{n})(\mathbf{x}-j\mathbf{n}) = x^2+n^2+j(\mathbf{nx}-\mathbf{xn}) = x^2+n^2-2j\mathbf{x}\wedge\mathbf{n},$$

and we obtain the outer product of real vectors that cannot be canceled: it is absent in $Z\underline{Z}+Z\underline{\mathbf{F}}+\underline{Z}\mathbf{F}$. Consequently, it must be $\underline{M}=\bar{M}$. We already found that $M\bar{M}=\bar{M}M$, but we can show this also from the requirement that the amplitude (any) belongs to the center of the algebra, which leads to commutativity:

$$M\underline{M}\in\mathbb{C}\Rightarrow M(M\underline{M}) = (M\underline{M})M = M(\underline{M}M)\Rightarrow M(M\underline{M}-\underline{M}M)=0,$$

due to associativity and distributivity. In a special case, the expression in parentheses could be nonzero because of the existence of *zero divisors* in the algebra; however, we need the commutativity for all elements M. Consequently, $M\underline{M}-\underline{M}M$ must be zero. The complex scalar $\sqrt{M\bar{M}}$ is called the *multivector amplitude* (MA in the text).

Using the MA, we can define the inverse of a multivector (if $M\bar{M}\neq 0$):

$$M^{-1}\equiv \bar{M}/M\bar{M}.$$

To find $1/M\bar{M}$, we use the ordinary complex numbers technique

$$\frac{1}{M\bar{M}} = \frac{(M\bar{M})^*}{M\bar{M}(M\bar{M})^*},\qquad M\bar{M}(M\bar{M})^*\in\mathbb{R},$$

where $*$ stands for complex conjugation, that is, $j\to -j$. The technique is the same, though the interpretation is not; specifically, the pseudoscalar j is the oriented unit volume, and it has an intuitive geometric interpretation.

Example What is $1/(1+j)$?

We have $1/(1+i)=(1-i)/2 \;\Rightarrow\; 1/(1+j)=(1-j)/2$.

Of course, this small *trick* is justified:

$$\frac{1}{1+j} = \frac{1-j}{(1+j)(1-j)} = \frac{1-j}{2}.$$

This technique is useful with computers, where ordinary complex numbers are an integral part of applications. We will see that this procedure is sometimes insufficient to find all possible solutions in geometric algebra. Namely, solutions for roots of complex numbers can be extended to complex vectors; a simple example is $\sqrt{1}=e_1$.

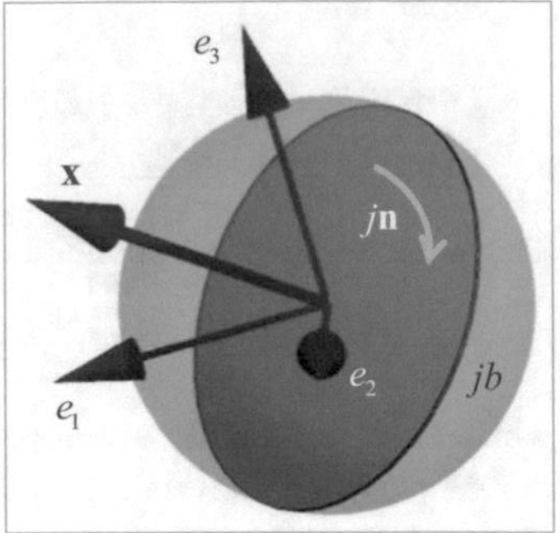

Fig. 2.1 A multivector in 3D: the **disk** represents an oriented bivector. The transparent **sphere** represents a pseudoscalar; an orientation is given by a color

An important concept is the *dual* of a multivector *M*, defined as (see Sect. 1.3.1 and Appendix 5.8)

$$M^{\Delta} \equiv Mj^{-1} = -jM$$

(here we use the character Δ instead of the more common $*$ to avoid confusion with complex conjugation). Note that with the dual operation, a real scalar becomes a pseudoscalar, and a vector becomes a bivector (and vice versa). ◆ An element $j\mathbf{n}$ is a bivector. We suggest that the reader express $j\mathbf{n}$ in an orthonormal basis and interpret it. Also, take any two vectors in an orthonormal basis and form their outer product. Then find the dual of the obtained bivector and check that this dual is formally just the cross product of your vectors. ◆ We should replace (redefine) the cross product as

$$\mathbf{x} \times \mathbf{y} \to -j\mathbf{x} \wedge \mathbf{y},$$

due to its invalid behavior under space inversion, as we discussed in Sect. 1.3.

In *Cl*3, from the general form of a multivector

$$M = t + \mathbf{x} + j\mathbf{n} + jb = Z + \mathbf{F},$$

we see that a multivector is essentially determined by two real numbers $(t,\ b)$, and two vectors $(\mathbf{x},\ \mathbf{n})$. We usually represent bivectors by oriented disks, while a sphere can represent the pseudoscalar, with two possible colors to represent the orientation. Thus, we have a nice and simple image that represents a multivector (Fig. 2.1), and this helps greatly, especially with interactive versions. For the reader, apart from an imagination, we suggest the application *GAViewer*.

2.1.1 The Square of a Complex Vector

Let us look at the properties of a complex scalar $\mathbf{F}^2 = x^2 - n^2 + 2j\mathbf{x} \cdot \mathbf{n}$. In particular, for orthogonal vectors $(\mathbf{x} \cdot \mathbf{n} = 0)$, we have $\mathbf{F}^2 \in \mathbb{R}$, where the values -1, 0, and 1 are of particular interest.

Recall that $j\mathbf{n}$ is a bivector that defines the plane orthogonal to the vector $\mathbf{n}$, which means that for $\mathbf{x} \cdot \mathbf{n} = 0$, the vector $\mathbf{x}$ belongs to that plane. This is an often-utilized

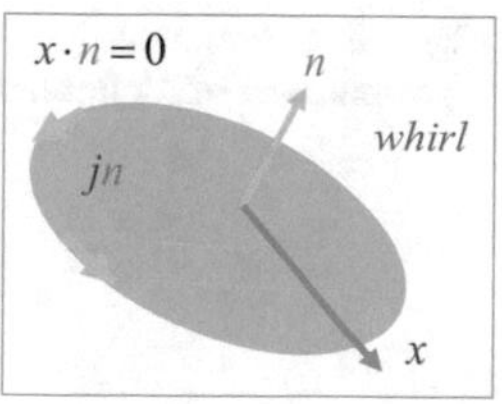

Fig. 2.2 Whirl: a special complex vector

situation (for instance, a complex vector of the electromagnetic field in an empty space), and consequently, it is important to imagine a clear picture. Note that in this case, the real value of $\mathbf{F}^2 = x^2 - n^2$ is determined by the lengths of vectors $\mathbf{x}$ and $\mathbf{n}$ (see Fig. 2.2). There is no special name for this kind of complex vector in the literature (probably?), and therefore, we suggest the name *whirl* (short for whirligig). We proceed with the three important cases of $\mathbf{F}^2$ (**the heart of 3D geometry**).

2.2 Nilpotents and Dual Numbers

1. $\mathbf{F}^2 = 0$

This means that such a complex vector is a *nilpotent* ($\mathbf{F} \neq 0$).

Theorem 2 *The general form of nilpotents in Cl3 is* $\mathbf{x} + j\mathbf{n}$, $x = n$, $\mathbf{x} \cdot \mathbf{n} = 0$.
Proof:

$$(Z + \mathbf{F})^2 = 0 = Z^2 + 2Z\mathbf{F} + \mathbf{F}^2 \Rightarrow Z = 0, \mathbf{F}^2 = 0 \Rightarrow x = n, \quad \mathbf{x} \cdot \mathbf{n} = 0.$$

Note that we frequently use the form $Z + \mathbf{F}$ to draw conclusions, and this is not a coincidence. It is a good practice to avoid the habits of some authors to express multivectors by components, which results in long and opaque expressions. Here the focus is on the structure of multivectors, and that structure reflects the geometric properties. A simple example of a nilpotent is $e_1 + je_3$ (check this, Fig. 2.3). Functions with a nilpotent as an argument are easy to find; using a series expansion, almost all terms just disappear. For example, from $\mathbf{N}^2 = 0$ follows $\exp(\mathbf{N}) = 1 + \mathbf{N}$ (see Sect. 2.4.5).

Nilpotents are welcome in physics. For example, an electromagnetic wave in vacuum is a nilpotent in the *Cl*3 formulation; a field is a complex vector $\mathbf{F} = \mathbf{E} + j\mathbf{B}$, $E = B$, $c = 1$; $\mathbf{E}$ and $\mathbf{B}$ are vectors of electric and magnetic fields. We can define the *direction of the nilpotent* $\mathbf{N} = \mathbf{x} + j\mathbf{n}$ as $\hat{\mathbf{k}} = -j\hat{\mathbf{x}} \wedge \hat{\mathbf{n}} = -j\hat{\mathbf{x}}\hat{\mathbf{n}}$, $\hat{\mathbf{k}}^2 = \hat{\mathbf{x}}^2 = \hat{\mathbf{n}}^2 = 1$, so we have

$$\hat{\mathbf{k}}\mathbf{N} = -\mathbf{N}\hat{\mathbf{k}} = \mathbf{N}, \quad (1 + \hat{\mathbf{k}})\mathbf{N} = 2\mathbf{N}.$$

These relations are not difficult to prove. Note that $\mathbf{N}\bar{\mathbf{N}} = 0$. Find $\mathbf{N}\mathbf{N}^\dagger$ and $\mathbf{N} * \mathbf{N}^\dagger$.

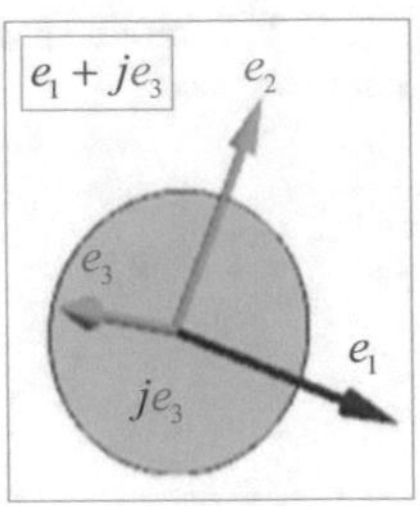

Fig. 2.3 A graphical representation of a nilpotent

2.2.1 Dual Numbers

Let us now mention the possibility of defining the *dual numbers*. The dual numbers are like complex numbers, with an "imaginary unit" that squares to zero. For the nilpotent $\mathbf{N} = \mathbf{x} + j\mathbf{n}$, we have $x = n,\ \mathbf{x} \cdot \mathbf{n} = 0$. Therefore, let's define a "unit nilpotent" (nilpotents have a zero MA; their magnitude is zero):

$$\varepsilon \equiv \mathbf{N}/x = \hat{\mathbf{x}} + j\hat{\mathbf{n}},\ \ \varepsilon^2 = 0.$$

Now we can define dual numbers as $\alpha + \beta\varepsilon,\ \alpha, \beta \in \mathbb{R}$. The addition of these numbers is similar to that of complex numbers, while for multiplication we have

$$(\alpha_1 + \beta_1\varepsilon)(\alpha_2 + \beta_2\varepsilon) = \alpha_1\alpha_2 + (\alpha_1\beta_2 + \alpha_2\beta_1)\varepsilon.$$

Therefore, it is a real number for $\alpha_1\beta_2 + \alpha_2\beta_1 = 0$. If $\alpha_1 = \alpha_2 = 0$, the product is zero, which distinguishes the dual and the complex numbers. For a dual number $z = \alpha + \beta\varepsilon$, we define the conjugation $\bar{z} = \alpha - \beta\varepsilon$ (note that it is just the Clifford involution), whence we have

$$z\bar{z} = (\alpha + \beta\varepsilon)(\alpha - \beta\varepsilon) = \alpha^2,$$

that is, the MA of a dual number is $|z| = \sqrt{z\bar{z}} = \alpha$ (it could be negative). Note that there is no dependence on β. For $|z| = \alpha \neq 0$, we have the polar form

$$z = \alpha + \beta\varepsilon = \alpha(1 + \varphi\varepsilon),\ \ \varphi = \beta/\alpha,$$

where φ is the *argument* of a dual number. ◆ Check that

$$(1 + \varphi\varepsilon)(1 - \varphi\varepsilon) = 1,\quad (1 + \varphi\varepsilon)^n = (1 + n\varphi\varepsilon),\ \ n \in \mathbb{N}.\ \ ◆$$

For polynomials, we have (check)

$$P(\alpha + \beta\varepsilon) = p_0 + p_1(\alpha + \beta\varepsilon) + \cdots + p_n(\alpha + \beta\varepsilon)^n = P(\alpha) + \beta P'(\alpha)\varepsilon,$$

where P' is the first derivative of the polynomial. This may be extended to analytic functions (see Sect. 1.13.1) or to maintain the *automatic differentiation* (AD). To see how it works, let's specify $P(x) = 1 + x^2$, and then, replacing x by $x + \varepsilon$, it follows that

$$P(x+\varepsilon) = 1 + (x+\varepsilon)^2 = x^2 + 1 + 2x\varepsilon = P(x) + P'(x)\varepsilon.$$

Thus, we have calculated $P(x)$ and $P'(x)$ simultaneously. Just take, for example, $x = 2$. You immediately have $5 + 4\varepsilon$, the function value (5), and the function slope value (4). This technique can be useful in game programming, satellite trajectory calculations, etc., where we might want to know the position and the slope at the same instant of time.

The division by dual numbers is defined as

$$\frac{\alpha+\beta\varepsilon}{\gamma+\delta\varepsilon} = \frac{(\alpha+\beta\varepsilon)(\gamma-\delta\varepsilon)}{(\gamma+\delta\varepsilon)(\gamma-\delta\varepsilon)} = \frac{(\alpha+\beta\varepsilon)(\gamma-\delta\varepsilon)}{\gamma^2}, \quad \gamma \neq 0.$$

In particular,

$$\frac{1}{1+\varphi\varepsilon} = \frac{1-\varphi\varepsilon}{(1+\varphi\varepsilon)(1-\varphi\varepsilon)} = 1 - \varphi\varepsilon \Rightarrow \left(\frac{1}{1+\varphi\varepsilon}\right)^n = (1-\varphi\varepsilon)^n = 1 - n\varphi\varepsilon,$$

where we see that de Moivre's formula is valid:

$$z^n = \alpha^n(1+\varphi\varepsilon)^n = \alpha^n(1+n\varphi\varepsilon), \quad n \in \mathbb{Z}.$$

2.2.2 Dual Numbers and Galilean Transformations

Dual numbers are of some interest in physics. For example, let's define the special dual number (*event*) $t + x\varepsilon$ and the *proper velocity* (*boost*) $u \equiv 1 + v\varepsilon$, $u\bar{u} = 1$, where the coordinates of time (t) and position (x) are introduced. The velocity $v = x/t = \varphi$ is the argument of the dual number. Now we have

$$(t + x\varepsilon)u = (t + x\varepsilon)(1 + v\varepsilon) = t + (x + vt)\varepsilon = t' + x'\varepsilon,$$

which means that $t' = t$ and $x' = x + vt$, so we have Galilean transformations. The velocity addition rule follows immediately:

$$u = 1 + v\varepsilon = u_1u_2 = (1 + v_1\varepsilon)(1 + v_2\varepsilon) = 1 + (v_1 + v_2)\varepsilon \Rightarrow v = v_1 + v_2.$$

There is also a possibility to formulate Galilean transformations using *dual quaternions*. Using dual quaternion units q_i with properties $q_iq_j = 0$, $i, j = 1,2,3$, and identifying vectors like $e_i \to q_i$, we get

$$(1 + vq_1)(t + xq_1 + yq_2 + zq_3) = t + (x + vt)q_1 + yq_2 + zq_3,$$

and we have Galilean transformations again. A simple way to accomplish this is to identify $q_i \to je_i = B_i$, using the outer product, since $B_i \wedge B_j = 0$ (see problem 34).

Apart from Lorentz transformations (hyperbolic numbers, vectors, and complex vectors; see Sect. 2.6.1) and rotors (complex numbers, bivectors), Galilean transformations (dual numbers, nilpotents) also have a common place in the geometry of $\mathfrak{R}^3$. Since all this is a part of a larger structure ($Cl3$), one can get the idea that Galilean transformations are not just the approximation of Lorentz transformations for the infinite speed of light; possibly, there is some deeper physical content, independent of the velocity. However, such an idea is simply due to our special choice of components of the dual number (t, x). Dual numbers like $t+x\varepsilon$ could be useful in nonrelativistic physics. Certainly, however, they are not in accordance with the special theory of relativity. In the chapter on special relativity (Sect. 2.7), it is shown that the Galilean transformations (also rotations and Lorentz transformations) follow from the simple symmetry assumptions about our world (*relativity principle*, *homogeneity*, and *isotropy*). If there is a deeper physics behind the formalism of dual numbers, then it certainly does not include explicit space-time events. However, what if we choose differently? For example, the electromagnetic wave in vacuum is a typical nilpotent. If we define $\xi=E=B$ and $\varepsilon=(\mathbf{E}+j\mathbf{B})/\xi$, one could investigate dual numbers like $\lambda+\xi\varepsilon$, $\lambda\in\mathbb{R}$; however, then there is a question: how to interpret λ. We already have such quantities in electromagnetism, connected to the *Lorenz gauge*, where we choose $\lambda=0$. Namely, we can define the fields $\mathbf{E}$ and $\mathbf{B}$ starting from the paravector of a potential field and taking the derivative (see Sect. 2.11):

$$\begin{aligned}\bar{\partial}A &= (\partial_t-\nabla)(V+\mathbf{A}) = \left(\dot{V}-\nabla\cdot\mathbf{A}\right)+\left(-\nabla V+\dot{\mathbf{A}}\right)-j(\nabla\times\mathbf{A})\\ &= \lambda+\mathbf{E}+j\mathbf{B},\quad \lambda=\dot{V}-\nabla\cdot\mathbf{A}.\end{aligned}$$

2.3 Idempotents and Hyperbolic Structure

2. $\mathbf{F}^2=1$

For $\mathbf{F}^2=1$, we can find a general form using the relation $\cosh^2\varphi-\sinh^2\varphi=1$, whence we have $\mathbf{F}\equiv\mathbf{f}=\mathbf{n}\cosh\varphi+j\mathbf{m}\sinh\varphi$, $\mathbf{n}^2=\mathbf{m}^2=1$, $\mathbf{n}\perp\mathbf{m}$, where $\mathbf{f}$ is a unit complex vector, for example, $\mathbf{f}=e_1\cosh\varphi+je_2\sinh\varphi$. Such a complex vector can be obtained using $\sqrt{\mathbf{F}^2}$. ◆ Check that the complex vector $\mathbf{f}\equiv\mathbf{F}/\sqrt{\mathbf{F}^2}$ possesses the required properties. Check that $p=(1+\mathbf{f})/2\Rightarrow p^2=p$. ◆ Consequently, we have an *idempotent*.

Theorem 3 *All idempotents in Cl3 have the form* $p=(1+\mathbf{f})/2$.
Proof:

$$(Z+\mathbf{F})^2=Z^2+2Z\mathbf{F}+\mathbf{F}^2=Z+\mathbf{F}\Rightarrow Z=1/2\Rightarrow\mathbf{F}^2=1/4\Rightarrow\mathbf{F}=\mathbf{f}/2.$$

Note the Z, $\mathbf{F}$ form again. The general form of idempotents is now

$$p=(1+\mathbf{n}\cosh\varphi+j\mathbf{m}\sinh\varphi)/2,\quad \mathbf{n}^2=\mathbf{m}^2=1,\quad \mathbf{n}\perp\mathbf{m}.$$

Idempotents like $(1+\mathbf{n})/2$, where $\mathbf{n}$ is a unit vector, are called *simple*.

Theorem 4 *Each idempotent in Cl3 can be expressed as the sum of a simple idempotent and a nilpotent.*

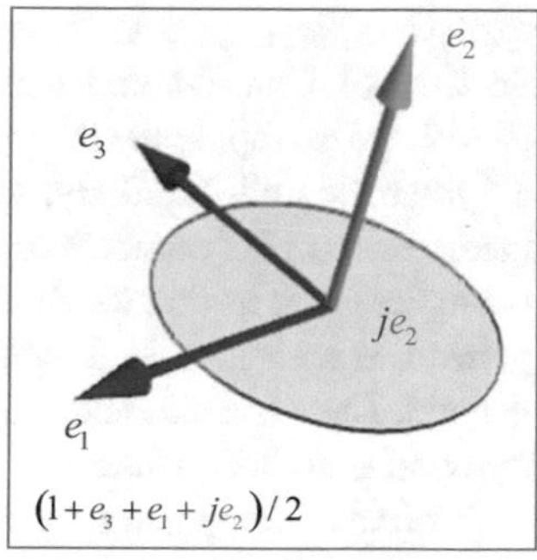

Fig. 2.4 An idempotent in Cl3 expressed as the sum of a simple idempotent and a nilpotent

Proof:

For the simple idempotent $p = (1+\hat{\mathbf{u}})/2$ and the nilpotent $\mathbf{N} = (\mathbf{x} + j\mathbf{y})/2$, $\mathbf{x} \cdot \mathbf{y} = 0$, $|\mathbf{x}| = |\mathbf{y}|$ (the factor 1/2 is just for convenience), we have

$$(p+\mathbf{N})^2 = p + p\mathbf{N} + \mathbf{N}p = p + \mathbf{N} + (\hat{\mathbf{u}}\mathbf{N} + \mathbf{N}\hat{\mathbf{u}})/2.$$

Therefore, we see that the statement is correct if $\hat{\mathbf{u}}\mathbf{N} = -\mathbf{N}\hat{\mathbf{u}}$, which means that $\hat{\mathbf{u}} = \pm\hat{\mathbf{k}}$, where $\hat{\mathbf{k}}$ is the direction of the nilpotent $\mathbf{N}$ (see Sect. 2.2). We have proved the theorem and found the conditions for the nilpotent.

Example (Fig. 2.4):

$p = (1+e_3)/2$ (*simple idempotent*),
$\mathbf{N} = (e_1 + je_2)/2$ (*nilpotent*),
$\hat{\mathbf{k}} = -je_1e_2 = e_3$ (*nilpotent direction*),
$(1 + e_3 + e_1 + je_2)/2$ (*idempotent*).

For a given $\mathbf{x}$, find $\mathbf{y}$.

2.4 Spectral Decomposition and Functions of Multivectors

2.4.1 Motivation

We are accustomed to functions of a real or complex variable. Now we can try to define functions of a multivector variable. We know that this is possible for simple elements, like bivectors, which square to a real number. In Sect. 1.9, we defined rotors as the exponential functions, like $\exp(\theta je_3/2)$, which means that there is a valid reason to look for other possibilities. The representation of analytic functions as series is well established, which immediately gives us the possibility to formally define functions like $\exp(\varphi e_3)$, $\varphi \in \mathbb{R}$. All even terms in the series give a real number, while the odd terms give a vector. Then collecting all real terms and all vector terms, we get the hyperbolic cosine and sine. There is an important question about the geometric meaning, if any, of such an expression. We will see in the text that $\exp(\varphi e_3)$, $\varphi \in \mathbb{R}$, defines a *boost* in the special theory of relativity. We can also consider functions of the form

$\exp(\varphi e_3 + j\theta e_2)$, φ, $\theta \in \mathbb{R}$, and find that they represent *Lorentz transformations* (see Sect. 2.7.6). Consider now a multivector of the form $\alpha + \beta e_3$ (*paravector*), where due to $e_3^2 = 1$, we can apply the theory of hyperbolic numbers immediately; e_3 behaves just like a hyperbolic unit. Similarly, we can use any element that squares to -1 and apply the theory of complex numbers or use nilpotents to apply the theory of *dual numbers*. These examples teach us that we should try to find a way to define functions of multivectors in general, at least in *Cl*3. However, the powers of multivectors are hard to manipulate in general. There is a nice text [16] about functions of multivectors in *Cl*3; however, we will try another approach here.

Consider idempotents $p_\pm$, like $(1 \pm e_3)/2$, with the special property $p_+ p_- = p_- p_+ = 0$ (*zero divisors*). Squaring the linear combination $m = \alpha p_+ + \beta p_-$, α, $\beta \in \mathbb{R}$, we get $m^2 = \alpha^2 p_+ + \beta^2 p_-$, or more generally,

$$m^n = \alpha^n p_+ + \beta^n p_-, \quad n \in \mathbb{N}.$$

This magical property gives us the opportunity to find functions of general multivectors; all we need is a way to decompose a multivector as a linear combination of idempotents with coefficients that are complex scalars (*spectral decomposition*), and we can do this in *Cl*3. ◆ Show that from $\alpha p_+ + \beta p_- = 0$ follows $\alpha = \beta = 0$. ◆

Of course, there remain a few things on which to comment. First, we have to reconsider our concept of a number and accept the possibility that numbers are something more general than we thought. We have real numbers in geometric algebra, but we also have oriented objects with magnitude, which define subspaces, like lines and planes. Second, if we define the function $\exp(\theta j)$, $\theta \in \mathbb{R}$, we can consider it to be like a complex phase, where the pseudoscalar j is the commutative imaginary unit. However, the pseudoscalar j is also an oriented volume, and we should try to understand the geometric meaning of such an expression. Here we will propose a technique to define functions; the question of possible geometric interpretations is beyond the scope of this text (except for simple cases). We can also anticipate possible questions about convergence of series with multivectors; however, recall that multivectors have finite magnitudes, which means that we should not bother with such questions here.

The reader should master the Sects. 2.1, 2.2, and 2.3 before reading this part. With a clear understanding of the structure of a multivector in *Cl*3, this chapter should be easy to understand.

2.4.2 Spectral Decomposition

We represent a general function by f (do not confuse with **f**).

Using elements from the text above, we define idempotents $u_\pm = (1 \pm \mathbf{f})/2$, $\mathbf{f}^2 = 1$, with the properties

$$u_+ + u_- = 1, \quad u_+ - u_- = \mathbf{f}, \quad u_+ u_- = u_- u_+ = 0, \quad u_\pm^2 = u_\pm, \quad \bar{u}_+ = u_-.$$

The idempotents $u_\pm$ do not make a full spectral basis in *Cl*3 (for details about the *spectral basis*, see Sect. 4.2, [43]). Also, we should write $\mathbf{f} = \mathbf{f}(M)$ and $u_\pm = u_\pm(M)$, but we omit this. We can express a general multivector with the complex vector $\mathbf{F}^2 \neq 0$ as (see the definition of the *unipodal numbers* in the Glossary).

$$M = Z + \mathbf{F} = Z + \sqrt{\mathbf{F}^2}\mathbf{f} \equiv Z + Z_f\mathbf{f},\ \ \mathbf{f}^2 = 1,\ \ Z_f = \sqrt{\mathbf{F}^2};$$

therefore, if we define complex scalars $M_\pm \equiv Z \pm Z_f$, we get the form

$$M = Z + Z_f\mathbf{f} = Z(u_+ + u_-) + Z_f(u_+ - u_-) = M_+u_+ + M_-u_-.$$

We say that we have the *spectral decomposition* of a multivector. The spectral decomposition gives us the **magical opportunity**

$$M^2 = (M_+u_+ + M_-u_-)^2 = M_+^2u_+ + M_-^2u_-,$$

which we can immediately generalize to any positive integer in the exponent, and then to negative integers as well if the inverse of a multivector exists. E29: Prove that in the spectral decomposition we have $M\bar{M} = M_+M_-$.

2.4.3 Series Expansion

For analytic functions, we can utilize series expansions to find

$$\mathrm{f}(M) = \mathrm{f}(M_+)u_+ + \mathrm{f}(M_-)u_- = \mathrm{f}_+ + \mathrm{f}_-\mathbf{f}$$

(**if** $\mathrm{f}(M_\pm)$ **exist**), where

$$\mathrm{f}_\pm \equiv (\mathrm{f}(M_+) \pm \mathrm{f}(M_-))/2$$

are functions that are symmetric and antisymmetric with respect to the substitutions $M_+ \leftrightarrow M_-$. Note that these substitutions arise from the Clifford conjugation $\mathrm{f}(\bar{M}) = \mathrm{f}(Z - Z_f\mathbf{f})$.

As an example, consider the series expansion for the exponential function (see Appendix 5.1):

$$\exp(x) = \sum_{k=0}^{\infty} \frac{x^k}{k!}.$$

Note that for exp(1), we get a satisfactory result after just a few terms in the series expansion (Fig. 2.5).

Important multivectors have finite magnitudes, which ensures the convergence under the geometric product (see the literature).

Now we can write

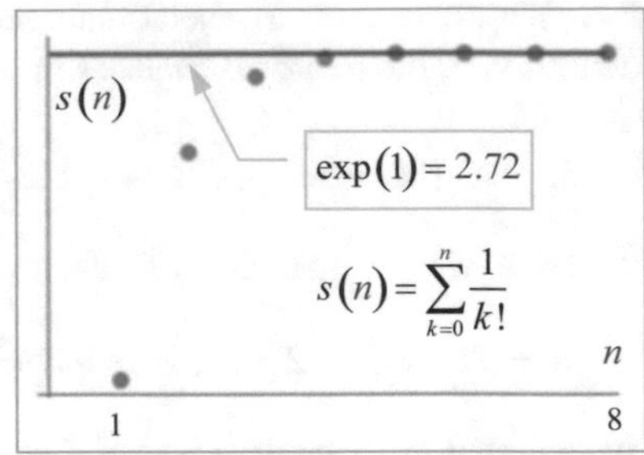

Fig. 2.5 The fast convergence of the exponential function

$$\begin{aligned}\exp(M) &= \sum_{k=0}^{\infty}\frac{(M_+u_+ + M_-u_-)^k}{k!} = u_+\sum_{k=0}^{\infty}\frac{(M_+)^k}{k!} + u_-\sum_{k=0}^{\infty}\frac{(M_-)^k}{k!} \\ &= u_+\exp(M_+) + u_-\exp(M_-)\end{aligned}.$$

Recall that we can use the theory of complex numbers to find f($M_\pm$): just switch $j \to i = \sqrt{-1}$, find the function of the complex variable, and switch again $i \to j$. As an example, let us find $\exp(1+j)$:

$$\exp(1+i) = e(\cos 1 + i\sin 1) \xrightarrow{i\to j} e(\cos 1 + j\sin 1).$$

Here we should be careful, especially with the square roots. Consider the multivector

$$M = \mathbf{F} = \mathbf{x} + j\mathbf{n},\quad |\mathbf{x}| = |\mathbf{n}|,\quad \mathbf{x}\cdot\mathbf{n} = 1/2,$$

for which we have $\mathbf{F}^2 = x^2 - n^2 + 2j\mathbf{x}\cdot\mathbf{n} = j$ and therefore $M_\pm = \pm\sqrt{j}$. For the square root of j we get

$$\sqrt{j} = \frac{1 \pm j}{\sqrt{2}};$$

however, from $\mathbf{F}^2 = j$, we could also write $\sqrt{j} = \mathbf{F}$. This result is new; however, it should not be surprising, because for vectors, we have $\mathbf{x}^2 = \alpha \in \mathbb{R}$ from the start. We can understand this as $|\mathbf{x}| = \sqrt{\alpha}$; however, with the geometric product, we have the other possibility

$$\mathbf{x}^2 = \mathbf{x}\mathbf{x} = |\mathbf{x}|^2 = \alpha \Rightarrow \mathbf{x} = \sqrt{\alpha},$$

specifically,

$$\begin{aligned}&\sqrt{1} = 1 \quad \text{(defined over real numbers)},\\ &\sqrt{1} = 1 \;\text{ or }\; \mathbf{f} \quad \text{(defined over multivectors in GA)},\end{aligned}$$

where **f** is a unit complex vector. Note that in this way, the square root becomes a function defined by means of the geometric product, and not only by means of the ordinary product of complex numbers. In geometric algebra, we deal with many different objects, all united in the same system; consequently, we should rethink our understanding of the concept of number. Points, lines, areas, volumes, and so on, are represented by real numbers, vectors, bivectors, trivectors, and so on, respectively. There is no fundamental difference between them; they are just subspaces in an n-dimensional vector space described by 2^n-dimensional linear space—geometric algebra. Such a unified understanding of geometric objects represented by oriented numbers may have far-reaching consequences.

2.4.4 Functions of Complex Vectors

For multivectors $M = \mathbf{F} = Z_f \mathbf{f}$ (complex vectors), we have $M_\pm = \pm Z_f$, whence follows

$$\mathrm{f}(\mathbf{F}) = \mathrm{f}(Z_f)u_+ + \mathrm{f}(-Z_f)u_-.$$

For even functions we have

$$\mathrm{f}(\mathbf{F}) = \mathrm{f}(Z_f)(u_+ + u_-) = \mathrm{f}(Z_f),$$

while for odd functions we get

$$\mathrm{f}(\mathbf{F}) = \mathrm{f}(Z_f)(u_+ - u_-) = \mathrm{f}(Z_f)\mathbf{f}.$$

2.4.5 Functions of Nilpotents

Multivectors of the form $M = Z + \mathbf{F},\ \mathbf{F}^2 = \mathbf{N}^2 = 0$ do not have a spectral decomposition. However, using the binomial expansion,

$$M^n = (Z + \mathbf{N})^n = Z^n + nZ^{n-1}\mathbf{N},$$

we have (the function expansion about Z)

$$\mathrm{f}(Z + \mathbf{N}) = \mathrm{f}(Z) + \frac{d\mathrm{f}(Z)}{dZ}\mathbf{N}.$$

For example, $d\exp(z)/dz = \exp(z)$, $\exp(\mathbf{N}) = \exp(0) + \exp(0)\mathbf{N} = 1 + \mathbf{N}$. However, we could obtain this result from the series expansion; almost all terms simply disappear. As we have shown above, this could help us to carry out *automatic differentiation*; for $\mathrm{f}(Z)$, just calculate $\mathrm{f}(Z+\mathbf{N})$. For example, specifying $\mathrm{P}(Z) = Z^2 - 1$, we have $\mathrm{P}(j + \mathbf{N}) = -2 + 2j\mathbf{N}$, where $\mathrm{P}(j) = -2$ and $P'(j) = 2j$. This is easy to implement on a computer.

2.4.6 Functions of Idempotents and Unit Complex Vectors

We can look at some special cases:

2

$$\mathrm{f}(u_\pm) = \mathrm{f}(1)u_\pm + \mathrm{f}(0)u_\mp, \quad (2.1)$$
$$\mathrm{f}(\mathbf{f}) = \mathrm{f}(u_+ - u_-) = \mathrm{f}(1)u_+ + \mathrm{f}(-1)u_-, \quad (2.2)$$
$$\mathrm{f}(j\mathbf{f}) = \mathrm{f}\left(ju_+ - ju_-\right) = \mathrm{f}(j)u_+ + \mathrm{f}(-j)u_-. \quad (2.3)$$

From the relation (2.1), we see that the logarithm of an idempotent is not defined, because log(0) is not defined. For the exponential function, we have

$$\exp(u_+) = u_+\exp(1) + u_-\exp(0) = \frac{e+1}{2} + \frac{e-1}{2}\mathbf{f},$$
$$\exp(\mathbf{f}) = \exp(u_+ - u_-) = eu_+ + e^{-1}u_- = \cosh 1 + \mathbf{f}\sinh 1,$$
$$\exp(je_3) = \exp\left(ju_+ - ju_-\right) = e^j u_+ + e^{-j}u_- = \cos 1 + je_3 \sin 1.$$

We have such relations in the special theory of relativity (boosts and rotations). Generally, from (2.2) and (2.3), we have ($Z \in \mathbb{C}$)

$$\mathrm{f}(Z\mathbf{f}) = \mathrm{f}(Z)u_+ + \mathrm{f}(-Z)u_- = \frac{\mathrm{f}(Z) + \mathrm{f}(-Z)}{2} + \frac{\mathrm{f}(Z) - \mathrm{f}(-Z)}{2}\mathbf{f}$$
$$= \mathrm{f}_+(Z) + \mathrm{f}_-(Z)\mathbf{f},$$

where $\mathrm{f}(Z) = \mathrm{f}_+(Z) + \mathrm{f}_-(Z)$ is the decomposition of the function into its even and odd parts. The reader should investigate different possibilities, special cases, etc. It is important to illuminate the geometric content in relations if possible.

2.4.7 Inverse Functions

For the inverse function and multivectors X, Y, we have (note that from $\alpha u_+ + \beta u_- = 0$ follows $\alpha, \beta = 0$, $\alpha, \beta \in \mathbb{C}$)

$$\mathrm{f}^{-1}(Y) = X \Rightarrow \mathrm{f}(X) = Y \Rightarrow \mathrm{f}(X_\pm) = Y_\pm \Rightarrow X_\pm = \mathrm{f}^{-1}(Y_\pm),$$

where $X_\pm$, $Y_\pm \in \mathbb{C}$ are the complex scalars from the spectral decomposition of the multivectors X, Y and we assume the existence of $\mathrm{f}(X_\pm)$ and $\mathrm{f}^{-1}(Y_\pm)$.

2.4.8 Functions of Lightlike Multivectors

If $M\bar{M} = 0$ (the *lightlike multivectors*, sometimes referred as *null multivectors*; see below), we have

$$Z^2 - \mathbf{F}^2 = Z^2 - Z_f^2 = 0 = \left(Z - Z_f\right)\left(Z + Z_f\right),$$

whence two options follow:

1. $Z_f = Z \Rightarrow M_+ = 2Z,\ M_- = 0 \Rightarrow \mathrm{f}(M) = \mathrm{f}(2Z)u_+ + \mathrm{f}(0)u_-$,
2. $Z_f = -Z \Rightarrow M_+ = 0,\ M_- = 2Z \Rightarrow \mathrm{f}(M) = \mathrm{f}(0)u_+ + \mathrm{f}(2Z)u_-$,

where, of course, the conditions of existence must be met (for example, the logarithm of a lightlike multivector is not defined). We will see such multivectors again, for instance in the special theory of relativity.

2.4.9 Elementary Functions

We will now give some examples of *elementary functions*.

The **inverse** of a multivector ($M\bar{M} \neq 0$) is found easily:

$$M^{-1} = \frac{1}{M_+u_+ + M_-u_-} = \frac{M_+u_- + M_-u_+}{(M_+u_+ + M_-u_-)(M_+u_- + M_-u_+)} = \frac{M_+u_- + M_-u_+}{M_+M_-} = \frac{u_+}{M_+} + \frac{u_-}{M_-},$$

with the powers

$$M^{-n} = \frac{1}{(M_+u_+ + M_-u_-)^n} = \frac{u_+}{(M_+)^n} + \frac{u_-}{(M_-)^n}, \quad n \in \mathbb{N}.$$

The **square root** is (see [16] for a different form)

$$\sqrt{M} = S = S_+u_+ + S_-u_- \Rightarrow M = M_+u_+ + M_-u_- = (S_+)^2u_+ + (S_-)^2u_- \Rightarrow S_\pm = \pm\sqrt{M_\pm},$$

or

$$M^{\pm 1/n} = S \Rightarrow S_\pm = (M_\pm)^{\pm 1/n}, \quad n \in \mathbb{N}.$$

For example, $M = e_1$, $M_\pm = \pm 1$,

$$\sqrt{e_1} = \sqrt{1}u_+ + \sqrt{-1}u_- = \frac{\sqrt{1} + \sqrt{-1}}{2} + \frac{\sqrt{1} - \sqrt{-1}}{2}e_1.$$

One possibility is $\sqrt{1} = \pm 1$, $\sqrt{-1} = \pm j$, $\sqrt{e_1} = \pm(\pm j + e_1)/\sqrt{\pm 2j}$. This is in accordance with [16]:

$$\sqrt{M} = \frac{M \pm |M|}{\sqrt{M + \bar{M} \pm 2|M|}} = \frac{Z + \mathbf{F} \pm \sqrt{Z^2 - \mathbf{F}^2}}{\sqrt{Z \pm 2\sqrt{Z^2 - \mathbf{F}^2}}}.$$

We can also use $\sqrt{1} = \pm e_1$, $\sqrt{-1} = \pm je_1$, which gives the same solution again, but also $\sqrt{1} = \pm\mathbf{f}$, $\sqrt{-1} = \pm j\mathbf{f}$, $\mathbf{f}^2 = 1$, which gives the same solution multiplied by $\mathbf{f}$. Finally, we have

$$\sqrt{\mathbf{e}_i} = \pm\mathbf{f}(\pm j + \mathbf{e}_i)/\sqrt{\pm 2j}.$$

Now we can write the solution from [16] as (note that $M\mathbf{f} = \mathbf{f}M$)

$$\sqrt{M} = \pm\mathbf{f}\frac{M \pm |M|}{\sqrt{M + \bar{M} \pm 2|M|}}, \quad \mathbf{f}^2 = 1.$$

The **exponential function** has the decomposition

$$\exp(M) = \exp(M_+)u_+ + \exp(M_-)u_-,$$

and consequently, the **logarithm** is obtained as

$$\begin{aligned}\log M = X \Rightarrow e^X = M = M_+u_+ + M_-u_- \\ = \exp(X_+)u_+ + \exp(X_-)u_- \Rightarrow X_\pm = \log M_\pm \Rightarrow \\ \log M = \log(M_+)u_+ + \log(M_-)u_-.\end{aligned}$$

With the definition $\mathbf{I} \equiv \mathbf{F}/|\mathbf{F}| = -j\mathbf{f}, \quad \mathbf{I}^2 = -1$, the logarithmic function has the form (see [16])

$$\log M = \log|M| + \varphi\mathbf{I}, \quad \varphi = \arctan(|\mathbf{F}|/Z);$$

however, we can show that these two formulas are equivalent:

$$\begin{aligned}&\log(M_+)u_+ + \log(M_-)u_- = \frac{\log(M_+) + \log(M_-)}{2} + \frac{\log(M_+) - \log(M_-)}{2}\mathbf{f} \\ &= \log|M| - j\mathbf{I}\log\left(\sqrt{(1 - j|\mathbf{F}|/Z)/(1 + j|\mathbf{F}|/Z)}\right) \\ &= \log|M| + \mathbf{I}\arctan(|\mathbf{F}|/Z) = \log|M| + \varphi\mathbf{I}.\end{aligned}$$

Example 1

$$\begin{aligned}&M = e_I, \quad u_\pm = (1 \pm e_I)/2, \quad e_1u_\pm = \pm u_\pm, \quad e_I = u_+ - u_- \Rightarrow \\ &\log e_I = u_+\log 1 + u_-\log(-1) = j\pi u_-. \\ &\log|M| + \varphi\mathbf{I} = \log(j) - je_1\arctan(\infty) = j\pi/2 - je_1\pi/2 = j\pi u_-.\end{aligned}$$

◆ Using the spectral decomposition, show that ${e_I}^{e_I} = e_I$. Note that formally, $({e_I}^{e_I})^2 = (e_1^2)^{e_1} = 1$. ◆

For the **trigonometric functions**, we can start from

$$\sin(Z + \mathbf{F}) = \sin Z\cos\mathbf{F} + \cos Z\sin\mathbf{F}$$

(decompose both sides by $u_\pm$ to prove this). Using $|\mathbf{F}| = \sqrt{\mathbf{F}\bar{\mathbf{F}}} = \sqrt{-\mathbf{F}^2} = j\sqrt{\mathbf{F}^2} \Rightarrow \mathbf{I} = \mathbf{F}/|\mathbf{F}| = -j\mathbf{f}$, we can treat $\mathbf{I}$ as an imaginary unit and write (again, decompose both sides by $u_\pm$ or use the series expansions to prove this)

$$\sin\mathbf{F} = \sin(\mathbf{I}|\mathbf{F}|) = \mathbf{I}\sinh|\mathbf{F}|, \quad \cos\mathbf{F} = \cos(I|\mathbf{F}|) = \cosh|\mathbf{F}|,$$

whence follows

$$\sin M = \sin Z \cosh|\mathbf{F}| + \mathbf{I}\cos Z \sinh|\mathbf{F}|.$$

The reader should prove the following formulas:

$$\begin{aligned}
\cos M &= \cos Z \cosh|\mathbf{F}| - \mathbf{I}\sin Z \sinh|\mathbf{F}|,\\
\sinh M &= \sinh Z \cos|\mathbf{F}| + \mathbf{I}\cosh Z \sin|\mathbf{F}|,\\
\cosh M &= \cosh Z \cos|\mathbf{F}| + \mathbf{I}\sinh Z \sin|\mathbf{F}|,\\
\exp(M) &= \cosh M + \sinh M, \quad \exp(jM) = \cos M + j\sin M,\\
\cosh^2 M &- \sinh^2 M = \cos^2 M + \sin^2 M - 1,\\
\operatorname{arcsinh} M &= \log\left(M + \sqrt{1+M^2}\right), \quad \operatorname{arcsinh}(jM) = j\arcsin M,\\
\operatorname{arccosh} M &= \log\left(M + \sqrt{M^2-1}\right), \quad \operatorname{arccosh}(jM) = j\arccos M,\\
\operatorname{arctanh} M &= \log\sqrt{\frac{1+M}{1-M}}, \quad \operatorname{arctanh}(jM) = j\arctan M.
\end{aligned}$$

For $M = \mathbf{N}$, $\mathbf{N}^2 = 0$, we have $Z = |\mathbf{F}| = 0$, $\sin\ \mathbf{N} = 0$, $\cos\ \mathbf{N} = 1$, etc.

2.4.10 Polynomial Equations

We can now take an example of the polynomial equation

$$M^2 - 1 = 0,$$

whose solutions are all multivectors that square to 1. Apart from $\pm e_i$, we could try

$$(Z + \mathbf{F})^2 - 1 = 0 \Rightarrow Z^2 + 2Z\mathbf{F} + \mathbf{F}^2 - 1 = 0 \Rightarrow Z = 0 \Rightarrow \mathbf{F}^2 = 1,$$

and we know the general solution. Using the *spectral decomposition*, we have

$$\begin{aligned}
&M^2 - 1 = (M_+u_+ + M_-u_-)^2 - (u_+ + u_-) = \left(M_+^2 - 1\right)u_+ + \left(M_-^2 - 1\right)u_- = 0 \Rightarrow\\
&M_+^2 - 1 = 0, \;\; M_-^2 - 1 = 0.
\end{aligned}$$

Therefore, we get two equations with complex numbers. Then, for instance, from $M_\pm = \pm 1$ it follows that $Z = 0$, $Z_f = 1$, and the solution is $M = \mathbf{f}$. This was just a little demonstration of possibilities; the reader should do the complete calculations. Note that in general, from f $(M) = 0$ follows the pair of equations in complex numbers $f(M_\pm) = 0$. If we find $M_\pm$, then any unit complex vector $\mathbf{f}$ will give a possible solution for M.

2.4.11 A New Concept of Numbers

We have already pointed out that *Cl*3 has complex and hyperbolic structures, the complex one due to *j* and other elements that square to -1, and the hyperbolic one due to elements that square to 1; unit vectors are hyperbolic, for example. There are also dual numbers here (using nilpotents). It is possible to formulate the special theory of relativity efficiently using ordinary hyperbolic (*double*, *split-complex*) numbers. Consequently, it should not be a surprise if it turns out that the STR is easy to formulate in *Cl*3 (see Sect. 2.7). A unit complex vector **f** is the most general element of the algebra with features of a hyperbolic unit. For two multivectors that have the same unit complex vector **f** (the same "direction"),

$$M_1 = z_1 + z_{1f}\mathbf{f}, \quad M_2 = z_2 + z_{2f}\mathbf{f},$$

we can define the square of the *distance of multivectors* as

$$\bar{M}_1 M_2 \equiv \left(z_1 - z_{1f}\mathbf{f}\right)\left(z_2 + z_{2f}\mathbf{f}\right) = z_1 z_2 - z_{1f} z_{2f} + \left(z_1 z_{2f} - z_2 z_{1f}\right)\mathbf{f} \equiv h_{\mathrm{i}} + h_{\mathrm{o}}\mathbf{f},$$

where h_{i} and h_{o} are the *hyperbolic inner* and the *hyperbolic outer* products. If $M_1 = M_2 = M$, we have the *multivector amplitude*, which means that the *distance of multivectors* is a generalization of the *multivector amplitude*. For $h_{\mathrm{o}} = 0$, we say that multivectors are *h*-parallel, while for $h_{\mathrm{i}} = 0$, they are *h*-orthogonal.

Lemma *Let $M_1 \neq 0$ and $M_2 \neq 0$. Then from $\bar{M}_1 M_2 = 0$ follows $M_1 M_2 \neq 0$ and vice versa.*
Proof:

$$M_1 M_2 = (M_{1+}u_+ + M_{1-}u_-)(M_{2+}u_+ + M_{2-}u_-) = M_{1+}M_{2+}u_+ + M_{1-}M_{2-}u_-,$$
$$\bar{M}_1 M_2 = (M_{1+}u_- + M_{1-}u_+)(M_{2+}u_+ + M_{2-}u_-) = M_{1-}M_{2+}u_+ + M_{1+}M_{2-}u_-,$$

and consequently, $M_{1-}M_{2+} = M_{1+}M_{2-} = 0$, which means that $M_{1-} = 0$ & $M_{2-} = 0$ or $M_{1+} = 0$ & $M_{2+} = 0$. However, both cases imply $M_1 M_2 \neq 0$. The proof of the reverse statement is similar.

2.5 What Is the Square Root of −1?

3. $\mathbf{F}^2 = -1$

Generally, this kind of a complex vector can be obtained by $\sqrt{-\mathbf{F}^2} = \sqrt{\mathbf{F}\bar{\mathbf{F}}} \equiv |\mathbf{F}|$, and we define $\mathbf{I} \equiv \mathbf{F}/|\mathbf{F}| = -j\mathbf{f}$, $\mathbf{I}^2 = -1$. The general form is

$$\mathbf{I} = \mathbf{n}\sinh\varphi + j\mathbf{m}\cosh\varphi, \quad \mathbf{n}^2 = \mathbf{m}^2 = 1, \quad \mathbf{n}\perp\mathbf{m}.$$

Note that we have a nontrivial solution for $\sqrt{-1}$. In order to substantiate this further, we can look for all possible solutions for $\sqrt{c + jd}$, so we need to solve the equation $M^2 = c + jd = z$. One solution is just the ordinary square root of a complex number (for $\mathbf{F} = 0$); however, more generally,

$$z = c + jd = (Z + \mathbf{F})^2 = Z^2 + 2Z\mathbf{F} + \mathbf{F}^2 \Rightarrow Z = 0 \Rightarrow \mathbf{F} = \sqrt{c + jd} = \mathbf{v} + j\mathbf{w},$$

where **v** and **w** are vectors, and therefore,

$$c + jd = v^2 - w^2 + 2j\mathbf{v} \cdot \mathbf{w},$$

and $c = v^2 - w^2, \quad d = 2 \ \mathbf{v} \cdot \mathbf{w}$. Amazingly, the square root of a complex number is a complex vector (and this we expect; the square of a complex vector is a complex scalar)! Note that the square root is defined in *Cl*3, by means of the geometric product. ◆ The reader should explore different possibilities. ◆ For $z = 1$, we have

$$v^2 - w^2 = 1, \quad d = 2 \ \mathbf{v} \cdot \mathbf{w} = 0,$$

which gives unit vectors or unit complex vectors that square to 1. For $z = -1$, we have

$$v^2 - w^2 = -1, \quad d = 2 \ \mathbf{v} \cdot \mathbf{w} = 0,$$

which gives unit complex vectors that square to -1. For $z = j$, we have

$$v^2 - w^2 = 0, \quad d = 2 \ \mathbf{v} \cdot \mathbf{w} = 1,$$

which gives interesting complex vectors (draw them). However, for $\mathbf{F} = 0$, we have, for example,

$$z = j = Z^2 = (a + jb)^2 = a^2 + 2abj - b^2 \Rightarrow a = b = \pm 1/\sqrt{2}.$$

2.6 Trigonometric Forms of Multivectors

Recall that for $\mathbf{F}^2 = 0$ we defined dual numbers $z = \alpha + \beta\varepsilon$, $\varepsilon^2 = 0$, $\alpha, \beta \in \mathbb{R}$ and that for $\alpha \neq 0$, we found the polar form

$$z = \alpha + \beta\varepsilon = \alpha(1 + \varphi\varepsilon), \quad \varphi = \beta/\alpha,$$

where φ is the *argument* of the dual number.

The elements **f** and **I** can be used to define trigonometric forms of general multivectors. To take advantage of the theory of *complex numbers*, we use **I**. Accordingly, we define the *argument of a multivector* as

$$\varphi = \arg M \equiv \arctan(|\mathbf{F}|/Z).$$

Now, using $|M| \equiv \sqrt{M\bar{M}}$, we have (with the conditions of existence)

$$\cos\varphi = Z/|M|, \quad \sin\varphi = |\mathbf{F}|/|M|,$$

which gives

$$M = Z + \mathbf{F} = |M|(\cos\varphi + \mathbf{I}\sin\varphi).$$

Recalling that $\mathbf{I}^2 = -1$, the generalized de Moivre's formula is valid:

$$M^n = |M|^n[\cos(n\varphi) + \mathbf{I}\sin(n\varphi)].$$

Note that we have the same form as for complex numbers. However, there is a substantial difference. The element $\mathbf{I}$ has a clear geometric meaning: it contains the properties that are determined by the vectors that define the complex vector part of a multivector. Using $\mathbf{F} = \mathbf{I}|\mathbf{F}|$ and our techniques, we get

$$e^M = e^{Z+\mathbf{F}} = e^Z e^{\mathbf{F}} = e^Z(\cos|\mathbf{F}| + \mathbf{I}\sin|\mathbf{F}|),$$

which is possible due to the commutativity of the complex scalar Z. The case $\mathbf{F}^2 = 0$ was discussed earlier.

2.6.1 Mathematics of the Special Theory of Relativity

To take the advantage of the theory of *hyperbolic numbers*, we use $\mathbf{f}$:

$$\begin{aligned} M &= Z + \mathbf{F} = Z + Z_f\mathbf{f} \\ &= \rho\left(\frac{Z}{\rho} + \frac{Z_f}{\rho}\mathbf{f}\right) = \rho(\cosh\varphi + \mathbf{f}\sinh\varphi), \quad \rho = \sqrt{M\bar{M}} = \sqrt{Z^2 - Z_f^2}. \end{aligned}$$

If $M\bar{M} = 0$, there is no polar form (*lightlike* multivectors), but then we have $M = Z(1 \pm \mathbf{f})$. Defining the "velocity" $\vartheta \equiv \tanh\varphi$, $\gamma = \cosh\varphi$, $\gamma\vartheta = \sinh\varphi$, it follows that

$$M = \rho(\cosh\varphi + \mathbf{f}\sinh\varphi) = \rho\gamma(1 + \vartheta\mathbf{f}), \quad \gamma^{-1} = \sqrt{1 - \vartheta^2}.$$

Defining the *proper velocity* $u = \gamma(1 + \vartheta\mathbf{f})$, $u\bar{u} = 1$, we get the *velocity addition rule*

$$\begin{aligned} &\gamma_1\gamma_2(1 + \vartheta_1\mathbf{f})(1 + \vartheta_2\mathbf{f}) = \gamma_1\gamma_2(1 + \vartheta_1\vartheta_2 + (\vartheta_1 + \vartheta_2)\mathbf{f}) \\ &= \gamma_1\gamma_2(1 + \vartheta_1\vartheta_2)(1 + \mathbf{f}(\vartheta_1\vartheta_2)/(1 + \vartheta_1\vartheta_2)) \Rightarrow \\ &\gamma = \gamma_1\gamma_2(1 + \vartheta_1\vartheta_2), \quad \vartheta = (\vartheta_1 + \vartheta_2)/(1 + \vartheta_1\vartheta_2), \end{aligned}$$

which is just like the velocity addition rule of the special theory of relativity. The proper velocity in a "rest reference system" ($\vartheta = 0$) is $u_0 = 1$. Consequently, we can transform to a new reference frame by $u_0 u = u$ or, as in the previous example, $u_0 u_1 u_2 = u_1 u_2$. These formulas represent geometric relations. They are more general than the formulas of the STR, for which we need just the real part of a multivector (*paravectors*; see the next chapter). Note that for $u_0 = 1$ we have $M \in \mathbb{C}$, which means that we get the general form of a multivector $M = Z + \mathbf{F}$ using the boost $u = \gamma(1 + \vartheta\mathbf{f})$. This is an astonishing property of the 3D Euclidean vector space, which follows from the geometric multiplication of vectors. It would be difficult to overestimate the importance and consequences of this result.

Using the spectral decomposition, we have

$$M = \rho\gamma(1 + \vartheta \mathbf{f}) = K_+ u_+ + K_- u_- \Rightarrow$$
$$K_\pm = \rho\gamma(1 \pm \vartheta) \equiv \rho k^{\pm 1}, \quad k = \sqrt{(1+\vartheta)/(1-\vartheta)},$$

where from $\exp(2\varphi) = (1 + \tanh \varphi)/(1 - \tanh \varphi)$, it follows that

$$\varphi = \log k,$$

and we have the generalized *Bondi factor* from the special theory of relativity. It follows that

$$(k_1 u_+ + u_-/k_1)(k_2 u_+ + u_-/k_2) = k_1 k_2 u_+ + u_-/(k_1 k_2) \Rightarrow k = k_1 k_2,$$

which is the exact formula from the special theory of relativity, equivalent to the velocity addition rule.

It is self-evident that the geometric product gave us the possibility of writing relativistic formulas **without the use of Minkowski space.** *If Einstein had known that…*

2.6.2 Everything Is a "Boost"

In the case of $\rho = 1$, for a complex vector $\mathbf{F} = \mathbf{v} + j\mathbf{w}$, we define $W = \sqrt{\mathbf{F}^2} \in \mathbb{C}$ (above, the quantity W was denoted by Z_f; note that it is also possible to write $\sqrt{\mathbf{F}^2} = \mathbf{F}$). For $\mathbf{F}^2 \neq 0$, we define $\mathbf{F}/W = \mathbf{f}$, $\mathbf{f}^2 = 1$, and $\mathbf{F}/\sqrt{-W} = \mathbf{I} = -j\mathbf{f}$, $\mathbf{I}^2 = -1$. From the exponential form $\exp(\varphi \mathbf{f})$, defining $\tanh\varphi = W$, we follow the formulas from the previous chapter with $\vartheta \to W$. Generally, we have a complex scalar $\varphi = \varphi_{\mathrm{Re}} + j\varphi_{\mathrm{Im}}$, which leads to

$$\exp(\varphi \mathbf{f}) = \exp(\varphi_{\mathrm{Re}} \mathbf{f}) \exp(j\varphi_{\mathrm{Im}} \mathbf{f}).$$

From $\mathbf{F} = \mathbf{v} + j\mathbf{w}$, we have $W = \sqrt{(\mathbf{v} + j\mathbf{w})^2} = \sqrt{v^2 - w^2 + 2j\mathbf{v}\cdot\mathbf{w}}$, and we will examine some special cases.

For $\mathbf{w} = 0$, we have $W = v = \tanh \varphi$, $\mathbf{f} = \hat{\mathbf{v}}$, $\gamma = \cosh\varphi = 1/\sqrt{1 - v^2} \in \mathbb{R}$ (why?), $v < 1$, and the proper velocity is $u = \gamma(1 + \mathbf{v})$, whence follow the well-known relations for boosts in restricted special relativity. Note that the limiting speed follows from the geometry of 3D Euclidean space (in GA).

For $\mathbf{v} = 0$, we have $W = \sqrt{(j\mathbf{w})^2} = \sqrt{-w^2} = jw$, $\mathbf{f} = j\mathbf{w}/jw = \hat{\mathbf{w}}$, $\gamma = 1/\sqrt{1 + w^2}$, $k = \sqrt{(1 + jw)/(1 - jw)}$, $\varphi = \log k = \arctan w$, $\exp(\varphi \hat{\mathbf{w}}) = \gamma(1 + jw\hat{\mathbf{w}})$, whence for successive "boosts" we have

$$\gamma = \gamma_1\gamma_2(1 - w_1 w_2), \quad w = (w_1 + w_2)/(1 + w_1 w_2).$$

In physics, $j\mathbf{w}$ could be interpreted as a quantity associated with spin (see Appendix 5.4, where we discuss the case in which both $\mathbf{v}$ and $\mathbf{w}$ are nonzero vectors).

For the well-known pure rotations $\exp(\theta j\hat{\mathbf{n}})$, we have $W = \tanh(\theta j) = j \tan \theta$, $\gamma = 1/\sqrt{1 + \tan^2\theta} = \cos\theta$, $k = \sqrt{(1 + j\tan\theta)/(1 - j\tan\theta)}$, $u_\pm = (1 + \hat{\mathbf{n}})/2$, whence

$$\exp(\theta j\hat{\mathbf{n}}) = \gamma(1 + j\hat{\mathbf{n}}\tan\theta) = \cos\theta + j\hat{\mathbf{n}}\sin\theta \quad (\textit{proper velocity}),$$
$$\gamma = \gamma_1\gamma_2(1 - \tan\theta_1 \tan\theta_2) \quad (\textit{gamma rule}),$$
$$\tan\theta = (\tan\theta_1 + \tan\theta_2)/(1 - \tan\theta_1\tan\theta_2) = \tan(\theta_1 + \theta_2) \quad (\textit{velocity addition}).$$

Can you recognize the trigonometric addition formulas?

2.7 The Special Theory of Relativity

The special theory of relativity (STR), in its classic form, is the theory of transformations of coordinates, where the concept of velocity is particularly important. Geometric algebra does not substantially depend on specific coordinates, which gives us the opportunity to consider general geometric relationships, not only relations between coordinates, which is certainly desirable; physical processes do not depend on the coordinate systems in which they are expressed. Unfortunately, many authors who use geometric algebra adore coordinates, making thus formulas opaque and blurring the geometric content. It is hard to get rid of old habits. There are many texts and comments about STR; there are many opponents as well, which often just demonstrate a lack of understanding of the theory. Therefore, for example, they say that Einstein *wrote nonsense because he uses the speed of a photon as c, as well as* $c \pm v$ in his formulas. Critics do not recognize the important and simple fact that the speed of a photon is c in any inertial reference system. However, if we want to find the time a photon needs to reach the wall of the rail car that is moving away from the photon (viewed from the rail system, the *collision time*) we must use $c \pm v$. Why? Because it is the *collision velocity* between the photon and the wall of the rail car in the rail system. Both the speed of the photon and the speed of the wall are measured in the same reference system; consequently, they are simply added, without the relativistic addition rule. It is quite another matter when we have a man in the rail car walking in the direction of movement of the train with velocity u relative to the train. The velocity of the man, as measured in the rail system, is $(v + u)/(1 + uv/c^2)$; however, here the velocity u is measured in the train system, while the velocity v (the velocity of the train) is measured in the rail system. Therefore, we use relativistic velocity addition formulas for velocities measured in different frames of reference. We cannot transform quantities from a single system of reference. Consequently, there are no formulas that arise from transformations (here the *Lorentz transformations*).

Before we proceed, it may be useful to clarify some terms. We say that the laws of physics need to be *covariant*, meaning that in different reference frames they have the same form. Consequently, the formula $A = B$ transforms to $A' = B'$. A physical quantity is a *constant* if it does not depend on coordinates, for example the number 3 or the charge of an electron. The speed of light is not a constant in that sense; it is an *invariant*. It means that it generally depends on coordinates ($c = |d\mathbf{r}/dt|$), but has the same value in any inertial reference frame. The speed of light is a constant of nature in the sense that it is the limiting speed. However, related to the Lorentz transformations, it is an invariant (scalar).

2.7.1 Postulates of the Special Theory of Relativity

There is a common misconception about the postulates of the special theory of relativity. Let the covariance postulate be first and the invariance of the speed of light postulate second. From the first postulate we have, for example, $v = dx/dt$ and $v' = dx'/dt'$. The second postulate is motivated primarily by Maxwell's electromagnetic theory, which predicts the invariance of the speed of light in inertial reference frames. However, it is important to note that we need only the first postulate to derive the Lorentz transformations (LT) (it is not hard to find references, so we highly recommend that the reader do so; see [36]). Once we have LT, the existence of the maximum speed (v_m), which is invariant, follows immediately, which means that we do not need the second postulate to have a maximum speed in the theory. Accordingly, we can use v_m instead of c in relativistic formulas. Einstein simply assumed that $v_m = c$, relying mostly on Maxwell's theory. However, the existence of a maximum speed does not necessarily mean that there must exist an object moving at such a speed. We believe that light is such an object. But we can imagine that the maximum speed is 1 mm/s larger than c. What experiment could show the difference? But if it were so, then a photon would have to have a mass, no matter how small it was. Then we could imagine a reference system that moves along with the photon, and therefore, the photon must be at rest in it. However, light is a wave as well, and we would see a wave that is not moving. The wave phase would be constant to us (maximum amplitude, for example); consequently, we would see no vibrations. Now, without a change in an electric field over time, there is no magnetic field, and therefore, we see an electrostatic field. However, there is no a charge distribution in an empty space that could create such a field (*Einstein*). Therefore, instead of v_m, we use c. However, that does not mean that the assumption of the invariance of the speed of light is necessary for the validity of STR. The first postulate is certainly deeply natural and typical for Einstein, who was among the first to stress the importance of symmetries in physics, and this is certainly a question of symmetry. True, it is easier to make sense of thought experiments and derive formulas using the postulate of the speed of light, as is done in almost all textbooks, so that students get the impression that such a theory cannot exist without the second postulate. Let us also mention that there are numerous tests that confirm STR and none (as far as is known to the author) that refute it, although many are trying to show things differently; they even make up stories about a "relativists' conspiracy." Let us mention two important facts. First, the quantum electromagnetic theory (QED) is deeply based on the special theory of relativity, and it is known that the predictions of QED are in unusually good agreement with experiments. Second, we have the opportunity to monitor what is happening at speeds comparable to the speed of light almost every day; that is, we have particle accelerators. They are built using the formulas of the special theory of relativity, and it is hard to imagine that they would operate if STR were not valid. In addition, the Global Positioning System (GPS), which we use daily, is based on results of both the special and general theories of relativity

2.7.2 Inertial Coordinate Systems and Reference Frames

There is an additional topic to discuss. Usually, an inertial coordinate system is defined in textbooks as an "unaccelerated system"; however, that implies *homogeneity*, in agreement with Newton's first law only, not all of Newton's laws. To include Newton's third law, we have to introduce the concept of *isotropy*. Why? Consider two protons at rest and let them move freely in vacuum. Then we expect that, due to the repulsion, the protons will move in opposite orientations.

However, we also expect that both protons have exactly the same kinematic properties, since all orientations in space are equal. Without that, we do not have Newton's third law. The isotropy is connected to the possibility of synchronizing clocks (two protons could be a possible tool). It is also natural to expect that the speed of light is equal in all possible orientations (although this is not so important here, since we will not use light in the derivation of LT). Then we have inertial coordinate systems (ICS) with the homogeneity and isotropy included. We call the class of inertial coordinate systems (rotated, translated) that are not moving relative to some inertial coordinate system the inertial reference frame (IRF). Now, with homogeneity and isotropy included, we do not need the speed of light postulate; symmetries are enough to obtain the Lorentz transformations. Thus, light loses its central role in the theory.

2.7.3 Derivation of the Lorentz Transformations from Symmetries

Due to the linearity, we expect transformations like (here v is the relative velocity between systems along the x-axis, measured in one of the systems)

$$\begin{pmatrix} x' \\ t' \end{pmatrix} = \begin{pmatrix} A & B \\ C & D \end{pmatrix} \begin{pmatrix} x \\ t \end{pmatrix},$$

where A, B, C, and D may depend on v, and $\Delta = AD - BC$ is the determinant of the transformation matrix. From $x' = \text{const}$, we have $\Delta x' = A\Delta x + B\Delta t = 0$, that is, $B = -vA$, where $v = \Delta x/\Delta t$. The inverse transformations are (you need the inverse of the transformation matrix)

$$\begin{pmatrix} x \\ t \end{pmatrix} = \frac{1}{\Delta} \begin{pmatrix} D & -B \\ -C & A \end{pmatrix} \begin{pmatrix} x' \\ t' \end{pmatrix},$$

and then from $x = \text{const}$, we have $B = -vD$, where $-v = \Delta x'/\Delta t'$ and consequently, $D = A$. If we replace v with $-v$ and exchange primes, these two transformations should be exchanged (due to isotropy); therefore, we have $\Delta = 1$ (note that this means that the transformations are orthogonal). We therefore have

$$C = \frac{A^2 - 1}{-vA}, \quad t' = A\left(t - \frac{A^2 - 1}{v^2 A^2} vx\right),$$

and consequently, after setting

$$\kappa = \frac{A^2 - 1}{v^2 A^2}, \quad A = \pm 1/\sqrt{1 - \kappa v^2},$$

and choosing A to be positive (why?), we get the transformations

$$\begin{pmatrix} x' \\ t' \end{pmatrix} = \frac{1}{\sqrt{1-\kappa v^2}} \begin{pmatrix} 1 & -v \\ -\kappa v & 1 \end{pmatrix} \begin{pmatrix} x \\ t \end{pmatrix}, \quad \begin{pmatrix} x \\ t \end{pmatrix} = \frac{1}{\sqrt{1-\kappa v^2}} \begin{pmatrix} 1 & v \\ \kappa v & 1 \end{pmatrix} \begin{pmatrix} x' \\ t' \end{pmatrix}.$$

Due to isotropy, we have $\kappa(\mathbf{v}) = \kappa(v)$, which means that $\kappa(v) = \kappa(-v)$. Application of two successive transformations with different velocities v_1 and v_2 should give a transformation matrix of the same form, which means that the diagonal elements must be the same, whence follows (check) $\kappa(v_1) = \kappa(v_2)$, that is, $\kappa(v) = \text{const}$. Using appropriate physical units, we get only three interesting possibilities for κ: $-1,\ 0,\ 1$. Does this look familiar? Note the following group properties of the transformations ($\mathrm{T}(v)$, I is the identity transformation):

1. $\mathrm{T}(0) = I$ (identity)
2. associativity (check)
3. $\mathrm{T}^{-1}(v) = \mathrm{T}(-v)$ (inverse)
4. $\mathrm{T}(v_1)\mathrm{T}(v_2) = \mathrm{T}(v)$ (closure)

For $\kappa = -1$, we have pure rotations in the (x, t)-plane (show this) by the angle arctan (v). For $\kappa = 0$, we have Galilean transformations, and for $\kappa = 1$ we have Lorentz transformations. Experiments in physics teach us that we have to use $\kappa = 1$; however, note that Galilean relativity is also a valid theory of relativity, since all of this is a consequence of our definition of the ICS. A direct consequence of the Lorentz transformations is the existence of a maximum speed, since $\sqrt{1-v^2}$ must be a real number.

Recall that we have already seen the numbers $-1,\ 0,\ 1$ here in the text, when we discussed properties of $\mathbf{F}^2 \in \mathbb{R}$ in *Cl*3.

2.7.4 Paravectors and Restricted Lorentz Transformations

Paravectors in *Cl*3, such as $t + \mathbf{x}$ (a multivector with grades 0 and 1), give a paravector again when squared (check it). Therefore, the magnitude of a paravector is to be defined differently. For complex, dual, and hyperbolic numbers, as well as for quaternions, we have a similar obstacle, so we use conjugations. In *Cl*3, we do not need any *ad hoc conjugation*, since we already have the *Clifford involution*, so we define

$$p\bar{p} = (t + \mathbf{x})(t - \mathbf{x}) = t^2 - x^2 \in \mathbb{R},$$

which is exactly the desired form of an invariant interval required in the special theory of relativity. Recall that the *Clifford involution* is a combination of the *grade* involution and the *reverse* involution. Therefore, we can try to interpret it geometrically in $\mathfrak{R}^3$. Specifically, the grade involution means a space inversion, while for the reverse involution we have seen that it is related to the fact that the Pauli matrices are Hermitian.

Note that with the Clifford involution, there is no need for a negative signature (Minkowski). According to the Minkowski formulation of STR, we can define the unit vector "in the time direction" $e_0,\ \ e_0^2 = 1$, and three space vectors $e_i,\ \ e_i^2 = -1$, which means that we have a 4D vector space with the negative signature $(1, -1, -1, -1)$. Such an approach is possible in geometric algebra as well; we have STA (*spacetime algebra, Hestenes*). However, everything we can do with STA, we can do in *Cl*3 as well, without the negative signature (see [10]). Those who argue that the negative signature is necessary in STR are possibly mistaken. Some authors write sentences like, *the principle of relativity forces us to consider the scalar product with a negative square of vectors*, forgetting that their definition of norm of elements prejudices such a result (R2). Yet it is possible to describe the geometry of a vector space using the formalism of a higher-dimensional vector space. Consequently, we can say that the Minkowski geometry formulation of STR is a 3D problem described in 4D. However, in *Cl*3, all we need is the geometric product and one involution. Time is not a fourth dimension in *Cl*3; it is just a real parameter (as in quantum mechanics). If there is a fourth dimension of time, how is it that we cannot move through the time as we move through space? There are other interesting arguments in favor of 3D space. For example, gravitational and electrostatic forces depend on the square of the distance. And what about the definition of velocity (we use it also in the theory of relativity): dx/dt? If there exists a time dimension, then time is a vector, which means that velocity is naturally a bivector, like a magnetic field, not a vector. It does not matter whether we use a proper time to define the four-velocity vector; the space velocity is still defined by the previous formula, up to a real factor. Minkowski gave us a nice mathematical theory; however, his conclusion about the fourth time dimension was a pure mathematical abstraction, widely accepted among physicists. At that time, the geometric ideas of Grassmann, Hamilton, and Clifford were largely suppressed. This forces us to question what Einstein would have chosen had he known this. At the beginning of the twentieth century, another important theory was developing, quantum mechanics, in which Pauli introduces his matrices to formulate the half spin, on which we have already commented. Dirac's matrices are also a representation of one Clifford algebra, but again, Dirac's theory has a nice formulation in *Cl*3 (see [5]), as well as the minimal standard model in *Cl*7 (see [44]), and so on. It is not baseless to question the merits of introducing time as a fourth physical dimension. A usual argument is one that Minkowski gave; however, this is not an argument: it is just the observation that in the special theory of relativity, an invariant interval is not $dt^2 + dx^2$; it is $dt^2 - dx^2$. However, we see that the invariant interval $dt^2 - dx^2$ is easy to obtain in *Cl*3, with completely natural requirements for a multiplication of vectors. Minkowski introduced a fourth dimension ad hoc. If his formalism were undoubtedly the only possible one with which to formulate the special theory of relativity, then there would be a solid basis for believing that indeed there must be a fourth dimension of time. However, without that condition, with the knowledge that there is a natural way to formulate the theory without the fourth dimension, it is difficult to avoid the impression that this widely accepted mantra of a fourth dimension does not have a solid foundation. According to some authors, one of the obstacles in the theory of quantum gravity is, possibly, the existence of a fourth dimension of time in the formalism. Here we develop a formalism using paravectors, which define 4D linear space, but time is identified as a real scalar; we say that time is a real parameter. It would be interesting to investigate whether there is an experiment that would unambiguously prove the existence of a fourth physical dimension of time. Probably there is no such experiment. Therefore, it is difficult to avoid the impression that physicists are tying up a ritual cat during their meditations (see the Zen story). However, the future will show, perhaps, that the time dimension does indeed exist, maybe more than one of them (if time exists). Whatever the case may be, it is not true that the Minkowski space is the only correct framework for the formulation of STR. Particularly, it is not true that in STR we must introduce vectors whose squares are negative.

We shall use a system of physical units in which is $c = 1$. In geometric algebra, we are combining different geometric objects that may have different physical units. Therefore, we always choose a system of units such that all coefficients of a multivector are reduced to the same physical unit (usually the length). We study geometric relationships, and that is the goal here. In an application to a particular situation (experiment), physical units are converted (analysis of physical units); consequently, there is no problem with physical units.

Starting from the invariant interval in STR $t^2 - x^2 = \tau^2$, where τ is the invariant *proper time* in the particle rest frame, it follows that

$$t^2 - x^2 = t^2\left(1 - v^2\right) = \tau^2 \Rightarrow t^2/\tau^2 = 1/\left(1 - v^2\right) = \gamma^2,$$

where γ is the well-known relativistic factor. Now, instead of the four-velocity vector, we define the *proper velocity* $u \equiv \gamma(1 + \mathbf{v})$, $u\bar{u} = 1$, a *paravector* that is simply $u_0 = 1$ in the rest frame. Notice that the proper velocity is not a list of coordinates, like a four-velocity vector. However, it plays the same role. Imagine that we want to analyze a body, initially at rest, in a new reference frame in which the body has velocity $\mathbf{v}$ (*boost*). The recipe is very simple: just take the geometric product of two proper velocities $u_0 \rightarrow u_0 u = u$. For a series of boosts, we have a series of transformations

$$u_0 \rightarrow u_0 u_1 \rightarrow u_0 u_1 u_2 = u_1 u_2.$$

Note how this is really easy to calculate and that from the form of the proper velocity paravector we immediately see the relativistic factor γ and the 3D velocity vector $\mathbf{v}$. For example, if we specify all velocity vectors to be parallel to e_1, it follows that

$$\begin{aligned} u &= \gamma_1(1 + v_1 e_1)\gamma_2(1 + v_2 e_1) = \gamma_1\gamma_2(1 + v_1 v_2 + (v_1 + v_2)e_1) \\ &= \gamma_1\gamma_2(1 + v_1 v_2)\left(1 + \frac{v_1 v_2}{1 + v_1 v_2} e_1\right). \end{aligned}$$

Therefore, from the form of the paravector, we immediately get

$$\gamma = \gamma_1\gamma_2(1 + v_1 v_2) = (u + \bar{u})/2 = \langle u \rangle_S, \quad \mathbf{v} = \frac{v_1 + v_2}{1 + v_1 v_2} e_1 = \frac{(u - \bar{u})/2}{\langle u \rangle_S} = \frac{\langle u \rangle_V}{\langle u \rangle_S},$$

the well-known result of the special theory of relativity (relativistic velocity addition). Notice how the geometric product makes the derivation of formulas easy, and as stated earlier, the formulas obtained are just special cases of general formulas in *Cl*3. From the polar form of a general multivector

$$M = \rho(\cosh\varphi + \mathbf{f}\sinh\varphi) = \rho\gamma(1 + \vartheta\mathbf{f}), \quad \gamma^{-1} = \sqrt{1 - \vartheta^2},$$

reducing to the real part of a multivector (*paravector*), we have

$$\gamma = \cosh\varphi,\quad \gamma v = \sinh\varphi,\quad u = \cosh\varphi + \hat{\mathbf{v}}\sinh\varphi = \cosh\varphi(1 + \hat{\mathbf{v}}\tanh\varphi) = \exp(\varphi\hat{\mathbf{v}}).$$

Note that $\gamma = t/\tau$ is now a real number (generally, it is a complex number), which gives the maximum speed. Using the spectral decomposition, it follows that

$$\gamma(1 + v\hat{\mathbf{v}}) = k_+u_+ + k_-u_- \Rightarrow k_\pm = \gamma(1 \pm v) = \cosh\varphi \pm \sinh\varphi,$$

whence, defining implicitly the factor k (*Bondi factor*), $\varphi \equiv \log k$, and recalling the definitions of the hyperbolic sine and cosine, we get

$$k^{\pm 1} = \cosh\varphi \pm \sinh\varphi,\quad k = \sqrt{(1+v)/(1-v)},\quad u = ku_+ + k^{-1}u_-.$$

Our earlier example with two "boosts" parallel to e_1 now has the form

$$u_1u_2 = (k_1u_+ + u_-/k_1)(k_2u_+ + u_-/k_2) = k_1k_2u_+ + u_-/(k_1k_2),$$

that is, the relativistic velocity addition rule is equivalent to the multiplication of the Bondi factors

$$k = k_1k_2.$$

Generally, if velocity vectors do not lie in the same direction, in expressions there will appear versors, like $\mathbf{v}_1\mathbf{v}_2$, which may seem like a complication, but they actually provide new opportunities for elegant research. For example, it is rather easy to derive the *Thomas precession* (see Sect. 3.1, [17]).

Example In the reference frame S_1, a starship has velocity v, while in the reference frame of the starship, another starship has velocity v, and so on, all in the same direction. Find the velocity v_n of the nth starship in S_1. Discuss a solution for $n \to \infty$.

Solution:

From $k_1 = \sqrt{(1+v)/(1-v)}$, it follows that

$$k_n = \sqrt{(1+v_n)/(1-v_n)} = k_1^n = \left(\sqrt{(1+v)/(1-v)}\right)^n,$$

from which we can find the required velocity v_n.

2.7.5 Paravectors and the Minkowski Metric

For two paravectors p and q, we define the scalar product

$$\langle p\bar{q}\rangle_S = (p\bar{q} + q\bar{p})/2,$$

whence follows the orthogonality condition $\langle p\bar{q}\rangle_S = 0$. The scalar product of $p = t_1 + \mathbf{x}_1$ and $q = t_2 + \mathbf{x}_2$ is $\langle p\bar{q}\rangle_S = t_1t_2 - \mathbf{x}_1 \cdot \mathbf{x}_2$. If we define $e_0 = 1$ and use indices $\mu = 0, 1, 2, 3$, we have

$$\left\langle e_\mu \bar{e}_\nu \right\rangle_S = \eta_{\mu\nu} = \begin{cases} 1, & \mu = \nu = 0, \\ -1, & \mu = \nu > 0, \\ 0, & \mu \neq \nu, \end{cases}$$

that is, we get the Minkowski metric. Now we can write ($\bar{e}_\mu \equiv e^\mu$)

$$\langle p\bar{q}\rangle_S = p^\mu q^\nu \left\langle e_\mu \bar{e}_\nu \right\rangle_S = p^\mu q^\nu \eta_{\mu\nu},$$

where we use the Einstein summation rule. The norm of a paravector is

$$\langle p\bar{p}\rangle_S = t^2 - x^2,$$

where $\langle p\bar{p}\rangle_S = 0$ is for a lightlike paravector (orthogonal to itself), $\langle p\bar{p}\rangle_S > 0$ is for timelike paravectors, and $\langle p\bar{p}\rangle_S < 0$ is for spacelike paravectors. For $\langle p\bar{q}\rangle_S = 0$, we have

$$t_1 t_2 - \mathbf{x}_1 \cdot \mathbf{x}_2 = t_1 t_2 - x_1 x_2 \cos\varphi = 0,$$

and then if $t_1^2 - x_1^2 > 0$ (timelike), we define $t_1 = kx_1$, $k > 0$, whence follows

$$kx_1 t_2 - x_1 x_2 \cos\varphi = 0 \Rightarrow x_2 = kt_2/\cos\varphi, \quad t_2^2 - x_2^2 = t_2^2\left(1 - k^2/\cos^2\varphi\right) < 0,$$

which means that orthogonality holds between different types of paravectors (except for lightlike paravectors).

William E. Baylis

◆ Show that for paravectors in $Cl3$, one has

2

$$p = \frac{1}{2} e_\mu \bar{p} e^\mu, \quad \bar{p} = -\frac{1}{2} e_\mu p e^\mu \quad \text{(summation)},$$

where $e^\mu = \bar{e}_\mu$. ◆

The geometric product of two paravectors is a paravector; therefore, paravectors in *Cl*3 make a 4-dimensional linear subspace of the algebra. Note that multivectors in *Cl*3 are like complex numbers, where components are real paravectors:

$$M = t + \mathbf{x} + j\mathbf{n} + jb = t + \mathbf{x} + (b + \mathbf{n})j.$$

2.7.6 The Restricted Lorentz Transformations

We are now ready to comment on the restricted Lorentz transformations (LT). Generally, LT consists of "boosts"B and rotors R. We can generally write (see Sect. 3.1, [27]) $L = BR$, $L\bar{L} = 1$ (the *unimodularity condition*). Here we can regard the *unimodularity* condition as the *definition* of the Lorentz transformations, which is well researched and justified. If we define (see above)

$$\begin{aligned} B &= \cosh(\varphi/2) + \hat{\mathbf{v}} \sinh(\varphi/2) = \exp(\varphi \hat{\mathbf{v}}/2), \\ R &= \cos(\theta/2) - j\hat{\mathbf{w}} \sin(\theta/2) = \exp(-j\hat{\mathbf{w}}\theta/2) \end{aligned}$$

(the unit vector $\hat{\mathbf{w}}$ defines the rotation axis), we can write the LT of some element, say a vector, as

$$p' = LpL^\dagger = BRpR^\dagger B^\dagger = BRpR^\dagger B,$$

where the element p is rotated, then boosted. There is also the possibility to write L as

$$L = \exp(\varphi \hat{\mathbf{v}}/2 - j\hat{\mathbf{w}}\theta/2) \neq \exp(\varphi \hat{\mathbf{v}}/2)\exp(-j\hat{\mathbf{w}}\theta/2),$$

where we have to be careful due to the general noncommutativity of vectors in the exponent (Appendix 5.1, [23]). However, it is always possible (not necessarily easy) to find vectors $\hat{\mathbf{v}}'$and $\hat{\mathbf{w}}'$ that satisfy

$$L = \exp(\varphi \hat{\mathbf{v}}/2 - j\hat{\mathbf{w}}\theta/2) = \exp(\varphi \hat{\mathbf{v}}'/2)\exp(-j\hat{\mathbf{w}}'\theta/2).$$

It is convenient in applications to resolve an element into commuting and anticommuting components with L and then take the advantage of the commutation properties. For further details, see Baylis's articles about APS (*the algebra of physical space*, *Cl*3). In [27], you can find a nice chapter about the special theory of relativity. We see that rotations are a natural part of LT; consequently, because of powerful rotor techniques, the geometric algebra formalism can provide many opportunities (see Sect. 3.1). Later in the text, we will discuss some powerful techniques with spinors (*eigenspinors*). Generally, $L = BR \Rightarrow L^\dagger = R^\dagger B$, from which it follows that

$$u = Lu_0 L^\dagger = BRR^\dagger B = \; B^2 \Rightarrow B = \sqrt{u}.$$

In addition,

$$\langle B \rangle_S = (B + \bar{B})/2, \;\; B\bar{B} = 1, \;\; u = \gamma(1 + \mathbf{v}),$$
$$4\langle B \rangle_S^2 = (B + \bar{B})^2 = u + 2 + \bar{u} = 2\gamma + 2 \Rightarrow 2\langle B \rangle_S = \sqrt{2(\gamma + 1)},$$

from which it follows that

$$B(B + \bar{B}) = 2B\langle B \rangle_S = B^2 + B\bar{B} = u + 1 \Rightarrow$$
$$B = \frac{u + 1}{\sqrt{2(\gamma + 1)}}.$$

For further analysis of Lorentz transformations, see Appendix 5.4.

2.7.7 Electromagnetic Fields in Geometric Algebra

Here we will not describe the entire EM theory in *Cl3* (see *Hestenes*, *Baylis*, *Chappell*, *Jancewicz*). Instead, we will comment on only a few ideas. In the geometric algebra formalism, for an electromagnetic (EM) wave in vacuum, we define

$$E = B, \mathbf{E} \perp \mathbf{B}, \;\; \mathbf{F} = \mathbf{E} + j\mathbf{B} \Rightarrow \mathbf{F}^2 = 0, \quad c = 1,$$

where the complex vector **F** is a *nilpotent*. Note that the term $j\mathbf{B}$ with a magnetic field is a bivector. It is useful to expand the magnetic field bivector in an orthonormal basis. Thus, if we start with the magnetic field vector

$$\mathbf{B} = B_1 e_1 + B_2 e_2 + B_3 e_3,$$

we get the bivector

$$j\mathbf{B} = j(B_1 e_1 + B_2 e_2 + B_3 e_3) = B_1 e_2 e_3 + B_2 e_3 e_1 + B_3 e_1 e_2.$$

◆ The reader should check this simple expression and try to create a mental image of it. In addition, we can represent bivectors using parallelograms; it is straightforward to see how to add them graphically (see the figure at the beginning of the Chap. 2). ◆ Although this may seem like just a neat trick, here we are going to try to show that the bivector, as a geometric object, is fully adequate for the description of a magnetic field. In fact, the physical properties of a magnetic field require it to be treated as a bivector. In any formalism that does not imply the existence of bivectors (such as the Gibbs vector formalism in the language of the scalar and cross products), problematic situations must necessarily occur. Here we will discuss the issue of Maxwell's theory and mirror symmetry as an example. If we use a coordinate approach, then in 3D we can define a richer structure by introducing tensors. Let us look at a quotation from the article R3:

> For the three-dimensional description, antisymmetric contravariant tensors are needed of ranks from zero to three (they are known as multivectors) as well as antisymmetric covariant

tensors of ranks from zero to three (called exterior forms). It makes altogether eight types of directed quantities.

Therefore, for example, axial vectors (like the cross product) become antisymmetric tensors of the second rank. This entirely geometric structure becomes simple and intuitive in geometric algebra, without the need for introducing any coordinates (here we often introduce a basis, but it is solely for easier understanding of the text, and it is not necessary). Due to the independence of the basis, it ceases to be important, for example, whether we work with the right or left coordinate system; the geometric product takes care of everything. For two vectors, we could have expressions like $\mathbf{ab} \pm \mathbf{ba}$, without the need to use any basis to conclude that it is a scalar or a bivector, in any dimension.

Maxwell's electromagnetic theory is the first physical theory that initially satisfied the postulates of the special theory of relativity. It is therefore no wonder that both theories fit in *Cl*3 perfectly. Let us look at some interesting facts related to the theory in the language of geometric algebra. For example, we can visualize solutions for an electromagnetic wave in vacuum (see [27]) by a simple and interesting picture, that is, a wave vector $\mathbf{k}$ is parallel to the direction vector of the nilpotent $\mathbf{F}$. Consequently, the solution (for $\mathbf{k} \parallel \mathbf{x}$) can be written as

$$\mathbf{F}_0 e^{\pm jkx} e^{j\omega t}, \quad \mathbf{F}_0 = \mathbf{E}_0 + j\mathbf{B}_0.$$

We can imagine the spatial part of the wave $\mathbf{F}_0 e^{\pm jkx}$ as a spatial periodic **steady** spiral, which extends along the direction of the wave's propagation. This "spiral" is a nilpotent, since it is proportional to $\mathbf{F}_0$. It turns out (see below) that rotation of a nilpotent around its directional axis can be achieved by multiplying it by a complex phase, such as $\exp(j\omega t)$. Consequently, we have a spiral in space that rotates around the axis of the direction of propagation of the wave. The bivector part $j\mathbf{B}$ defines the plane orthogonal to $\mathbf{B}$. Consequently, the vector $\mathbf{E}$ belongs to that plane. The bivector $j\mathbf{B}$ provides an opportunity for consideration of electromagnetic phenomena more completely and more elegantly than an axial vector $\mathbf{B}$. See also [35], p. 63.

2.7.8 Problems with the Cross Product

Let us look at some properties of EM fields in vacuum. Maxwell's equations are completely mirror symmetric in the language of geometric algebra, as are their solutions as well. When we use the cross product, we immediately need to introduce the right-hand rule, and we see that we have the left-hand rule in the mirror. If we superimpose the two images (the original and the one in the mirror, Fig. 2.6), they do not match: the vectors are standing incorrectly. However, if the vectors (axial) of the magnetic field are replaced by bivectors, the images match exactly. Moreover, of course, we do not need the right-hand rule: the geometric product takes care of everything, and the right-hand rule is an internal part of the geometric algebra, it is contained in the definition of the unit pseudoscalar. It is important to understand that such problems occur due to the strange properties of the cross product. In fact, one could fill a book with such examples. For those who have thought a long time in the language of the vectorial right-hand rule, it may be difficult to

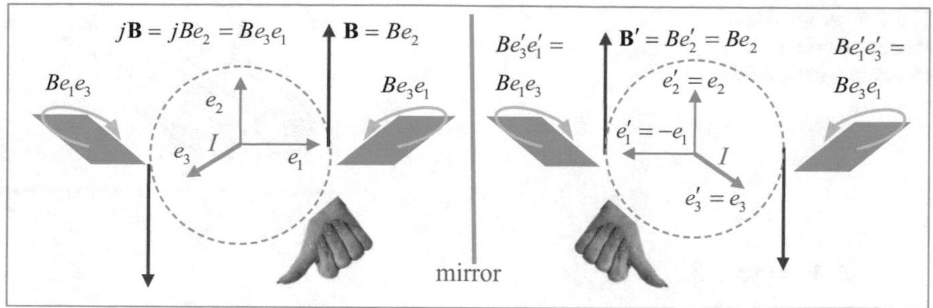

Fig. 2.6 Maxwell's equations are completely mirror-symmetric in the language of geometric algebra

accept a new paradigm. We are accustomed to imagining arrows; therefore, we have yet to develop an intuition for objects that are not vectors or scalars. Geometry is the language of physics more than we dreamed.

The geometric intuition of human beings is very powerful, and we should teach elements of geometric algebra at the high-school level. However, various games with oriented objects could be introduced already in elementary school.

2.7.9 Complex Vectors Are Powerful

For the electromagnetic field $\mathbf{F} = \mathbf{E} + j\mathbf{B}$ in vacuum, the vector $\mathbf{E}$ belongs to the plane defined by the bivector $j\mathbf{B}$ ($\mathbf{F}$ is a *whirl* in vacuum). The area of the circle that represents the bivector is B; hence its radius must be $\sqrt{B/\pi} \approx 0.56\sqrt{B}$. The direction and possible orientations of the wave propagation are plotted in Fig. 2.7.

This image is rotated about the wave propagation axis by an angle dependent on the position $(\mathbf{k} \cdot \mathbf{x})$, which gives a static image in space. This complete spatial image rotates in time, depending on the frequency of the wave (ω). ◆ E30: The reader can check this by taking a simple nilpotent $e_1 + je_2$ and multiplying it by a complex phase $\exp(j\varphi)$. Immediately we get the matrix of the rotation around the z-axis. ◆ We can rotate the whole picture elegantly; consequently, this particular example is not special in any way.

For any complex vector $\mathbf{F} = \mathbf{v} + j\mathbf{w}$ in $Cl3$, we have

$$\mathbf{F}\mathbf{F}^{\dagger} = (\mathbf{v} + j\mathbf{w})(\mathbf{v} - j\mathbf{w}) = v^2 + w^2 - 2j\mathbf{v} \wedge \mathbf{w}.$$

Consequently, if we use the complex vector of the electromagnetic field in vacuum (nilpotent) $\mathbf{F} = \mathbf{E} + jc\mathbf{B}$, where for a moment we use the SI system of units, we get

$$\mathbf{F}\mathbf{F}^{\dagger} = E^2 + c^2B^2 - 2cj\mathbf{E} \wedge \mathbf{B}.$$

Now we have

Fig. 2.7 A graphical representation of the electromagnetic field

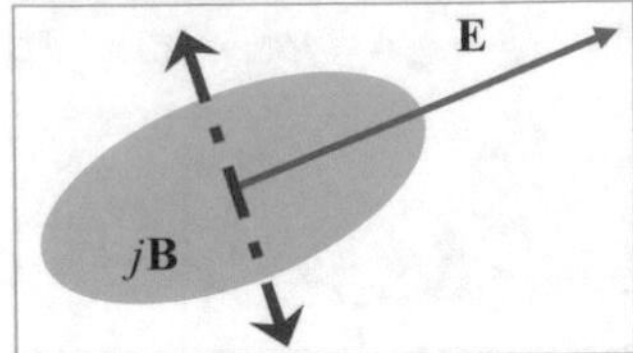

$$\frac{1}{2}\varepsilon_0 c\mathbf{F}\mathbf{F}^{\dagger} = c\xi + \mathbf{S},$$

with

$$\xi = \frac{1}{2}\varepsilon_0\left(E^2 + c^2B^2\right), \quad \mathbf{S} = -j\mathbf{E}\wedge\mathbf{B}/\mu_0.$$

Here ξ is the energy density, while $\mathbf{S}$ is the energy–current density (energy flow), also known as the *Poynting vector* (note that here we have a polar vector, not an axial one). Therefore, the Poynting vector is proportional to the nilpotent direction vector. Note also that in general,

$$\mathbf{F}^2 = E^2 - c^2B^2 + 2jc\mathbf{E}\cdot\mathbf{B}$$

is a complex scalar, which we can use to classify fields (it is zero in vacuum). Note that for a static electric field we have $\mathbf{F}^2 = E^2$, while for a static magnetic field we have a negative value $\mathbf{F}^2 = -c^2B^2$. An electromagnetic field can have the property $\mathbf{E}\cdot\mathbf{B} \neq 0$, for instance in an inhomogeneous medium.

There is a nice book [9] about electromagnetism; we recommend it. Many articles on geometric algebra cover this subject as well. See also Appendices 5.5 and 5.6.

2.8 Eigenspinors

Let us look at one rather elegant and powerful way to describe a motion of particles (see [5], [6], and [7]). Imagine the laboratory reference frame and the frame that is fixed to the particle in motion under the influence of, say, an electromagnetic field. Here we will consider paravectors and restricted Lorentz transformations only. At any instant of time, we can find a transformation that transforms elements from the lab frame to the inertial frame of reference that coincides with the particle movement (*comoving frame*). The proper velocity in the lab frame is $u_0 \equiv e_0 = 1$. Thus for that very instant of time, we can get the proper velocity of the particle as

$$u = \Lambda e_0 \Lambda^{\dagger} = \Lambda\Lambda^{\dagger},$$

where Λ is the Lorentz transformation, called the *eigenspinor* (Fig. 2.8) due to the special choice of the reference frame. Let us mention in passing that in applying such a transformation to orthonormal basis vectors $u_\mu = \Lambda\ e_\mu\ \Lambda^{\dagger}$, $\mu = 0, 1, 2, 3$, we get a

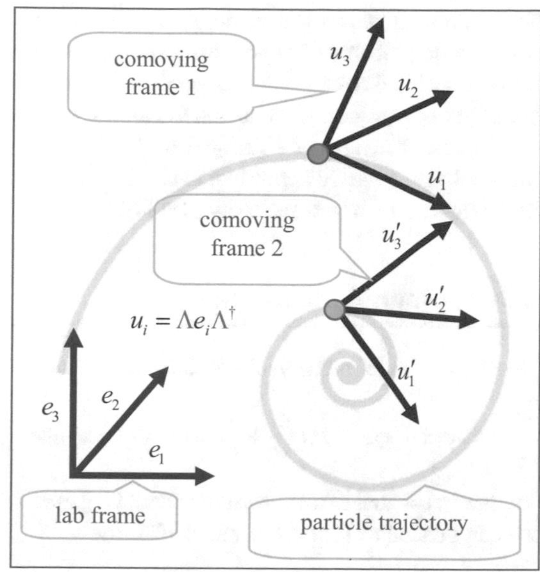

Fig. 2.8 Eigenspinors are powerful as a tool for dynamical problems

so-called *Frenet tetrad*. Recall that for Lorentz transformations we have $\bar{\Lambda}\Lambda = 1$ (*unimodularity*). If an eigenspinor Λ is known at every instant of time, we have all the information needed to describe the particle's movement. Eigenspinors are changing in time, and consequently, we need the first time derivative

$$\dot{\Lambda} = \dot{\Lambda}\bar{\Lambda}\Lambda = \mathbf{\Omega}\Lambda/2, \quad \mathbf{\Omega} \equiv 2\dot{\Lambda}\bar{\Lambda}.$$

This all seems like a trivial relation; however, it is not. We have (due to the unimodularity)

$$\frac{d}{dt}(\Lambda\bar{\Lambda}) = 0 = \dot{\Lambda}\bar{\Lambda} + \Lambda\bar{\dot{\Lambda}},$$

whence, using $\overline{\dot{\Lambda}\bar{\Lambda}} = \Lambda\bar{\dot{\Lambda}} = -\dot{\Lambda}\bar{\Lambda}$, we see that $\mathbf{\Omega}$ is a complex vector (therefore, it is in the bold format). For the first time derivative of the proper velocity, we have

$$\dot{u} = \dot{\Lambda}e_0\Lambda^\dagger + \Lambda e_0\dot{\Lambda}^\dagger = \dot{\Lambda}\bar{\Lambda}\Lambda e_0\Lambda^\dagger + \left(\dot{\Lambda}\bar{\Lambda}\Lambda e_0\Lambda^\dagger\right)^\dagger == \frac{\mathbf{\Omega}u + (\mathbf{\Omega}u)^\dagger}{2} = \langle\mathbf{\Omega}u\rangle_R,$$

which is a paravector. The Lorentz force (see [9]) in $Cl3$ is defined as

$$\dot{p} = m\dot{u} = e\langle\mathbf{F}u\rangle_R, \mathbf{F} = \mathbf{E} + j\mathbf{B},$$

whence we see how $\mathbf{\Omega}$ just defined acquires a physical meaning: it is proportional to the complex vector of the electromagnetic field $\mathbf{F}$. It is surprising how the electromagnetic theory can be simply and naturally formulated in $Cl3$. This is not an isolated example; the geometric product makes the geometry of our 3D world a natural framework for physics.

Someone who knows geometric algebra well, but knows nothing about electromagnetism, would probably discover an electromagnetic field as a purely geometric object. The Gibbs scalar and cross products, and then the whole apparatus of theoretical physics with coordinates, matrices, tensors, and the like, very much blurred the whole picture.

This brief review of eigenspinors should point out the powerful and elegant techniques that are widely applicable in electromagnetism and quantum mechanics, but also in the general theory of relativity (see [8]).

2.9 Spinorial Equations

Having a particle radius vector in 2D, we can write

$$\mathbf{r} = e_1 x + e_2 y = e_1 r\left(\frac{x}{r} + \frac{y}{r} e_1 e_2\right) = e_1 r \exp(\varphi e_1 e_2).$$

We have seen that this expression cannot be generalized to higher dimensions; however, we can generalize it in the "sandwich" form

$$\mathbf{r} = U e_1 U^\dagger, \quad U \propto \sqrt{r}.$$

What have we written? In 2D, U is a complex number with the imaginary unit $e_1 e_2$. However, it generally can be treated as a *spinor* (for the definition of spinors, see Sect. 4.4 or the literature). Note that starting with the unit vector e_1, we can get any vector in the plane defined by the bivector $e_1 e_2$, where the special choice of the vector and the bivector is not important. These relations are easy to generalize to higher dimensions. If the spinor U depends on time, we have all the dynamics contained in the spinor. This is a powerful technique for describing various types of movement (see below). It turns out, as a rule, that equations are much easier to solve if they are expressed in terms of U instead of $\mathbf{r}$.

Note that for a complex number U we have a complex conjugate $U^\dagger$. The modulus of $\mathbf{r}$ is just $r = \sqrt{\mathbf{r}^2}$, i.e., $r = UU^\dagger$. The time rate of the vector $\mathbf{r}$ is the first derivative of $\mathbf{r} = U e_1 U^\dagger$; however, in 2D, for simplicity, we can take the derivative of $U^2 e_1$(why?):

$$\dot{\mathbf{r}} = 2\dot{U} U e_1 \Rightarrow \dot{\mathbf{r}} e_1 = 2\dot{U} U \Rightarrow \dot{\mathbf{r}} e_1 U^\dagger = 2r\dot{U}.$$

Introducing a new variable s,

$$\frac{d}{ds} = r\frac{d}{dt}, \quad \frac{dt}{ds} = r,$$

and using $dU/ds = U'$, we get the new equation for U

$$2U' = \dot{\mathbf{r}} U e_1,$$

or differentiating once more,

$$2U'' = r\ddot{\mathbf{r}} U e_1 + \dot{\mathbf{r}} U' e_1 = \left(\ddot{\mathbf{r}}\mathbf{r} + \dot{\mathbf{r}}^2/2\right) U. \quad (*)$$

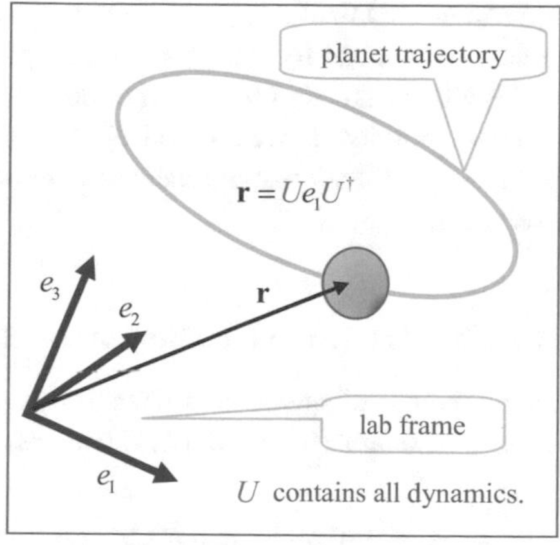

Fig. 2.9 Eigenspinors and the Kepler problem

2.9.1 Kepler's Problem

As an example, let us look at the motion of a body under the action of a central force (*Kepler's problem*, Fig. 2.9):

$$\mu\,\ddot{\mathbf{r}} = -k\mathbf{r}r^{-3},$$

where μ is the *reduced mass* and k is a constant. The equation (*) for U now becomes

$$U'' = \frac{1}{2\mu}U\left(\frac{\mu\dot{\mathbf{r}}^2}{2} - \frac{k}{r}\right) = \frac{E}{2\mu}U \Rightarrow U'' = \kappa U, \quad \kappa = \text{const},$$

where we have introduced the total energy E. This is a well-known and relatively simple equation, which for bound states ($E < 0$) takes the form of the equation for a harmonic oscillator (recall the connection between an ellipse and the harmonic functions). The advantages of this approach are numerous, including ease of solving specific equations and better stability of solutions (no singularities at $r = 0$). Note that the equation is linear, which has great advantages in the perturbation approach (better stability).

2.10 *Cl3* and Quantum Mechanics

There is a common belief that there is no possibility to describe the special theory of relativity without using 4D Minkowski space (see Sect. 2.7.4). It is also common to say that in quantum mechanics we have to use the imaginary unit. Here we will try to show that the reason for the appearance of the imaginary unit in quantum mechanics lies in the fact that the imaginary unit has the ability to rotate, although, as we have seen, with limited possibilities, without a clear geometric interpretation. In geometric algebra, we have rotors, and we can say that there are infinitely many orientated imaginary units, all with an obvious geometric interpretation. Thus, instead of using complex vector spaces, we can use 3D Euclidean space with the geometric product; there is no need for complex numbers. Not bad for our good old 3D space!

We have already shown that orthonormal basis vectors in *Cl*3 can be represented by the Pauli matrices (see Sect. 4.1). Now we will develop this idea a little further, in order to get a sense of the quantum mechanics expressed in the formalism of geometric algebra (see [21]).

2.10.1 The Wave Function of the Electron. Spinors

In the "standard" quantum mechanics formulation, a wave function of an electron has the form of a *spinor* (see Sect. 4.4; $\mathbb{C}$ denotes the standard complex numbers here):

$$|\psi\rangle = \alpha|\uparrow\rangle + \beta|\downarrow\rangle, \ \alpha, \beta \in \mathbb{C}, \quad |\uparrow\rangle = \begin{pmatrix} 1 \\ 0 \end{pmatrix}, \quad |\downarrow\rangle = \begin{pmatrix} 0 \\ 1 \end{pmatrix},$$

where $|\uparrow\rangle$ and $|\downarrow\rangle$ are the pure *up* and *down* states of the spin, written in Dirac *bra-ket* notation. Note that two complex numbers have four real parameters. We can represent such a quantum state in the form of a complex column (row) vector

$$|\psi\rangle = \begin{pmatrix} \alpha \\ \beta \end{pmatrix}, \quad \langle\psi| = |\psi\rangle^{\dagger} = (\alpha^* \ \ \beta^*).$$

Here the character † stands for the *Hermitian adjoint* (see Sect. 1.1.4). We also define *spin operators* (operators are objects that act on vectors; spinors are complex vectors here) in the form $\hat{s}_k \equiv \hat{\sigma}_k \hbar/2$, where $\hat{\sigma}_k$ are the Pauli matrices, and h is Planck's constant ($\hbar = h/2\pi$). If we set the direction of the z-axis in the direction of the up state $|\uparrow\rangle$, we should have

$$\hat{s}_3 \propto \begin{pmatrix} 1 & 0 \\ 0 & -1 \end{pmatrix} = \hat{\sigma}_3$$

(check its effect on the pure states). Now we can look for *observables* (an observable is a dynamic variable that can be measured; note the "sandwich" form; see Appendix 5.3.3) in the form

$$s_k = \hbar n_k/2 = \langle\psi|\hat{s}_k|\psi\rangle, \quad n_k = \langle\psi|\hat{\sigma}_k|\psi\rangle,$$

where the components are given as (a matrix multiplication, check this)

$$n_1 = \alpha\beta^* + \alpha^*\beta, \quad n_2 = i(\alpha\beta^* - \alpha^*\beta), \quad n_3 = \alpha\alpha^* - \beta\beta^*. \quad (2.4)$$

We also have (check)

$$|\mathbf{n}|^2 = \langle\psi|\psi\rangle^2 = \left(|\alpha|^2 + |\beta|^2\right)^2,$$

where we can take advantage of this relation to normalize the vector **n** (to get |**n**| = 1). Introducing *spherical coordinates*, for the unit vector **n** we can write

$$n_1 = \sin\theta\cos\varphi, \quad n_2 = \sin\theta\sin\varphi, \quad n_3 = \cos\theta, \quad (2.5)$$

whence by comparison of (2.4) and (2.5), after some algebra (a nice exercise; see Sect. 4.4), follows

$$\alpha = \cos(\theta/2)e^{i\gamma}, \quad \beta = \sin(\theta/2)e^{i\delta}, \quad \delta - \gamma = \varphi,$$

which gives the spinor expressed in spherical coordinates:

$$|\psi\rangle = \begin{pmatrix} \cos(\theta/2)e^{-i\varphi/2} \\ \sin(\theta/2)e^{i\varphi/2} \end{pmatrix} \exp(i(\gamma+\delta)/2) = \begin{pmatrix} \cos(\theta/2)e^{-i\varphi/2} \\ \sin(\theta/2)e^{i\varphi/2} \end{pmatrix} \exp(-i\alpha/2),$$

$$\alpha = -(\gamma + \delta),$$

where we can neglect the overall phase $\exp(-i\alpha/2)$ due to the sandwich form.

2.10.2 Spinors in *Cl*3

We see the dependence on half angles, suggesting a link to rotors in *Cl*3. Introducing now the designation $e_i \to \sigma_i$, common in *Cl*3, there follows the following connection with rotors (E31: Show this):

$$\mathbf{n} = \sum_{k=1}^{3} n_k\sigma_k = \sin\theta(\sigma_1\cos\varphi + \sigma_2\sin\varphi) + \sigma_3\cos\theta \equiv R\sigma_3R^\dagger,$$

$$R = \exp(-j\varphi\sigma_3/2)\exp(j\theta\sigma_2/2).$$

Expanding both the spinor $|\psi\rangle$ and the rotor R, we can note the following quantities:

$$a^0 = \cos(\varphi/2)\cos(\theta/2), \quad a^1 = \sin(\varphi/2)\sin(\theta/2),$$
$$a^2 = -\cos(\varphi/2)\sin(\theta/2), \quad a^3 = -\sin(\varphi/2)\cos(\theta/2).$$

If we want to introduce a spinor in *Cl*3, which, by analogy, we denote by ψ, we will seek a general form of ψ by analogy (the analogy is denoted by $\leftrightarrow$):

$$|\psi\rangle = \begin{pmatrix} a^0 + ia^3 \\ -a^2 + ia^1 \end{pmatrix} \leftrightarrow \psi = a^0 + a^k j\sigma_k,$$

where the summation over k is understood. We see immediately that $|\uparrow\rangle \leftrightarrow 1$, $|\downarrow\rangle \leftrightarrow j\sigma_2$, while from (2.4) it follows that appropriate vectors of the observable have components (0, 0, ±1). Note that ψ is just a quaternion (in GA).

2.10.3 Analogies for the Action of Operators

We can find analogies for the action of operators, and we begin with the operator $\hat{\sigma}_i$:

$$\hat{\sigma}_i|\psi\rangle \leftrightarrow \sigma_i\psi\sigma_3 = \sigma_i(a^0 + a^k j\sigma_k)\sigma_3 = a^0\sigma_i\sigma_3 + j\sigma_i(a^k\sigma_k)\sigma_3,$$

where σ_3 is included to ensure membership in the even part of the algebra. Namely, in this way, we have the terms $j\sigma_i\sigma_k\sigma_3$, where for $i = 3 \neq k$ we get $-j\sigma_k$, for $i = k$ we get $j\sigma_3$, and ± 1 otherwise. The choice of σ_3 is, of course, a consequence of the initial choice of the z-axis and does not affect the generality of the expressions. The choice of the z-axis usually has a physical background, for example, the direction of an external magnetic field. ◆ What do we get if we multiply all three Pauli matrices? ◆ We can establish an analogy with multiplication by the imaginary unit i, treating the pseudoscalar j as an operator:

$$i|\psi\rangle \leftrightarrow j\psi\sigma_3 = \psi j\sigma_3.$$

It is indicative that we have a multiplication by the bivector $j\sigma_3$, since we expect this; the vector σ_3 is thus invariant under rotations in the $j\sigma_3$-plane (*phase invariance*). Note the role of the pseudoscalar and bivectors here, which, unlike the ordinary imaginary unit, immediately give a clear geometric meaning to the quantities in the theory. This is definitely a good motivation for the study of quantum mechanics in this new language. Instead of unintuitive vectors and matrices over complex numbers, we have elements of geometric algebra, which always introduce clarity.

2.10.4 Observables in the Pauli Theory

Let us look now at observables in the Pauli theory (for more details, see [21]). The *inner product* in standard quantum mechanics is defined as

$$\langle\psi|\phi\rangle = (\psi_1^*, \psi_2^*)\begin{pmatrix}\phi_1\\ \phi_2\end{pmatrix} = \psi_1^*\phi_1 + \psi_2^*\phi_2.$$

This product is a complex number; its real part is (check the analogy by a direct calculation)

$$\text{Re}\,\langle\psi|\phi\rangle \leftrightarrow \langle\psi^\dagger\phi\rangle$$

($\dagger$ is for the reverse involution here), for example,

$$\text{Re}\,\langle\psi|\psi\rangle \leftrightarrow \langle\psi^\dagger\psi\rangle = \langle(a^0 - a^k j\sigma_k)(a^0 + a^k j\sigma_k)\rangle = \sum_{k=0}^{3} a^k a^k.$$

We also have (check)

$$\langle\psi|\phi\rangle = \text{Re}\,\langle\psi|\phi\rangle - i\,\text{Re}\,\langle\psi|i\phi\rangle,$$

whence, using the analogy $i|\psi\rangle \leftrightarrow \psi j\sigma_3$, we can find a new analogy (**Caution**: do not confuse $\langle a|b\rangle$, the inner product, with $\langle ab\rangle$, the grade 0):

$$\langle\psi|\phi\rangle \leftrightarrow \langle\psi^\dagger\phi\rangle - \langle\psi^\dagger\phi j\sigma_3\rangle j\sigma_3. \quad (**)$$

Here we have $\langle\psi^\dagger\phi\rangle$, the grade 0 of the product $\psi^\dagger\phi$, as well as $-\langle\psi^\dagger\phi j\sigma_3\rangle j\sigma_3$, the projection of the product $\psi^\dagger\phi$ on the plane $j\sigma_3$ (see Sect. 1.11.3).

2.10.5 The Expected Value of the Spin

Let us look for the expected value of the spin $\langle\psi|\hat{s}_k|\psi\rangle$. From $\hat{\sigma}_i|\psi\rangle \leftrightarrow \sigma_i\psi\sigma_3$ and (**), we have (check, just replace $|\phi\rangle$ by $\hat{\sigma}_k|\psi\rangle$)

$$\langle\psi|\hat{\sigma}_k|\psi\rangle \leftrightarrow \langle\psi^\dagger\sigma_k\psi\sigma_3\rangle - \langle\psi^\dagger\sigma_k\psi j\rangle j\sigma_3.$$

If we take advantage of the reverse involution, it follows that

$$\left(\psi^\dagger\sigma_k\psi j\right)^\dagger = j^\dagger\psi^\dagger\sigma_k\psi = -\psi^\dagger\sigma_k\psi j,$$

which means that $\psi^\dagger\sigma_k\psi j$ is imaginary (there is no grade 0 or 1). Therefore, it must be $\langle\psi^\dagger\sigma_k\psi j\rangle = 0$, so the expected value of the spin is real (which we expect, since the $\hat{\sigma}_k$ are Hermitian operators, giving the real expected values). The element $\psi^\dagger\sigma_k\psi$ has odd grades only (check), and it is equal to its reverse; therefore, it is a vector. Thus, we can define the *spin vector* as

$$\mathbf{s} \equiv \frac{1}{2}\hbar\psi\sigma_3\psi^\dagger.$$

The expected value is now

$$\langle\psi|\hat{s}_k|\psi\rangle \leftrightarrow \frac{1}{2}\hbar\langle\psi^\dagger\sigma_k\psi\sigma_3\rangle = \frac{1}{2}\hbar\langle\sigma_k\psi\sigma_3\psi^\dagger\rangle = \sigma_k \cdot \mathbf{s}.$$

This expression is different from what we are accustomed to in quantum mechanics. Instead of calculating the expected value of the operator, here we have a simple projection of the spin vector onto the desired direction in space. This immediately raises the question of the coexistence of all three components of the spin vector. The problem does not really exist; the reader is referred to the article R4.

2.10.6 Spinors Are Rotors with Dilatation

We can use our form of spinors to introduce the real scalar $\rho = \psi\psi^\dagger$, and then, if we define

$$R = \rho^{-1/2}\psi,$$

it follows that $RR^\dagger = 1$, which means that we have a rotor. In view of that, spinors (here, not necessarily in general) are just rotors with dilatation, and the spin vector is

$$\mathbf{s} = \frac{1}{2}\hbar\rho R\sigma_3 R^{\dagger}.$$

It follows that the form of the expected value $\sigma_k \cdot \mathbf{s}$ is just an instruction for the rotation of the fixed vector σ_3 in the direction of the spin vector, followed by a dilatation and taking an appropriate component. Note again the clear geometric meaning, which is not as easy to achieve in quantum theory as it is commonly formulated.

For the relationship between the Schrödinger equation and the Pauli theory, see R1.

2.10.7 Half Spin is Due to the 3D Geometry

Let us imagine now that we want to rotate the spin vector. Therefore, let us introduce a transformation $\mathbf{s} \to R_0 \mathbf{s} R_0^{\dagger}$. In doing so, the spinor must transform as $\psi \to R_0\psi$ (E32: Show this), which is often taken as a way to identify an object as a spinor. ◆ E33: Taking into account these properties of transformations, show that spinors change sign after rotation by 2π. ◆ This result is also clear in "ordinary" quantum theory. However, here we see that there is nothing "quantum" in this phenomenon; it is actually the property of our 3D space (of rotors, actually) and it appears in classical mechanics, too, when formulated in geometric algebra. This is probably a significant conclusion. One could say that we have a good reason to reexamine the fundamentals (and philosophy, if you like) of quantum theory. Again, all this is just due to the new product of vectors. For possible implications in the special theory of relativity, see Appendix 5.4.

2.11 Differentiation and Integration

Here we will comment on this area only briefly. The reader is referred to the literature (for example, [13], [21], [25], and [35]). Geometric algebra contains the powerful apparatus of differential and integral calculus. It should be no surprise that here we have a significant improvement over the traditional approach. In particular, there is a *fundamental theorem of calculus* that combines and expands many well-known theorems of classical integral calculus. In addition, all elements of the algebra (including full multivectors) can be included in the calculus on an equal footing. Thus, we can differentiate in the "direction" of a multivector, a bivector for example. This is, actually, a very nice feature! It is common to use the term *geometric calculus* for calculus in the language of geometric algebra. What essentially distinguishes the classical calculus from the geometric calculus is reflected in several elements. First, in the geometric calculus we use various "operators" containing elements of the algebra, which, due to the property of noncommutativity, offer us new opportunities. Let us look at an example of such an "operator"

$$\nabla = e^k \partial_k, \quad \partial_k \equiv \frac{\partial}{\partial x^k},$$

where the Einstein summation convention is understood. Here we have introduced the vectors of a reciprocal basis again, which in orthonormal frames are equal to the basis vectors. This is convenient here for the Einstein summation convention and for possible generalizations. Notice that $\nabla \equiv e^k \partial_k$ has grade 1, i.e., acts as a vector. In *Cl*3, we often use the operator

$$\partial = \partial_t + \nabla,$$

which has the form of a *paravector*. We can differentiate from the left and from the right; however, in this we have the geometric product with the basis vectors, which is noncommutative. Therefore, it is customary to write, for example, $\dot{\mathbf{r}}\dot{\nabla}$, or $\dot{\mathbf{r}}\mathbf{a}\dot{\nabla}$, where the dots indicate that the operator ∇ acts on $\mathbf{r}$, with the order of the unit basis vectors in the geometric products preserved. The derivative operator ∂ can be inverted. For example, Maxwell's equations can be written, without considerable effort, in the form ($J = \rho - \mathbf{J}, \rho$ is a charge density, $\mathbf{J}$ is a current vector)

$$\partial \mathbf{F} = J. \quad (*)$$

◆ Take all grades of (*). ◆ This makes it possible to find the inverse of the operator ∂ using the Green's functions. The simple mathematical form of equation (*) is not just a neat trick; in fact, it provides features that without the geometric product would not exist (or it would be difficult to achieve them). Moreover, we can choose to solve problems in electromagnetism using the *eigenspinors* formalism (see Sect. 2.8), which is even easier than solving equation (*) directly. For interesting examples of the power of geometric algebra in electromagnetism, see [2] or [11].

Second, in the integral calculus, we encounter objects like $dx\,dy$ (see *measure*, *differential forms,* etc.). In geometric algebra, such objects have an orientation (like blades), which gives many new possibilities. For example, the unification of all the important theorems of classical integral calculus (Stokes, Gauss, Green, Cauchy, ...) into one is a great and inspiring achievement of geometric algebra (the *fundamental theorem of geometric calculus*).

2.12 Geometric Models (The Conformal Model)

It is known that the geometry of a 3D vector space can be formulated by embedding it in spaces of higher dimension. Geometric algebra is an ideal framework for investigations of such models. Here we will discuss, only briefly, one of them, the *conformal model* (CGA), developed and patented by *Hestenes*. The idea of the conformal model is that the n-dimensional Euclidean vector space is modeled in $(n+2)$-dimensional Minkowski space. For $n = 3$, apart from the usual unit vectors, let us introduce two more: e, $e^2 = 1$, and $\bar{e}$, $\bar{e}^2 = -1$, to have the basis $\{e, e_1, e_2, e_3, \bar{e}\}$. This is an orthonormal basis of the 5D Minkowski vector space $\mathfrak{R}^{4,1}$. The two added unit vectors define the 2D Minkowski subspace $\mathfrak{R}^{1,1}$, in which we will introduce the new basis $\{o, \infty\}$:

$$o = (e + \bar{e})/2, \quad \infty = \bar{e} - e$$

(the character ∞ represents the point at infinity, while the character o represents the origin). The factor ½ does not play an important role here and may be dropped, but then the rest of the formulas will have a somewhat different form. ◆ E34: Show that o and ∞ are nilpotents and that $e \wedge \bar{e} = o \wedge \infty, o \cdot \infty = \infty \cdot o = -1, o \cdot o = \infty \cdot \infty = 0$. ◆ Now we have the new basis of CGA

$$\{o, e_1, e_2, e_3, \infty\},$$

in which geometric elements, such as lines and circles, have a simple form.

2.12.1 Points as Null Vectors

If we have two points in $\mathfrak{R}^3$ and two 3D vectors $\mathbf{p}$ and $\mathbf{q}$ starting at the common origin and ending up at our two points, the squared distance between the points is given by $(\mathbf{p} - \mathbf{q}) \cdot (\mathbf{p} - \mathbf{q})$. The idea is to find vectors p and q in CGA whose inner product will give us the distance of 3D points (up to a factor), i.e., $p \cdot q \propto (\mathbf{p} - \mathbf{q}) \cdot (\mathbf{p} - \mathbf{q})$. In this case, it should be $p \cdot p = 0$, since the distance is zero. Therefore, such vectors are called *null vectors*. Thus, points are represented by *null vectors* in CGA (in GA, we usually use the term null vector for $nn = 0$). For the 3D vector $\mathbf{p}$, it can be shown that the corresponding null vector is given by

$$p = o + \mathbf{p} + \mathbf{p}^2 \infty / 2,$$

where $p \cdot p = 0$ (E35: Check this and find $p \cdot q$). In CGA, points can have a *weight*; however, here we will not deal with it, except to note that the weight has a geometric meaning; for example, the weight can show the way in which a straight line and a plane intersect.

2.12.2 Geometric Objects

Vectors of CGA that are not points (are not null vectors) can represent a variety of geometric elements. As an example, take the vector $\pi = \mathbf{n} + \lambda\, \infty$, $\lambda \in \mathbb{R}$. If we want to find all points x that belong to such an object, we have to write the condition $x \cdot \pi = 0$, which means that the distance between a point represented by x and a point from π is zero. This is a simple and powerful idea. Now we have

$$x \cdot \pi = \left(o + \mathbf{x} + \mathbf{x}^2 \infty / 2\right) \cdot (\mathbf{n} + \lambda \infty) = \mathbf{x} \cdot \mathbf{n} - \lambda = 0,$$

and we have obtained the equation of the plane perpendicular to the vector $\mathbf{n}$ with distance from the origin λ / n. Now we can take the object π and manipulate it easily, relate it to other objects, rotate, translate, etc. It is particularly important that transformations of elements can be implemented using a single formalism. Thus, for example, the same formalism operates on rotations and translations. If we recall that three points can define a circle in 3D, we could appreciate the fact that a circle in this model can be easily obtained: just take the outer product of the null vectors of these points. If one of the three points is ∞, we get a straight line. It cannot be easier than that. The interested reader can find the theory beautifully presented in [23], where you can take advantage of the software *GAViewer*, which accompanies the book. Everything is available free on the Internet.

Applications

Miroslav Josipović

M. Josipović, *Geometric Multiplication of Vectors*,
Compact Textbooks in Mathematics,
https://doi.org/10.1007/978-3-030-01756-9_3

Robot is playing a trumpet

3.1 Two Boosts

Consider two consecutive boosts in arbitrary directions and let us try to express them as a product of a boost and a rotation. For the hyperbolic sine and cosine we use the abbreviations c and s, while for the sine and cosine we use C and S. We have

---

The original version of this chapter was revised. A correction to this chapter can be found at https://doi.org/10.1007/978-3-030-01756-9_7.

Electronic supplementary material The online version of this chapter (https://doi.org/10.1007/978-3-030-01756-9_3) contains supplementary material, which is available to authorized users.

3

$$e^{\mathbf{v}_1\varphi_1/2}e^{\mathbf{v}_2\varphi_2/2} = (\cosh(\varphi_1/2) + \mathbf{v}_1 \sinh(\varphi_1/2))(\cosh(\varphi_2/2) + \mathbf{v}_2 \sinh(\varphi_2/2))$$
$$= c_1c_2 + c_2s_1\mathbf{v}_1 + c_1s_2\mathbf{v}_2 + s_1s_2\mathbf{v}_1\mathbf{v}_2 = c_1c_2 + c_2s_1\mathbf{v}_1 + c_1s_2\mathbf{v}_2$$
$$+ s_1s_2\mathbf{v}_1 \cdot \mathbf{v}_2 + s_1s_2\mathbf{v}_1 \wedge \mathbf{v}_2.$$

Introducing the angle $\mathbf{v}_1 \cdot \mathbf{v}_2 = \cos\vartheta$ and decomposing the vector $\mathbf{v}_2$ (we will get geometric products instead of outer ones), we get

$$\mathbf{x} = \mathbf{v}_{2\parallel} = \mathbf{v}_1(\mathbf{v}_1 \cdot \mathbf{v}_2) = \mathbf{v}_1 \cos\vartheta, \quad \mathbf{y} = \frac{\mathbf{v}_{2\perp}}{\sin\vartheta} = \frac{\mathbf{v}_2 - \mathbf{v}_{2\parallel}}{\sin\vartheta} = \frac{\mathbf{v}_2 - \mathbf{v}_1\cos\vartheta}{\sin\vartheta},$$
$$\mathbf{y}^2 = 1.$$

Now we have

$$c_1c_2 + s_1s_2\cos\vartheta + c_2s_1\mathbf{v}_1 + c_1s_2(\mathbf{x} + \mathbf{y}\sin\vartheta) + s_1s_2\mathbf{v}_1 \wedge (\mathbf{x} + \mathbf{y}\sin\vartheta)$$
$$= c_1c_2 + s_1s_2\cos\vartheta + (c_2s_1 + c_1s_2\cos\vartheta)\mathbf{v}_1 + c_1s_2\sin\vartheta\mathbf{y} + s_1s_2\sin\vartheta\mathbf{v}_1\mathbf{y}.$$

Expanding the product of the boost and the rotation, we get

$$e^{\mathbf{v}\varphi/2}e^{-j\mathbf{n}\theta/2} = (\cosh(\varphi/2) + \mathbf{v}\sinh(\varphi/2))(\cos(\theta/2) - j\mathbf{n}\sin(\theta/2))$$
$$= cC + Cs\mathbf{v} - cSj\mathbf{n} - sSj\mathbf{v}\mathbf{n}.$$

Note that in the last term we have the pseudoscalar $sSj\mathbf{v} \cdot \mathbf{n}$. Consequently, it must be $\mathbf{v} \cdot \mathbf{n} = 0$. We have our first important conclusion: the boost direction and the rotation axis must be orthogonal. The element $j\mathbf{v}\mathbf{n} = j(\mathbf{v} \wedge \mathbf{n}) = j(j\mathbf{v} \times \mathbf{n}) = -\mathbf{v} \times \mathbf{n}$ (here we understand the cross product as defined in *Cl*3, that is, we have a polar vector; see Sect. 1.3) is a vector. Therefore, the only bivector is $j\mathbf{n}$, suggesting that the unit vector $\mathbf{n}$ is parallel to the vector $\mathbf{v}_1 \times \mathbf{v}_2$. By comparison, we have

$$cC + Cs\mathbf{v} - sSj\mathbf{v}\mathbf{n} - cSj\mathbf{n}$$
$$= c_1c_2 + s_1s_2\cos\vartheta + (c_2s_1 + c_1s_2\cos\alpha)\mathbf{v}_1 + \mathbf{y}c_1s_2\sin\vartheta + \mathbf{v}_1 \wedge \mathbf{y}s_1s_2 sin\vartheta,$$

whence follows

$$cC = c_1c_2 + s_1s_2\mathbf{v}_1 \cdot \mathbf{v}_2 = c_1c_2 + s_1s_2\cos\vartheta, \quad (3.1)$$
$$Cs\mathbf{v} - sSj\mathbf{v}\mathbf{n} = (c_2s_1 + c_1s_2\cos\vartheta)\mathbf{v}_1 + \mathbf{y}c_1s_2\sin\vartheta, \quad (3.2)$$
$$cSj\mathbf{n} = -\mathbf{v}_1\mathbf{y}s_1s_2\sin\vartheta. \quad (3.3)$$

Choosing + in $\mathbf{v}_1\mathbf{y}\mathbf{n} = \pm j$, we see that relation (3.3) gives

$$cS = -s_1s_2\sin\vartheta,$$

from which follows (see the *Thomas precession* in the literature)

$$\frac{cS}{cC} = \tan(\theta/2) = -\frac{s_1s_2\sin\vartheta}{c_1c_2 + s_1s_2\cos\vartheta},$$

giving also $S = \sin(\theta/2)$ and $C = \cos(\theta/2)$. From (3.1), we have c, then also $s = \sqrt{c^2 - 1}$ and φ. For the unit vector $\mathbf{n}$ we have

$$\mathbf{n} = -j\mathbf{v}_1\mathbf{y} = \mathbf{v}_1 \times \mathbf{y} = \frac{\mathbf{v}_1 \times \mathbf{v}_2}{\sin \vartheta}.$$

We choose the sign for $\mathbf{n}$ in such a way that three unit vectors $\mathbf{v}_1$, $\mathbf{y}$, and $\mathbf{n}$ make a dextral system. If we place the boost vectors in the plane e_1e_2, we get $\mathbf{n} = \pm e_3$. From (3.2), we get

$$\begin{aligned}
&\mathbf{v}(1 - \tan(\theta/2)j\mathbf{n}) = s^{-1}C^{-1}(c_2s_1 + c_1s_2\cos\vartheta)\mathbf{v}_1 + \mathbf{y}s^{-1}C^{-1}c_1s_2\sin\vartheta \Rightarrow \\
&\mathbf{v} = \frac{(c_2s_1 + c_1s_2\cos\vartheta)\mathbf{v}_1 + \mathbf{y}c_1s_2\sin\vartheta}{sC(1 - \tan(\theta/2)j\mathbf{n})} \Rightarrow \\
&\mathbf{v} = \frac{(c_2s_1 + c_1s_2\cos\vartheta)\mathbf{v}_1 + \mathbf{y}c_1s_2\sin\vartheta}{sC[1 + \tan^2(\theta/2)]}(1 + \tan(\theta/2)j\mathbf{n}).
\end{aligned}$$

The reader can expand the last relation using

$$\mathbf{v}_1\mathbf{y}\mathbf{n} = j, \quad j\mathbf{v}_1\mathbf{n} = \mathbf{v}_1\mathbf{y}\mathbf{n}\mathbf{v}_1\mathbf{n} = \mathbf{y}, \quad j\mathbf{y}\mathbf{n} = \mathbf{v}_1\mathbf{y}\mathbf{n}\mathbf{y}\mathbf{n} = -\mathbf{v}_1.$$

For orthogonal boosts we can choose boost directions $\mathbf{v}_1$, $\mathbf{v}_2 \to e_1$, e_2. Then

$$\begin{aligned}
&cC = c_1c_2, \quad \mathbf{n} = e_3, \quad \tan(\theta/2) = -\tanh(\varphi_1/2)\tanh(\varphi_2/2), \\
&\mathbf{v} = \frac{c_2s_1e_1 + c_1s_2e_2}{sC[1 + \tan^2(\theta/2)]}(1 + \tan(\theta/2)e_1e_2).
\end{aligned}$$

We see that the formulas are rather complicated. However, it is a nice exercise to manage them. ◆ What happens if φ_2 is small? ◆ As an example, we will take boosts in the e_1 and e_2 directions, with $\gamma = 7 = \cosh\varphi_1 = \cosh\varphi_2$. Then

$$\begin{aligned}
&\cosh(\varphi_1/2) = 2 \Rightarrow \sinh(\varphi_1/2) = \sqrt{3}, \\
&e^{e_1\varphi_1/2}e^{e_2\varphi_1/2} = \left(2 + e_1\sqrt{3}\right)\left(2 + e_2\sqrt{3}\right) = 4 + 2\sqrt{3}(e_1 + e_2) + 3e_1e_2, \quad (*) \\
&e^{\mathbf{v}\varphi/2}e^{-j\mathbf{n}\theta/2} = (\cosh(\varphi/2) + \mathbf{v}\sinh(\varphi/2))(\cos(\theta/2) - j\mathbf{n}\sin(\theta/2)) \\
&\qquad = cC + Cs\mathbf{v} - cSj\mathbf{n} - sSj\mathbf{v}\mathbf{n}, \quad \mathbf{v}\cdot\mathbf{n} = 0, \\
&cC + Cs\mathbf{v} - sSj\mathbf{v}\mathbf{n} - cSj\mathbf{n} = 4 + 2\sqrt{3}(e_1 + e_2) + 3je_3, \\
&cC = 4, \\
&Cs\mathbf{v} - sSj\mathbf{v}\mathbf{n} = 2\sqrt{3}(e_1 + e_2), \\
&cSj\mathbf{n} = -3je_3.
\end{aligned}$$

Now we have $cS\mathbf{n} = -3e_3 \Rightarrow cS = \pm 3$, choosing $cS = -3$ to have $\mathbf{n} = e_3$. Solving, we get

$$\begin{aligned}
&\frac{cS}{cC} = \tan(\theta/2) = -\frac{3}{4}, \\
&S = \sin(\theta/2) = -3/5, \quad \theta \approx -0.4\pi, \\
&C = \cos(\theta/2) = 4/5,
\end{aligned}$$

$$
\begin{aligned}
&c = \cosh(\varphi/2) = 5, \quad \gamma = 49, \quad \beta = 0.9998, \\
&s = \sinh(\varphi/2) = \sqrt{c^2 - 1} = \sqrt{24} = 2\sqrt{6}, \\
&\mathbf{v}(4 + 3je_3) = 5(e_1 + e_2)/\sqrt{2}, \\
&1/(4 + 3je_3) = (4 - 3je_3)/25, \\
&\mathbf{v} = \frac{(e_1 + e_2)}{5\sqrt{2}}(4 - 3e_1e_2) = \frac{1}{5\sqrt{2}}(4e_1 - 3e_2 + 4e_2 + 3e_1) = \frac{1}{5\sqrt{2}}(7e_1 + e_2) \\
&= \frac{7e_1 + e_2}{\sqrt{50}},
\end{aligned}
$$

and finally

$$
\exp(\mathbf{v}\varphi/2)\exp(-j\mathbf{n}\theta/2) = \left(5 + \frac{7e_1 + e_2}{\sqrt{50}} 2\sqrt{6}\right)\left(\frac{4}{5} + je_3\frac{3}{5}\right),
$$

in agreement with (*).

3.2 Two Rotors in 3D

Two rotors in *Cl*3 give a third rotor

$$
R = R_2R_1 \Rightarrow RR^\dagger = R_2R_1R_1^\dagger R_2^\dagger = 1.
$$

How can we show this explicitly and find the rotor R (i.e., its bivector)? Using the abbreviations $c_i = \cos(\theta_i/2)$ and $s_i = \sin(\theta_i/2)$, we have

$$
\begin{aligned}
R_2R_1 &= \left(c_2 - \hat{B}_2s_2\right)\left(c_1 - \hat{B}_1s_1\right) = c_1c_2 - c_2s_1\hat{B}_1 - c_1s_2\hat{B}_2 + s_1s_2\hat{B}_2\hat{B}_1 \\
&= c_1c_2 - c_2s_1\hat{B}_1 - c_1s_2\hat{B}_2 + s_1s_2\left\langle\hat{B}_2\hat{B}_1\right\rangle - s_1s_2\left\langle\hat{B}_1\hat{B}_2\right\rangle_2
\end{aligned}
$$

(note that bivectors anticommute). We want to find the rotor

$$
R = \cos(\theta/2) - \hat{B}\sin(\theta/2),
$$

and therefore, introducing $\mathbf{n}_i = -j\hat{B}_i$, where the rotation axes in 3D satisfy $\mathbf{n}_1 \cdot \mathbf{n}_2 = \cos\delta$, by comparison, we have

$$
\begin{aligned}
&\cos(\theta/2) = c_1c_2 + s_1s_2\cos\delta, \\
&\hat{B}\sin(\theta/2) = c_2s_1\hat{B}_1 + c_1s_2\hat{B}_2 + s_1s_2\left\langle\hat{B}_1\hat{B}_2\right\rangle_2.
\end{aligned}
$$

Using the decomposition of the vector $\mathbf{n}_2$ as

$$
\mathbf{n}_2 = \mathbf{n}_{2\parallel} + \mathbf{n}_{2\perp}, \quad \mathbf{n}_{2\parallel} = \mathbf{n}_1\cos\delta, \quad \mathbf{n}_{2\perp} = \mathbf{u}\sin\delta, \quad \mathbf{n}_1 \perp \mathbf{u},
$$

we have $\mathbf{n}_1 \times \mathbf{u} = -j\mathbf{n}_1\mathbf{u}$, $\mathbf{n}_1\mathbf{u}(-j\mathbf{n}_1\mathbf{u}) = -j\mathbf{n}_1\mathbf{u}\mathbf{n}_1\mathbf{u} = j$. For the unit bivectors $\hat{B}_i = j\mathbf{n}_i$, we have

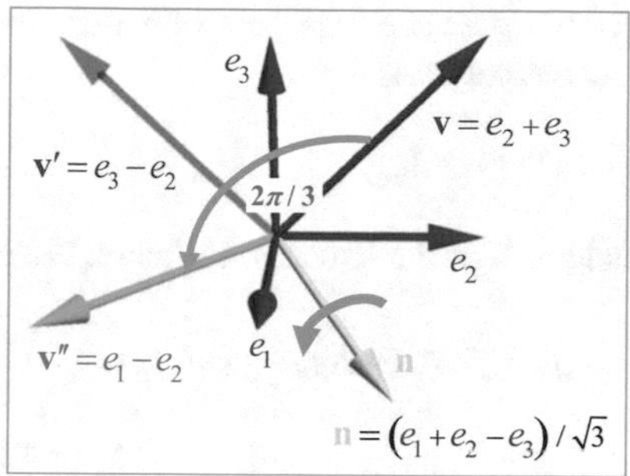

Fig. 3.1 Two rotors

$$\langle \hat{B}_1 \hat{B}_2 \rangle_2 = -\mathbf{n}_1 \wedge \mathbf{n}_2 = -\mathbf{n}_1 \wedge \mathbf{u} \sin\delta = -j\mathbf{n}_1 \times \mathbf{u} \sin\delta,$$

and comparison gives (you can cancel j)

$$j\mathbf{n} \sin(\theta/2) = c_2 s_1 j\mathbf{n}_1 + c_1 s_2 j\mathbf{n}_1 \cos\delta + c_1 s_2 j\mathbf{u} \sin\delta - s_1 s_2 j\mathbf{n}_1 \times \mathbf{u} \sin\delta.$$

Now a direct calculation gives the consistency with $\sin^2(\theta/2) + \cos^2(\theta/2) = 1$, and the rotor is

$$R = c_1 c_2 + s_1 s_2 \cos\delta - j((c_2 s_1 + c_1 s_2 \cos\delta)\mathbf{n}_1 + \mathbf{u} c_1 s_2 \sin\delta - \mathbf{n}_1 \times \mathbf{u} s_1 s_2 \sin\delta).$$

For $\mathbf{n}_1 \cdot \mathbf{n}_2 = 0$, we have $R = c_1 c_2 - j(c_2 s_1 \mathbf{n}_1 + c_1 s_2 \mathbf{n}_2 - s_1 s_2 \mathbf{n}_1 \times \mathbf{n}_2)$. ◆ What about the case $\mathbf{n}_2 = -\mathbf{n}_1$? ◆

The mathematician Olinde Rodrigues discovered these formulas before Hamilton's discovery of the quaternions.

Consider two rotors, first with the rotation axis e_1 and second with the rotation axis e_2, both by the angle $\pi/2$ (see Sect. 1.9.8). Then we have

$$\theta = 2\pi/3, \quad \mathbf{n} = (e_1 + e_2 - e_3)/\sqrt{3}.$$

In Fig. 3.1, we have the vector $\mathbf{v} = e_2 + e_3$ rotated around e_1 to the vector $\mathbf{v}' = e_3 - e_2$, then the vector $\mathbf{v}'$ rotated to $\mathbf{v}''$ around e_2. The vectors $\mathbf{v}$ and $\mathbf{v}''$ both lie in the plane defined by the bivector $j\mathbf{n}$. Find the bivector $B = \mathbf{v} \wedge \mathbf{v}''$ and then the unit bivector $\hat{B}$.

3.3 Reciprocal Frames

Here we indicate the Einstein summation rule using Greek letters for indices.

As a basis in an n-dimensional vector space, we can choose any set of n *linearly independent* vectors $f_1, f_2, \cdots, f_n$. Then we define the pseudoscalar (I is the unit pseudoscalar)

$$I_n = f_1 \wedge f_2 \wedge \cdots \wedge f_n = \Omega I, \quad \Omega \in \mathbb{R}, \quad I_n^{-1} = \pm\Omega^{-1} I,$$

where we have made use of earlier results for pseudoscalars. It is useful to define the reciprocal basis

$$f^i \cdot f_j \equiv \delta_{ij},$$

where δ_{ij} is the Kronecker symbol. From the antisymmetry of the outer product, we have

$$\begin{aligned} f_1 \wedge f_2 \wedge \cdots \wedge f_j \wedge \cdots \wedge f_n &= (-1)^{j-1} f_j \wedge \left(f_1 \wedge f_2 \wedge \cdots \wedge \breve{f}_j \wedge \cdots \wedge f_n\right) \\ &= (-1)^{j-1} f_j \wedge \breve{I}_j, \end{aligned}$$

where $\breve{f}_j$ means that f_j is missing from the product. From the general formula

$$a \cdot \left(A_j I\right) = a \wedge A_j I,$$

taking that $A_j = \breve{I}_j = f_1 \wedge f_2 \wedge \cdots \wedge \breve{f}_j \wedge \cdots \wedge f_n$, it follows that

$$f_i \cdot \left(A_j I\right) = \left(f_i \wedge A_j\right) I = 0, \quad i \neq j.$$

This gives the possibility to define

$$f^i \equiv (-1)^{i-1} f_1 \wedge f_2 \wedge \cdots \wedge \breve{f}_i \wedge \cdots \wedge f_n I_n^{-1},$$

or in abbreviated form,

$$f^i = (-1)^{i-1} \breve{I}_i I_n^{-1}.$$

For example,

$$f_1 \cdot f^1 = f_1 \cdot \left(f_2 \wedge \cdots \wedge f_n I_n^{-1}\right) = f_1 \wedge f_2 \wedge \cdots \wedge f_n I_n^{-1} = 1,$$

or

$$f_1 \cdot f^1 = f_1 \cdot \left(\breve{I}_1 I_n^{-1}\right) = \left(f_1 \wedge \breve{I}_1\right) I_n^{-1} = I_n I_n^{-1} = 1.$$

Note that in orthogonal frames we have $f^i \propto f_i$.

A vector can be specified using both bases $a = \mathrm{a}^\mu f_\mu = \mathrm{a}_\mu f^\mu$, where the components are

$$\begin{aligned} \mathrm{a}^i &= f^i \cdot a \quad (\textit{contravariant}), \\ \mathrm{a}_i &= f_i \cdot a \quad (\textit{covariant}). \end{aligned}$$

In orthogonal frames, we have the obvious relations $f_i \wedge f^i = 0$. However, in nonorthogonal frames we generally have $f_i \wedge f^i \neq 0$ (no summation). Interestingly, the sum $f_\mu \wedge f^\mu$ is always zero:

$$f_\mu \wedge f^\mu = \sum_i (-1)^{i-1} f_i \wedge \left(f_1 \wedge f_2 \wedge \cdots \wedge \breve{f}_i \wedge \cdots \wedge f_n I_n^{-1}\right) = 0;$$

see below. Consequently, it follows that

$$f_\mu f^\mu = f_\mu \cdot f^\mu = n.$$

From the relation $f_\mu f^\mu \cdot f_j = f_j$, we get

$$f_\mu f^\mu \cdot (f_j \wedge f_k) = f_\mu (f^\mu \cdot f_j f_k - f^\mu \cdot f_k f_j) = f_j f_k - f_k f_j = 2 f_j \wedge f_k,$$

and generally, for a blade of grade k,

$$\begin{aligned}
&f_\mu f^\mu \cdot A_k = k A_k,\\
&f_\mu f^\mu \wedge A_k = f_\mu f^\mu A_k - f_\mu f^\mu \cdot A_k = n A_k - k A_k = (n-k) A_k,\\
&f^i \wedge A_k \neq 0 \Rightarrow f^i \wedge A_k = (-1)^k A_k \wedge f^i,\\
&f^i \cdot A_k = -(-1)^k A_k \cdot f^i,\\
&f_\mu A_k \cdot f^\mu + f_\mu A_k \wedge f^\mu = f_\mu (A_k \cdot f^\mu + A_k \wedge f^\mu) =\\
&f_\mu A_k f^\mu = -(-1)^k k A_k + (-1)^k (n-k) A_k = (-1)^k (n-2k) A_k = (-1)^k (n-2k) A_k.
\end{aligned}$$

Using the left contractions rule, it follows that

$$(f^p \wedge f^q \wedge \cdots \wedge f^s) \rfloor (f_i \wedge f_j \wedge \cdots \wedge f_k) = \delta_{is} \cdots \delta_{jq} \delta_{kp}$$

(note the order of indices). Now we can express each multivector as a sum:

$$M = \sum_{i<j<\cdots<k} M_{ij\cdots k} f^i \wedge f^j \wedge \cdots \wedge f^k, \quad M_{ij\cdots k} = (f^k \wedge \cdots \wedge f^j \wedge f^i) \cdot M,$$

where the $M_{ij\cdots k}$ are totally antisymmetric in all indices. For orthogonal coordinates, we have

$$I_n = f_1 f_2 \cdots f_n, \quad f^i = (-1)^{i-1} \frac{f_1 f_2 \cdots \breve{f}_i \cdots f_n}{f_1 f_2 \cdots f_n} = (-1)^{i-1} f_i^{-1} \propto f_i.$$

3.3.1 An Example of a Reciprocal Frame

As an example of a nonorthogonal frame in 3D, consider the frame (Fig. 3.2)

$$\begin{aligned}
&f_1 = e_1, \quad f_2 = e_2, \quad f_3 = e_1 + e_2 + e_3,\\
&I_3 = e_1 \wedge e_2 \wedge (e_1 + e_2 + e_3) = j, \quad I_3^{-1} = -j,
\end{aligned}$$

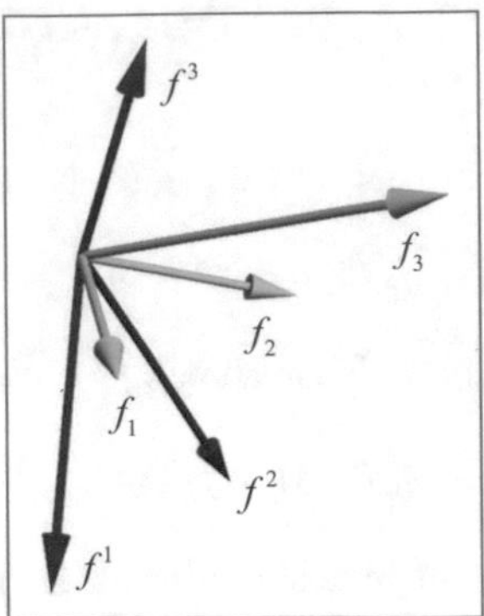

Fig. 3.2 The example of a reciprocal frame

$$
\begin{aligned}
f^1 &= f_2 \wedge f_3 I_3^{-1} = -je_2 \wedge (e_1 + e_2 + e_3) \\
&= -j(e_2 e_1 + e_2 e_3) = e_1 - e_3, \\
f^2 &= -f_1 \wedge f_3 I_3^{-1} = je_1 \wedge (e_1 + e_2 + e_3) \\
&= j(e_1 e_2 + e_1 e_3) = e_2 - e_3, \\
f^3 &= f_1 \wedge f_2 I_3^{-1} = -je_1 e_2 = e_3.
\end{aligned}
$$

Note that

$$
\begin{aligned}
f_\mu \wedge f^\mu &= e_1 \wedge (e_1 - e_3) + e_2 \wedge (e_2 - e_3) + (e_1 + e_2 + e_3) \wedge e_3 \\
&= -e_1 \wedge e_3 - e_2 \wedge e_3 + e_1 \wedge e_3 + e_2 \wedge e_3 = 0.
\end{aligned}
$$

Now we will try to clarify some properties of a reciprocal basis. Using the reciprocal basis e^i, $e^i \cdot e_j = \delta_{ij}$, we can express vectors as $a = \mathrm{a}^\mu e_\mu = \mathrm{a}_\mu e^\mu$, $\mathrm{a}_\mu \in \mathbb{R}$ (summation). Then it follows that

$$e^i \cdot a = \mathrm{a}^\mu e^i \cdot e_\mu = \mathrm{a}^i.$$

For the bivector $a \wedge b$, we have

$$e_\mu e^\mu \cdot (a \wedge b) = e_\mu (e^\mu \cdot ab - e^\mu \cdot ba) = e_\mu (\mathrm{a}^\mu b - \mathrm{b}^\mu a) = ab - ba = 2a \wedge b.$$

Using induction, for a multivector of grade r we can write (show this)

$$e_\mu e^\mu \cdot A_r = r A_r.$$

Expanding the basis vector $e_k = e_\mu \cdot e_k e^\mu$, we have

$$e^\mu e_\mu = e^\mu (e_\nu \cdot e_\mu e^\nu) = e_\nu \cdot e_\mu e^\mu e^\nu,$$

meaning that only the symmetric parts of $e^\mu e^\nu$ contribute to the sum ($e_\nu \cdot e_\mu$ is symmetric). Therefore, we get

$$e^{\mu}e_{\mu} = e^{\mu} \cdot e_{\mu} + e^{\mu} \wedge e_{\mu} = e^{\mu} \cdot e_{\mu} = n, \quad e^{\mu} \wedge e_{\mu} = 0$$

(n is the dimension of the vector space). This result leads to

$$e_{\mu}e^{\mu} \wedge A_r = e_{\mu}(e^{\mu}A_r - e^{\mu} \cdot A_r) = e_{\mu}e^{\mu}A_r - e_{\mu}e^{\mu} \cdot A_r = (n - r)A_r.$$

3.4 Rigid-Body Dynamics

In Sect. 1.5, we defined mixed products of a vector with a bivector. Let us see this now in a different way, since the properties of mixed products with bivectors will be useful here.

Consider the product aB, where a is a vector and B is a blade (it can be factored as an outer product of two vectors). It is always possible to decompose the vector a into a component in the bivector plane (a_{in}) and one out of the bivector plane (a_{out}, the dashed blue perpendicular) (Fig. 3.3). We can also write the bivector B as the geometric product $B = a_{\text{in}}b$, where the vector b belongs to the bivector plane, and it is orthogonal to a_{in}. Then we have

$$aB = aa_{\text{in}}b = (a_{\text{in}} + a_{\text{out}})a_{\text{in}}b = a_{\text{in}}a_{\text{in}}b + a_{\text{out}}a_{\text{in}}b, \quad (3.4)$$
$$Ba = a_{\text{in}}ba = a_{\text{in}}b(a_{\text{in}} + a_{\text{out}}) = a_{\text{in}}ba_{\text{in}} + a_{\text{in}}ba_{\text{out}}. \quad (3.5)$$

The term $a_{\text{in}}^2 b$ is a vector, while the term $a_{\text{out}}a_{\text{in}}b$ is a pseudoscalar. The first term is anticommutative ($a_{\text{in}}a_{\text{in}}b = -a_{\text{in}}ba_{\text{in}}$) and grade-lowering, whereas the second term is commutative ($a_{\text{out}}a_{\text{in}}b = -a_{\text{in}}a_{\text{out}}b = a_{\text{in}}ba_{\text{out}}$) and grade-raising. Taking the sum and the difference of (3.4) and (3.5), we can define

$$a \cdot B \equiv a_{\text{in}}a_{\text{in}}b = (aB - Ba)/2,$$
$$a \wedge B \equiv a_{\text{out}}a_{\text{in}}b = (aB + Ba)/2.$$

Here **the inner product means the lowest grade of the product** aB, whereas **the outer product means the highest grade of the product** aB. Finally, we have the familiar relation

$$aB = a \cdot B + a \wedge B.$$

Note that we have an extended definition of the inner and outer products, which is in accordance with the definition for vectors. Note also that the inner product is anticommutative here, while the outer product is commutative. The inner product is in accordance with the definition of the left contraction; just note that $a_{\text{in}}^2 b$ is the vector in the

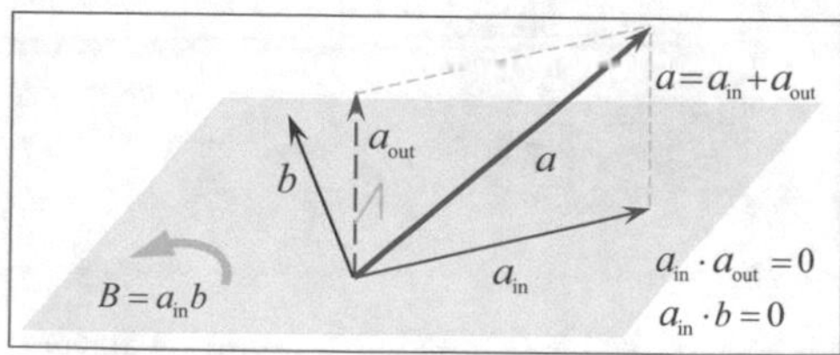

Fig. 3.3 Mixed products of a vector with a bivector

plane defined by B, perpendicular to the projection of the vector a on B. ◆ Find a_{in}, a_{out}, and b. ◆

Next, consider the exponential function $\exp(\alpha x)$, $\alpha = \text{const}$, with the property

$$(\exp(\alpha x))' \equiv \frac{d}{dx}\exp(\alpha x) = \alpha \exp(\alpha x).$$

If we have an equation of type $f'(x) = \alpha f(x)$, we can conclude that $f(x) = C_1 \exp(\alpha x) + C_2$, where the C_k are constants that can be found if $f(0)$ and $f'(0)$ are given.

The *angular momentum* of a particle with *momentum* $\mathbf{p}$ and *position vector* $\mathbf{r}$ is usually defined using the cross product $\mathbf{L} = \mathbf{r} \times \mathbf{p}$, which means that it is an axial vector. In geometric algebra, axial vectors become bivectors; therefore, we define the angular momentum as $L = \mathbf{r} \wedge \mathbf{p}$ (Fig. 3.4), which immediately gives an interpretation: the position vector is sweeping out an oriented plane. In this way, the angular momentum can be defined also in 2D, or any dimension. There is no need for a "vector orthogonal to the plane." The first derivative gives

$$\dot{L} = \dot{\mathbf{r}} \wedge \mathbf{p} + \mathbf{r} \wedge \dot{\mathbf{p}} = \mathbf{r} \wedge \dot{\mathbf{p}},$$

since $\mathbf{p} \propto \dot{\mathbf{r}}$. For the *central forces* ($\mathbf{F} \propto \mathbf{r}$), we have (Newton's second law) $\mathbf{F} = \dot{\mathbf{p}} \propto \mathbf{r}$, from which it follows that $\dot{L} = 0 \Rightarrow L = \text{const}$. Denoting the position vector of the center of mass by $\mathbf{r}$, and the position vector of a body point by $\mathbf{x}$ (its starting point is at the center of mass, Fig. 3.5), we see that the center of mass moves along a trajectory, while the

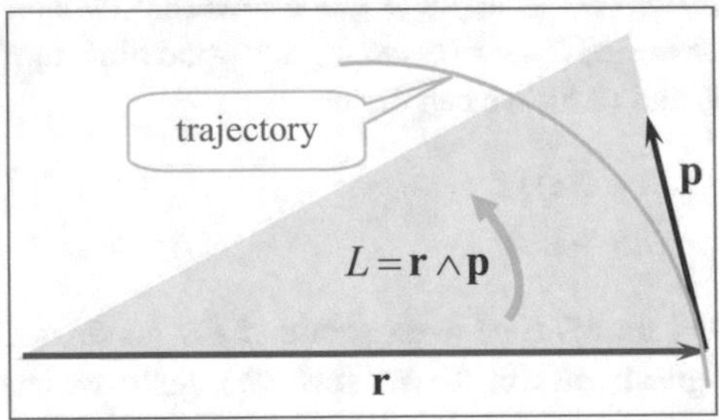

Fig. 3.4 The angular momentum definition

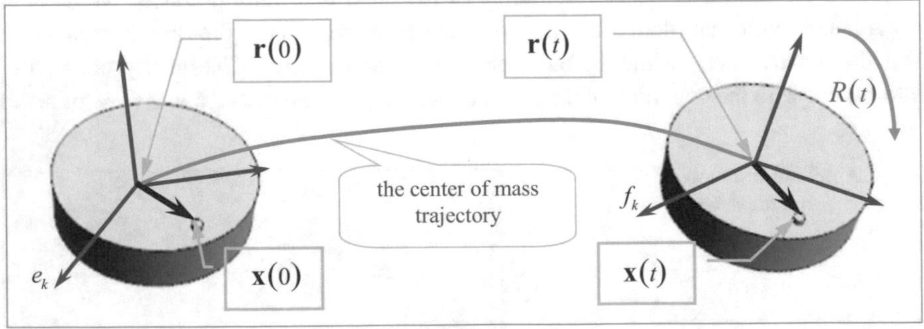

Fig. 3.5 A rigid body rotation in geometric algebra

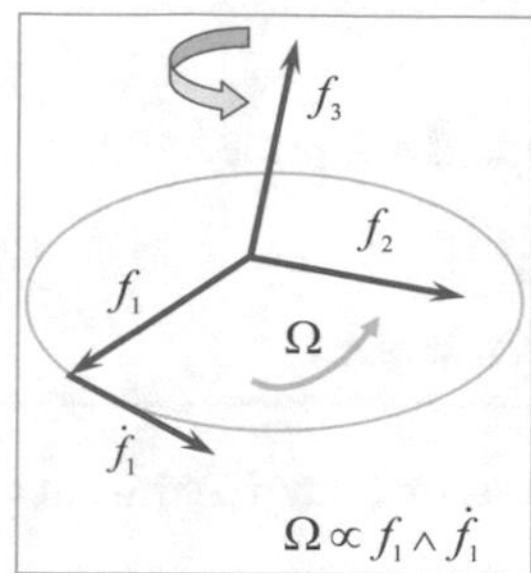

Fig. 3.6 The images of orthogonal unit vectors under a rotation

position vector **x** generally rotates around its initial position. Using a time-dependent rotor, we can write

$$\mathbf{x}(t) = \mathbf{r}(t) + R(t)\mathbf{x}(0)R^{\dagger}(t).$$

All rotational motion is contained in the time-dependent rotor $R(t)$. We can imagine that the basis unit vector e_k is rotated to the unit basis vector f_k:

$$f_k(t) = R(t)e_k R^{\dagger}(t).$$

It is customary to define the angular velocity vector ω (in the standard treatment we use the cross product) (Fig. 3.6)

$$\dot{f}_k = -j\omega \wedge f_k = (-j\omega) \cdot f_k \equiv -\mathbf{\Omega} \cdot f_k,$$

where we have introduced the angular velocity bivector $\mathbf{\Omega} \equiv j\omega$. For a rotation around f_3, we have $\mathbf{\Omega} = jf_3|\omega| \propto f_1 \wedge \dot{f}_1$, where $\dot{f}_1$ is orthogonal to both f_1 and f_3. The first derivative is now

$$\dot{f}_k = \dot{R}e_k R^{\dagger} + \mathrm{Re}_k\dot{R}^{\dagger} = \dot{R}R^{\dagger}\,\mathrm{Re}_k R^{\dagger} + \mathrm{Re}_k R^{\dagger}R\dot{R}^{\dagger} = \dot{R}R^{\dagger}f_k + f_k R\dot{R}^{\dagger}.$$

Due to $RR^{\dagger} = 1$, we have $\dot{R}R^{\dagger} = -R\dot{R}^{\dagger} = -\left(\dot{R}R^{\dagger}\right)^{\dagger}$, which means that $\dot{R}R^{\dagger}$ is imaginary, and since it is even, it must be a pure bivector. Consequently,

$$\dot{f}_k = \dot{R}R^{\dagger}f_k - f_k\dot{R}R^{\dagger} = 2\left(\dot{R}R^{\dagger}\right) \cdot f_k,$$

which leads to $\mathbf{\Omega} = -2\dot{R}R^{\dagger}$, meaning that all dynamics reduces to the equation

$$\dot{R} = -\mathbf{\Omega}R/2.$$

Such equations are easier to solve than their matrix counterparts. When formulated in the language of geometric algebra, they appear in many branches of physics (see Sect. 2.8).

If $\mathbf{\Omega} = \text{const}$, we immediately have

$$R = R(0)\exp(-\boldsymbol{\Omega}t/2),$$

a rotor with constant frequency in the $\boldsymbol{\Omega}$-plane.

3.5 *Cl*2 and Complex Numbers

Consider a product of two unit vectors, say $\hat{\mathbf{v}}$ and $\hat{\mathbf{w}}$, in *Cl*2,

$$U = \hat{\mathbf{v}}\hat{\mathbf{w}} = \cos\varphi + I\sin\varphi = \exp(I\varphi),$$
$$U^* = \hat{\mathbf{w}}\hat{\mathbf{v}} = \cos\varphi - I\sin\varphi = \exp(-I\varphi), \quad I = e_1e_2, \quad UU^* = 1,$$

where U and U^* are the unit complex number and its complex conjugate. We use the designation U^* for the complex conjugate (as usual), but note that this is just the reverse involution. We see that the geometric product of two unit vectors has an obvious interpretation in *Cl*2: it is a (noncommutative) complex number. In general, we can use any two vectors, say $\mathbf{v}$ and $\mathbf{w}$, and create a complex number

$$z = \mathbf{vw} = |\mathbf{v}||\mathbf{w}|(\cos\varphi + I\sin\varphi) = |z|\exp(I\varphi),$$

which is just a scaled version of a unit complex number. Complex numbers rotate objects in the e_1e_2-plane:

$$Ue_1 = (\cos\varphi + I\sin\varphi)e_1 = e_1\cos\varphi - e_2\sin\varphi,$$
$$e_1U = e_1(\cos\varphi + I\sin\varphi) = e_1\cos\varphi + e_2\sin\varphi,$$
$$U^*e_1 = (\cos\varphi - I\sin\varphi)e_1 = e_1\cos\varphi + e_2\sin\varphi,$$

where it is clear that the noncommutativity becomes an advantage; we can choose the orientation of a rotation, and the complex conjugate of a complex number rotates vectors in the opposite direction. We can also write the composition of rotations in the e_1e_2-plane as

$$U_1U_2 = \exp(I\varphi_1)\exp(I\varphi_2) = \exp(I(\varphi_1 + \varphi_2)),$$

where it is clear that rotations are commutative in 2D (can you see the trigonometric addition formulas?). Note that the composition of two rotations is just the product of two complex numbers, which can be represented as a **sum** of two arcs (Fig. 3.7). For a general complex number, we can write

$$z = |z|U,$$

which rotates (U) and dilates ($|z|$) vectors, that is, acts as a **spinor**. In fact, complex numbers can be treated as spinors (the designation U suggests the group U(1)); we just need the sandwich form

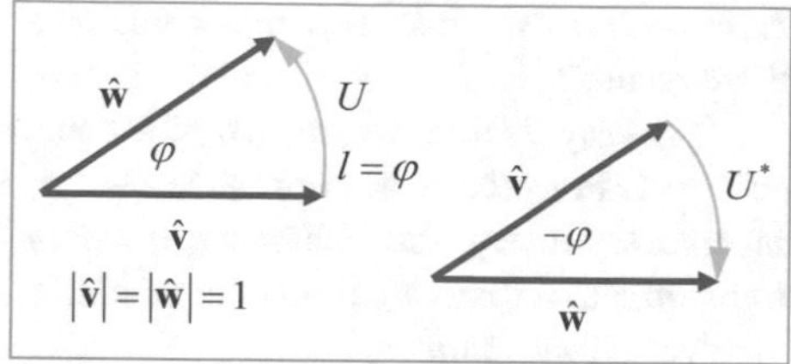

Fig. 3.7 An oriented arc of length (l) equal to the angle (in radians) represents a complex number

$$\exp(I\varphi)\mathbf{v} = \exp(I\varphi/2)\exp(I\varphi/2)\mathbf{v} = \exp(I\varphi/2)\mathbf{v}\exp(-I\varphi/2),$$

where we see that $\exp(I2\pi/2) = \exp(I\pi) = -1$, just what we expect from spinors.

Given two vectors, say $\mathbf{v}$ and $\mathbf{w}$, it is easy to find the complex number that transforms $\mathbf{v}$ to $\mathbf{w}$:

$$z\mathbf{v} = \mathbf{w}\mathbf{v}^{-1}\mathbf{v} = \mathbf{w},$$

since the inverse of a vector is a vector, and the modulus is

$$|z| = \sqrt{zz^*} = \sqrt{\mathbf{w}\mathbf{v}^{-1}\mathbf{v}^{-1}\mathbf{w}} = |\mathbf{w}|/|\mathbf{v}|.$$

These formulas are simple, and there is no need for rotation matrices, which is a great advantage. As an example, to transform the vector e_1 to the vector $e_1 + e_2$, we use

$$z = (e_1 + e_2)e_1^{-1} = (e_1 + e_2)e_1 = 1 - I = \sqrt{2}\exp(-I\pi/4),$$

and we see that the vector e_1 transforms to $ze_1 = (e_1 + e_2)e_1e_1 = e_1 + e_2$, which means that it is dilated by $|z| = \sqrt{2}$, while the unit complex number $\exp(-I\pi/4)$ rotates it by the angle $\pi/4$.

We can apply various transformations to complex numbers. For example, for $z = (e_1 + e_2)e_1 = 1 - I$, we can ask what happens in the transformation $z \to e_1 z e_1$,

$$z \to e_1(e_1 + e_2)e_1e_1 = 1 + I = z^*,$$

or, since I anticommutes with all vectors in the e_1e_2-plane,

$$z = \mathbf{v}\mathbf{w} \to \hat{\mathbf{u}}\mathbf{v}\mathbf{w}\hat{\mathbf{u}} = |\mathbf{v}||\mathbf{w}|\hat{\mathbf{u}}\exp(I\varphi)\hat{\mathbf{u}} = |\mathbf{v}||\mathbf{w}|\exp(-I\varphi) = z^*.$$

On the other hand, we can write $\hat{\mathbf{u}}\mathbf{v}\mathbf{w}\hat{\mathbf{u}} = \hat{\mathbf{u}}\mathbf{v}\hat{\mathbf{u}}\hat{\mathbf{u}}\mathbf{w}\hat{\mathbf{u}}$, and therefore, the vectors $\mathbf{v}$ and $\mathbf{w}$are reflected. As a simple example of a reflected vector, we have

$$e_1 + e_2 \to e_1(e_1 + e_2)e_1 = e_1 - e_2,$$

where the vector $e_1 + e_2$ is reflected across the vector e_1. Reflections across a unit vector produce complex conjugation of complex numbers. For the vector $\mathbf{v}$, we have

$IvI = -IIv = v$, while for complex numbers, we have $IzI = IIz = -z$. Can you interpret these results?

It is easy to find the inverse of a complex number z; we have $z^{-1} = z^*/zz^*$, with $zz^{-1} = 1$. For vectors, we also have $aa^{-1} = 1$. However, we can define another type of inverse, say inv(a), such that $a\,\mathrm{inv}(a) = r^2$, $r \in \mathbb{R}$. In fact, inv(a) can be understood as a transformation that transforms the interior of a circle of the radius r to its exterior, and vice versa (*inversion*).

Note that all our relations are easy to generalize: we can choose any plane, in any dimension, define the unit bivector in that plane, define complex numbers as products of vectors, etc. For example, $I = e_3e_4$ will be a nice anticommutative imaginary unit in the 4D Euclidean vector space.

3.5.1 Inversion

Consider a 2D Euclidean plane. The *inversion* of radius $r \in \mathbb{R}$ is the (geometric) transformation that maps a vector **v** to

$$\mathbf{v}' = r^2\mathbf{v}^{-1}.$$

For $r = 1$, we have just a vector inverse. This *inversion* has many interesting properties (see R5) (Fig. 3.8).

We can consider two vectors

$$\mathbf{v}' = r^2\mathbf{v}^{-1}, \quad \mathbf{w}' = r^2\mathbf{w}^{-1}$$

and define the complex number $z = \mathbf{vw}$. Then inversion gives

$$z' = \mathbf{v}'\mathbf{w}' = r^4 z|z|^{-2}.$$

We can study properties of inversion on sets of points, such as circles and triangles (see the reference R5). However, now we have the powerful language of geometric algebra. ◆ Consider two concentric circles placed anywhere in the exterior of a circle. Find their images in the interior. You can also try to do this using the CGA model. ◆

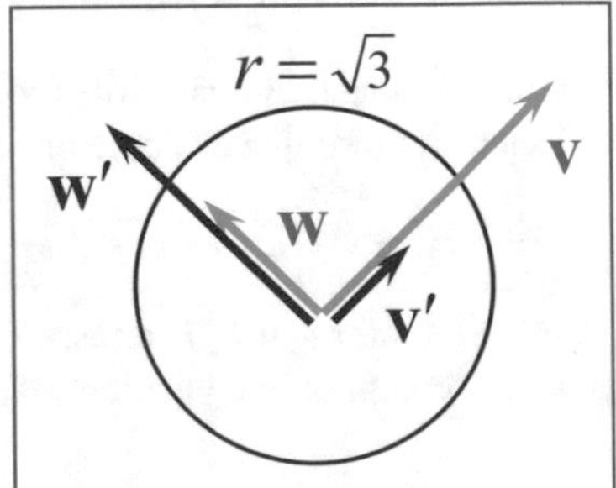

Fig. 3.8 The inversion of vectors

3.6 Generalization of Real and Complex Products of Complex Numbers

For two complex numbers z_1 and z_2, we can define (see R6) the *real product* as

$$z_1 \cdot z_2 = \left(z_1^* z_2 + z_1 z_2^*\right)/2$$

and the *complex product* as

$$z_1 \times z_2 = \left(z_1^* z_2 - z_1 z_2^*\right)/2$$

(do not confuse this with the inner and the cross product; see Sect. 4.4.5). We can try to define similar products for multivectors, where the role of complex conjugation can be taken by any involution. In doing that, we must worry about the involutions' properties (*automorphism*). Simply stated, the geometric product is not commutative, and we have a "problem" of the order of multivectors in products. It appears that there is a way to define the "real" and the "complex" products, namely, we can start from some involution $\mathrm{i}(M)$ and define

$$2M_1 \odot M_2 = \mathrm{i}(M_1)M_2 + \mathrm{i}[\mathrm{i}(M_1)M_2],$$
$$2M_1 \otimes M_2 = \mathrm{i}(M_1)M_2 - \mathrm{i}[\mathrm{i}(M_1)M_2].$$

Here the operator $\otimes$ is for the "complex" product, not for a commutator (as in the text). Note that the property of automorphism (antiautomorphism) is satisfied automatically. As an example, in *Cl*3, for the Clifford conjugation and a general multivector $Z + \mathbf{F}$, we get

$$M \odot M = Z^2 - \mathbf{F}^2, \quad M \otimes M = 0$$

Note that in general, the grades of products that change sign under an involution disappear with $\odot$ and remain with $\otimes$ (they can also disappear if $M_1 = M_2$). Therefore, if we take, for example, the reverse involution, we immediately know that $M_1 \odot M_2$ is a paravector, etc. The reader should consult the references for ordinary complex numbers and try to investigate the properties of the products in *Cl*3.

3.7 The Complex Geometric Product and Fractals

Starting from the 2D vector $r = xe_1 + ye_2$, we have a complex number

$$z = e_1 r = x + y e_1 e_2 = x + yI,$$

and a complex conjugate

$$z^* = re_1 = x - yI,$$

with the commutation property

$$e_1 z = z^* e_1.$$

For an ordinary complex number $q = x + i\ y$, we have

$$q^2 = x^2 - \ y^2 + 2ixy,$$

and the same is true for z,

$$z^2 = x^2 - \ y^2 + 2Ixy.$$

Consider the product

$$re_1r = re_1e_1e_1r = z^*e_1z = e_1z^2$$

or $e_1re_1r = z^2$. This gives us the opportunity to define the *complex geometric product* (a, b are vectors, e is a unit vector)

$$a \odot b = aeb,$$

or alternatively,

$$a \odot b = eaeb.$$

Such a product gives us the opportunity to generalize the theory of fractals easily to any dimension (see R7). For example, in the definition of the *Mandelbrot set* and the *Julia set*, functions of special type occur. For c, $z \in \mathbb{C}$ we define

$$\mathrm{f}(z, c) = z^2 + c,$$

which in GA takes the form

$$\mathrm{f}(v, c) = v \odot v + c,$$

where v is a vector. The important fact is that we can take vectors in higher dimensions, for example in 3D, and thus get fractals in higher dimensions quite easily.

3.8 Multiplication of Blades and Programming

The main problem of computing with geometric algebra is that we have computers based on the coordinate approach, which means that we can manipulate binary numbers or lists of numbers easily. Although oriented blades can be expressed as lists of numbers, it would be desirable to have computers that know how to work directly with oriented geometric objects, so we could declare objects as blades, and processors would know the multiplication rules.

There are various ways to implement geometric algebra on computers, but in essence, we have to choose how to represent blades and then find a way to calculate efficiently. For example, we can use multiplication tables, which means that products are not calculated using a processor, but are taken directly from RAM. Without multiplication tables, processors must work hard.

There is plenty of literature on the subject, for example, R8. Nice programming examples can be found in [23].

3.8.1 An Interesting Mathematica Implementation

There is a nice implementation clifford.m (Aragón-Camarasa et al.; see in R9) that can be used to play with various (not high) dimensions and signatures.

The blades of the Clifford basis can be written as $(m_1, m_2, \ldots, m_n)$, where $m_i \in \{0, 1\}$, $i = 1, 2, \ldots, n$, are the exponents in the products

$$\prod_{i=1}^{n} e_i^{m_i}.$$

For example, in $Cl3$ we have

$$e_1 e_3 = e_1^1 e_2^0 e_3^1.$$

Due to linearity, multiplication of two multivectors amounts to multiplications of blades, giving the formal expressions (m_i and r_i are the exponents for the factor blades)

$$(m_1, m_2, \ldots, m_n)(r_1, r_2, \ldots, r_n).$$

Note that the unit vector e_1, with exponents $r_1 \in \{0, 1\}$, has to make positional replacements with the blade $(m_2, \ldots, m_n)$, giving contributions to the overall sign

$$r_1 m_n + r_1 m_{n-1} + \cdots = r_1 \sum_{j=2}^{n} m_j.$$

The overall sign due to the all positional exchanges will be $(-1)^s$, where

$$s = \sum_{1 \le i < j \le n} r_i m_j$$

(prove this). Now we have factors

$$e_i^{r_i + m_i},$$

which means that due to the possibility of the case $r_i = m_i = 1$, we have to sum modulo 2, giving (for $r_i = m_i = 1$) e_i^2, the signature of the e_i. As an example, let's multiply $e_1 e_4$ and $e_3 e_4$ in $\mathfrak{R}^{3,1}$. We have lists $(1, 0, 0, 1)$ and $(0, 0, 1, 1)$, giving $s = 1$, and the result

3

$$(-1)^1\langle e_4, e_4\rangle e_1e_3 = e_1e_3 \to (1,0,1,0).$$

This technique is used to develop the implementation clifford.m; see R9.

3.8.2 Bits and Logical Operations

There is also a possibility to express elements of the Clifford basis in bits. In *Cl*3, we have

1	e_1	e_2	e_3	e_1e_2	e_1e_3	e_2e_3	$e_1e_2e_3$
000	001	010	100	011	101	110	111

Now we can use bitwise XOR to get products. The output of XOR is easy; it is 1 only if the arguments are different. For example, for e_1e_2 we have

$$001\mathrm{XOR}010 = 011.$$

◆ Make the multiplication table in *Cl*3 using this technique. ◆ Of course, as in the previous case, we have to take care of antisymmetry (see Sect. 1.6.2) and the signature. Note that there is a lot of work for a processor, so one can also consider the use of multiplication tables, which may be represented in a compact form.

3.8.3 Multiplication Tables

Multiplication tables could be memory expensive; they have 2^{2n} elements. Therefore, it is a good idea to find the easiest way to write elements in the table. One possibility is to write just a sign and a position in the Clifford basis. For example, if we multiply e_2 and e_1, we get $-e_1e_2$, which we can write as -7 in *Cl*3. This means the element at position 7 in the Clifford basis list, with a minus sign. The whole process of multiplying $(\alpha_1 e_2)$ $(\alpha_2 e_1)$, $\alpha_1, \alpha_2 \in \mathbb{R}$, looks like

$$((\alpha_1, 3), (\alpha_2, 2)) \to (-\alpha_1\alpha_2, 7).$$

It just takes reading (from fast RAM) the element at position $(3,2)$ in the multiplication table and multiplying real numbers. To decide which method to use, one should be familiar with computer performance. It is also important to reduce the memory that multivectors take. In our example, we have enough information in the list $((\alpha_1, 3),$ $(\alpha_2, 2))$, and we do not need to specify other positions with coefficients zero: they just take up space in memory. It is also advisable to predefine types of multivectors. For example, a list of real numbers with type such as $(-3, 2, 4, \{1\})$ can be interpreted as the vector (grade 1)

$$-3e_1 + 2e_2 + 4e_3.$$

3.8.4 Implementation of *Cl*3 via Spectral Decomposition

It is possible to express elements of the Clifford basis using spectral decomposition and the Pauli matrices. Thus, the multiplication rules follow from the properties of the simple idempotents ($u_{\pm} = (1 \pm e_3)/2$), given in the following multiplication Table 3.1 (see Sect. 4.2.2):

We can use this table to implement *Cl*3 on a computer, but note that the coefficients are complex numbers (with the "imaginary unit" j). Using complex lists, like $(\alpha_1, \alpha_2, \alpha_3, \alpha_4)$, and positions of elements in the Table 3.1, we can multiply easily. Furthermore, every complex number can be treated as a list, too, like $\alpha_1 \rightarrow$ (Re (α_1), Im (α_1)).

For example, here are transformations of e_1e_3 and j to the spectral basis (u_+, u_-, e_1u_+, e_1u_-):

$$e_1e_3 = e_1(u_+ - u_-) \rightarrow ((0,0),(0,0),(1,0),(-1,0)),$$
$$j = j(u_+ + u_-) \rightarrow ((0,1),(0,1),(0,0),(0,0)).$$

◆ Find the spectral basis representation for all elements of the Clifford basis in *Cl*3. ◆

After calculations, we just need transformations to get back to the Clifford basis, like

$$((0,0),(0,0),(0,0),(0,1)) \rightarrow je_1u_- = \frac{je_1}{2} - \frac{je_1e_3}{2} \rightarrow \left(0,0,-\frac{1}{2},0,\frac{1}{2},0,0,0\right).$$

The table of positions is easy to create, just replace, for example, e_1u_- with 4. ◆ Find involutions of a multivector in the spectral basis. ◆

There is such an implementation in *Mathematica*, named Cl3spectral.nb, created by the author as an auxiliary tool for reading this text. The implementation is free to use, and can be downloaded as Cl3spectral.zip from L1.

Table 3.1 The multiplication table of *Cl*3 in the spectral basis

GP	u_+	u_-	e_1u_+	e_1u_-
u_+	u_+	0	0	e_1u_-
u_-	0	u_-	e_1u_+	0
e_1u_+	e_1u_+	0	0	u_-
e_1u_-	0	e_1u_-	u_+	0

Geometric Algebra and Matrices

Miroslav Josipović

M. Josipović, *Geometric Multiplication of Vectors*,
Compact Textbooks in Mathematics,
https://doi.org/10.1007/978-3-030-01756-9_4

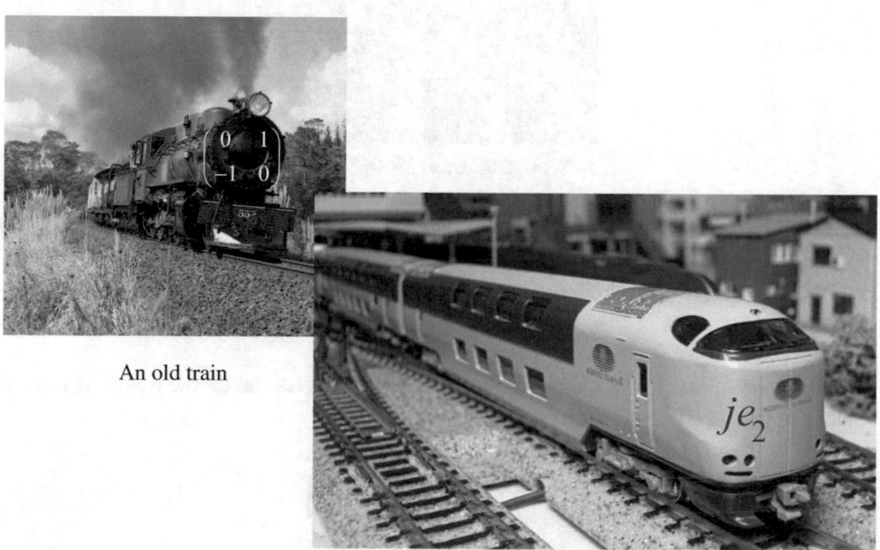

An old train

A modern train

The original version of this chapter was revised. A correction to this chapter can be found at https://doi.org/10.1007/978-3-030-01756-9_7

Electronic supplementary material The online version of this chapter (https://doi.org/10.1007/978-3-030-01756-9_4) contains supplementary material, which is available to authorized users.

4.1 The Pauli Matrices

4.1.1 Famous Representation of Vectors

Wolfgang Pauli (1900–1958)

We know how to multiply unit vectors; therefore, we can try to find some objects to *represent* them. For vectors in *Cl*3, we need three parameters (three coordinates), and the geometric product is noncommutative. Consequently, we can choose matrices; 2×2 complex matrices will be enough. Matrices are a common choice for various representations.

The unit vectors give 1 when squared. Therefore, we can relate the multiplication of vectors to matrix multiplication to get

$$\begin{pmatrix} a_{11} & a_{12} \\ a_{21} & a_{22} \end{pmatrix}^2 = \begin{pmatrix} a_{11}^2 + a_{12}a_{21} & a_{12}(a_{11} + a_{22}) \\ a_{21}(a_{11} + a_{22}) & a_{22}^2 + a_{12}a_{21} \end{pmatrix} = \begin{pmatrix} 1 & 0 \\ 0 & 1 \end{pmatrix},$$

from which we can find two simple possibilities. First,

$$a_{12} = a_{21} = 0, \quad a_{11} = \pm a_{22} = 1,$$

which gives (up to a sign; the unit matrix is ruled out)

$$\begin{pmatrix} 1 & 0 \\ 0 & -1 \end{pmatrix} = \hat{\sigma}_3.$$

Now we have one traceless diagonal matrix. The unit vectors anticommute, and therefore, other matrices cannot be diagonal, since diagonal matrices commute. We have a second possibility,

$$
\begin{aligned}
a_{11} &= a_{22} = 0, \\
a_{12}a_{21} &= 1,
\end{aligned}
$$

which gives two simple solutions (up to a sign):

$$
\begin{aligned}
a_{12} = a_{21} = 1 &\rightarrow \begin{pmatrix} 0 & 1 \\ 1 & 0 \end{pmatrix} = \hat{\sigma}_1, \\
a_{12} = -a_{21} = i &\rightarrow \begin{pmatrix} 0 & -i \\ i & 0 \end{pmatrix} = \hat{\sigma}_2,
\end{aligned}
$$

two more traceless matrices. Interestingly, it was enough to demand that the square of noncommutative objects (matrices) be 1. Note that due to the noncommutativity of the vector multiplication, we immediately get a tight relationship between vectors and the Pauli matrices.

We see that complex 2 × 2 matrices can represent multivectors from *Cl*3. However, we can find other representations, too. Specifying the matrix

$$
\varepsilon = \begin{pmatrix} 0 & 1 \\ -1 & 0 \end{pmatrix},
$$

we have $\varepsilon^2 = -I$ (I is 2 × 2 unit matrix), and ε represents the bivector je_2 (check). Starting with a complex 2 × 2 matrix, we can introduce a simple rule: we replace every real number α by αI, while we replace the imaginary unit i by ε. Thus we get 4 × 4 block matrices. Apply this rule to the Pauli matrices and show that the new matrices also represent the unit vectors.

4.1.2 Some Properties of the Pauli Matrices

Let us look at the Pauli matrices

$$
\begin{aligned}
&I = \hat{\sigma}_0 = \begin{pmatrix} 1 & 0 \\ 0 & 1 \end{pmatrix}, \quad \hat{\sigma}_1 = \begin{pmatrix} 0 & 1 \\ 1 & 0 \end{pmatrix}, \quad \hat{\sigma}_2 = \begin{pmatrix} 0 & -i \\ i & 0 \end{pmatrix}, \quad \hat{\sigma}_3 = \begin{pmatrix} 1 & 0 \\ 0 & -1 \end{pmatrix}, \\
&\hat{\sigma}_i^2 = 1, \quad \hat{\sigma}_i\hat{\sigma}_j + \hat{\sigma}_j\hat{\sigma}_i = 2\delta_{ij}, \quad i,j = 1,2,3, \quad (*)
\end{aligned}
$$

and some of their properties. Starting from vectors, like $a = \sum_{i=1}^{3} a_i e_i$, we can create linear combinations of the Pauli matrices:

$$
[a] = \sum_{i=1}^{3} a_i\hat{\sigma}_i, \quad [b] = \sum_{i=1}^{3} b_i\hat{\sigma}_i, \quad a_i, b_i \in \mathbb{R}.
$$

Note that only $\hat{\sigma}_3$ has nonzero elements on the diagonal, which makes it easy to find

$$[a] = \begin{pmatrix} \mathrm{a}_3 & \mathrm{a}_1 - i\mathrm{a}_2 \\ \mathrm{a}_1 + i\mathrm{a}_2 & -\mathrm{a}_3 \end{pmatrix},$$

which is a 2×2 matrix representation of the vector a. Now using (*) or calculating directly, we have

$$[a]^2 = [a][a] = I\sum_{i=1}^{3} \mathrm{a}_i^2 = Ia^2,$$

so $[a]$ behaves like a vector. In addition, we have

$$\frac{[a][b] + [b][a]}{2} = I\sum_{i=1}^{3} \mathrm{a}_i \mathrm{b}_i = Ia \cdot b,$$

which means that the Pauli matrices could be interpreted as unit vectors (we have the scalar product of vectors a and b). Of course, this means that products of the unit vectors e_i should be anticommutative (as for the Pauli matrices). If we find the antisymmetric part

$$([a][b] - [b][a])/2,$$

we can obtain coefficients of the cross product (E36: Show this). From $\hat{\sigma}_i\hat{\sigma}_j = -\hat{\sigma}_j\hat{\sigma}_i$, it follows that

$$(\hat{\sigma}_i\hat{\sigma}_j)^2 = \hat{\sigma}_i\hat{\sigma}_j\hat{\sigma}_i\hat{\sigma}_j = -\hat{\sigma}_i\hat{\sigma}_i\hat{\sigma}_j\hat{\sigma}_j = -I.$$

Therefore, we have objects with a negative square, and they are not vectors, obviously. This means that the matrix $\hat{\sigma}_i\hat{\sigma}_j$ does not represent a vector, and here we have a problem in the geometric interpretation. However, with the unit vectors, we have e_ie_j, which clearly gives us a geometric meaning (an oriented parallelogram), defining the plane along the way. Similarly, $\hat{\sigma}_1\hat{\sigma}_2\hat{\sigma}_3$ is just the product of the unit matrix and the imaginary unit, but $e_1e_2e_3$ is the oriented volume that squares to -1 and commutes with all vectors, which means that we have an "imaginary unit" again, however with a clear geometric interpretation this time. The geometric meaning is completely hidden with general 2×2 complex matrices. Finally, if we look for a 2D matrix representation of *Cl*3, we get the Pauli matrices as a solution (see Sect. 4.1.1). The very existence of the matrix representation proves that *Cl*3 is a well-defined algebra.

Consider a 2×2 complex matrix

$$A = \begin{pmatrix} a_{11} & a_{12} \\ a_{21} & a_{22} \end{pmatrix}.$$

The trace of the matrix A is the sum of the diagonal elements $\mathrm{tr}A = a_{11} + a_{22}$, while the determinant is $\det A = a_{11}a_{22} - a_{12}a_{21}$. We can define complex numbers (matrix components) as

$$a^\mu \equiv \mathrm{tr}\left(A\hat{\sigma}_\mu\right)/2, \quad \mu = 0, 1, 2, 3.$$

◆ Find a^μ. Show that (we have summation)

$$A = a^\mu \hat{\sigma}_\mu = \begin{pmatrix} a^0 + a^3 & a^1 - ia^2 \\ a^1 + ia^2 & a^0 - a^3 \end{pmatrix}, \quad i = \sqrt{-1},$$

$$a^0 = \mathrm{tr}\, A/2,$$

$$\det A = \left(a^0\right)^2 - \sum_{k=1}^{3} \left(a^k\right)^2.$$

To save time and perhaps frustration with matrices, use a computer. ◆ In formulas based on the Pauli matrices, like the previous ones, indices depend on the special choice of indices of the Pauli matrices. Formally, we can define a complex vector $\vec{a} = (a^1, a^2, a^3)$ and write $\det A = (a^0)^2 - \vec{a}^2$.

4.1.3 The Pauli Matrices and Relativity

Starting with a 4-vector $k = (k_0, k_1, k_2, k_3)$, define the Hermitian matrix

$$K = \begin{pmatrix} k_0 + k_3 & k_1 - ik_2 \\ k_1 + ik_2 & k_0 - k_3 \end{pmatrix}.$$

◆ Can you find a unimodular Hermitian matrix U such that $K' = UKU^\dagger$ († denotes the Hermitian adjoint) gives the correct Lorentz transformation for a boost in the 1-direction (see Sect. 4.4)? ◆ You can find the answer, as well as many other interesting facts, including historical ones, in the MIT course L2.

In the future, we hope that geometric algebra will be a common language. However, for the time being, we should understand the relationship between the old and new languages. Moreover, one can learn that geometric algebra really simplifies, unites, and generalizes the language of mathematics.

4.2 Spectral Basis and Matrices

We have already met the concept of the spectral basis, and here we will work out the algebras *Cl*2 and *Cl*3 in greater detail. We recommend the articles by Sobczyk.

4.2.1 *Cl*2

In the Euclidean 2D vector space, we can define an orthonormal basis (e_1, e_2), giving the *Cl*2 algebra with the Clifford basis $(1, e_1, e_2, I)$, $I = e_1 e_2$. Then a general multivector has the form

$$m = \alpha + x_1 e_1 + x_2 e_2 + \beta I, \quad \alpha, x_1, x_2, \beta \in \mathbb{R}.$$

We also define the special involutions me_1, $e_1 m$, and $e_1 m e_1$:

$$\begin{aligned} me_1 &= x_1 + \alpha e_1 - \beta e_2 - x_2 I, \\ e_1 m &= x_1 + \alpha e_1 + \beta e_2 + x_2 I, \\ e_1 m e_1 &= \alpha + x_1 e_1 - x_2 e_2 - \beta I, \end{aligned}$$

which will be useful in a moment. In addition, we define the idempotents $u_\pm = (1 \pm e_2)/2$ (note the vector e_2), with the properties

$$u_\pm^2 = u_\pm, \quad u_+ u_- = u_- u_+ = 0, \quad e_2 u_\pm = \pm u_\pm, \quad e_1 u_\pm = u_\mp e_1.$$

The transformation $X \to u_+ X u_+$, applied to the Clifford basis, gives

$$(1, e_1, e_2, I) \to (u_+, 0, u_+, 0).$$

Note that we have the simple relations

$$1 = u_+ + u_-, \quad e_1 = e_1(u_+ + u_-), \quad e_2 = u_+ - u_-, \quad e_1 e_2 = e_1(u_+ - u_-),$$

which means that the elements from the Clifford basis (and hence all multivectors) can be expressed as linear combinations of elements u_+, u_-, $e_1 u_+$, $e_1 u_-$, which we call the *spectral basis*. These relations have the matrix form

$$\begin{pmatrix} 1 \\ e_1 \\ e_2 \\ I \end{pmatrix} = \begin{pmatrix} 1 & 0 & 0 & 1 \\ 0 & 1 & 1 & 0 \\ 1 & 0 & 0 & -1 \\ 0 & 1 & -1 & 0 \end{pmatrix} \begin{pmatrix} u_+ \\ e_1 u_+ \\ e_1 u_- \\ u_- \end{pmatrix}.$$

Then, inverting the transformation matrix, we get

$$\begin{pmatrix} u_+ \\ e_1 u_+ \\ e_1 u_- \\ u_- \end{pmatrix} = \frac{1}{2} \begin{pmatrix} 1 & 0 & 1 & 0 \\ 0 & 1 & 0 & 1 \\ 0 & 1 & 0 & -1 \\ 1 & 0 & -1 & 0 \end{pmatrix} \begin{pmatrix} 1 \\ e_1 \\ e_2 \\ I \end{pmatrix}.$$

Note that the inverse matrix is easy to write directly, without inverting the transformation matrix. If we specify coefficients $\alpha^\mu \in \mathbb{R}$, $\mu = 0, 1, 2, 3$, we get

$$\begin{pmatrix} 1 & 0 & 0 & 1 \\ 0 & 1 & 1 & 0 \\ 1 & 0 & 0 & -1 \\ 0 & 1 & -1 & 0 \end{pmatrix} \begin{pmatrix} \alpha^0 \\ \alpha^1 \\ \alpha^2 \\ \alpha^3 \end{pmatrix} = \begin{pmatrix} \alpha^0 + \alpha^3 \\ \alpha^1 + \alpha^2 \\ \alpha^0 - \alpha^3 \\ \alpha^1 - \alpha^2 \end{pmatrix},$$

that is, the coefficients of a multivector in the spectral basis. ◆ Find the eigenvectors and eigenvalues of the transformation matrix. ◆ Similarly, if we specify coefficients α, x_1, x_2, and β in the Clifford basis, we get

$$\begin{pmatrix} 1 & 0 & 1 & 0 \\ 0 & 1 & 0 & 1 \\ 0 & 1 & 0 & -1 \\ 1 & 0 & -1 & 0 \end{pmatrix} \begin{pmatrix} \alpha \\ x_1 \\ x_2 \\ \beta \end{pmatrix} = \begin{pmatrix} \alpha + x_2 \\ x_1 + \beta \\ x_1 - \beta \\ \alpha - x_2 \end{pmatrix},$$

the coefficients of a multivector in the spectral basis. Note that the coefficients are real. Now, using matrix multiplication, we can write ($(1,\ e_1)$ is a row matrix)

$$(1,\ e_1)u_+ \begin{pmatrix} 1 \\ e_1 \end{pmatrix} = (u_+,\ e_1 u_+) \begin{pmatrix} 1 \\ e_1 \end{pmatrix} = u_+ + e_1 u_+ e_1 = u_+ + u_- = 1.$$

Therefore, for the multivector m we can write

$$m = (1,\ e_1)u_+ \begin{pmatrix} 1 \\ e_1 \end{pmatrix} m (1,\ e_1)u_+ \begin{pmatrix} 1 \\ e_1 \end{pmatrix},$$

where we have the matrix

$$\begin{pmatrix} 1 \\ e_1 \end{pmatrix} m (1,\ e_1) = \begin{pmatrix} 1 \\ e_1 \end{pmatrix} (m,\ m e_1) = \begin{pmatrix} m & m e_1 \\ e_1 m & e_1 m e_1 \end{pmatrix}.$$

Note that the matrix elements are just the previously defined involutions. Taking $m = u_+$, we get the matrix

$$\begin{pmatrix} u_+ & u_+ e_1 \\ e_1 u_+ & e_1 u_+ e_1 \end{pmatrix} = \begin{pmatrix} u_+ & e_1 u_- \\ e_1 u_+ & u_- \end{pmatrix},$$

where the matrix elements form the *spectral basis* in *Cl*2. Now we have

$$m = (1,\ e_1)\ u_+ \begin{pmatrix} m & m e_1 \\ e_1 m & e_1 m e_1 \end{pmatrix} u_+ \begin{pmatrix} 1 \\ e_1 \end{pmatrix},$$

$$u_+ \begin{pmatrix} m & m e_1 \\ e_1 m & e_1 m e_1 \end{pmatrix} u_+ = \begin{pmatrix} u_+ m u_+ & u_+ m e_1 u_+ \\ u_+ e_1 m u_+ & u_+ e_1 m e_1 u_+ \end{pmatrix} = u_+ \begin{pmatrix} \alpha + x_2 & x_1 - \beta \\ x_1 + \beta & \alpha - x_2 \end{pmatrix},$$

and finally

$$m = (1,\ e_1)u_+ \begin{pmatrix} \alpha + x_2 & x_1 - \beta \\ x_1 + \beta & \alpha - x_2 \end{pmatrix} \begin{pmatrix} 1 \\ e_1 \end{pmatrix}.$$

Now we have the (real) matrix of the multivector m

$$[m] = \begin{pmatrix} \alpha + x_2 & x_1 - \beta \\ x_1 + \beta & \alpha - x_2 \end{pmatrix},$$

which gives

$$[e_1] = \begin{pmatrix} 0 & 1 \\ 1 & 0 \end{pmatrix}, \quad [e_2] = \begin{pmatrix} 1 & 0 \\ 0 & -1 \end{pmatrix}, \quad [I] = \begin{pmatrix} 0 & -1 \\ 1 & 0 \end{pmatrix}.$$

It is easy to check that

$$[e_1]^2 = [e_2]^2 = \begin{pmatrix} 1 & 0 \\ 0 & 1 \end{pmatrix}, \quad [e_1][e_2] = [I], \quad [e_1][e_2] = -[e_2][e_1].$$

We have found a 2D matrix representation of the basis vectors in *Cl*2. Note that $[e_1]$, $[e_2]$, and $i[I]$ are the Pauli matrices (see Sect. 4.1).

Defining $e_0 = 1$, we can write a multivector in *Cl*2 in the form

$$m = \alpha^\mu e_\mu, \quad \alpha^\mu \in \mathbb{C}, \quad \mu = 0, 1, 2,$$

where summation over μ is understood ($I = e_1 e_2$ is the "imaginary unit" here). Then it is straightforward to find

$$\alpha = \mathrm{Re}\,(\alpha^0), \quad x_1 = \mathrm{Re}\,(\alpha^1) + \mathrm{Im}(\alpha^2), \quad x_2 = \mathrm{Re}\,(\alpha^2) - \mathrm{Im}(\alpha^1), \quad \beta = \mathrm{Im}(\alpha^0).$$

We see that $\alpha^0 = \alpha + \beta I$, and then using

$$e_1 = \frac{e_1 + Ie_2}{2}, \quad e_2 = \frac{e_2 - Ie_1}{2},$$

it follows that

$$\alpha^1 = \frac{x_1 - x_2 I}{2}, \quad \alpha^2 = \frac{x_2 + x_1 I}{2}.$$

We can write $m = \alpha + x_1 e_1 + x_2 e_2 + \beta I = \alpha + \mathbf{F}$, $\mathbf{F}^2 = x_1^2 + x_2^2 - \beta^2 = Z_f^2$, and then we have (show this)

$$\det[m] = m\bar{m}, \quad [m]^{-1} = [m]/(m\bar{m}).$$

The eigenvalues of $[m]$ are $e_\pm = \alpha \mp Z_f = m_\pm$, with the eigenvectors (column vectors)

$$v_\pm = \left(\frac{x_2 \mp Z_f}{x_1 + \beta}, 1 \right)^T,$$

which means that $[m]v_\pm = m_\mp v_\pm$.

Note that we can define $\mathbf{f} = \mathbf{F}/Z_f$, $\mathbf{f}^2 = 1$, $u_\pm = (1 \pm \mathbf{f})/2$, giving the possibility to express m in the form $m = m_+ u_+ + m_- u_-$, from which follows $mu_\pm = m_\pm u_\pm$. Note also that we have

$$v_{\pm} = \left(\frac{[m]_{11} - \alpha \mp Z_f}{[m]_{21}}, 1 \right)^T = \left(\frac{[m]_{11} - m_{\pm}}{[m]_{21}}, 1 \right)^T .$$

The reader should find the matrices of [(1 ± **f**)/2] to establish a relationship with the eigenvectors $v_{\pm}$.

4.2.2 *Cl*3

Consider now *Cl*3. Assuming that the previous chapter has been mastered well, this one will be easy to comprehend.

Starting from a general multivector $M = \alpha + x + jn + j\beta$, we can show (see problem 19) that $M = a_1 + ja_2$, where $a_i = \alpha_i + x_{i1}e_1 + x_{i2}e_2 + \beta_i I$ are multivectors from *Cl*2. This suggests that we can express *M* using the spectral basis from *Cl*2. We can repeat all the previous procedures to obtain

$$\begin{pmatrix} u_+Mu_+ & u_+Me_1u_+ \\ u_+e_1Mu_+ & u_+e_1Me_1u_+ \end{pmatrix}, \quad M = \alpha + \sum_{i=1}^{3} x_i e_i + j \sum_{i=1}^{3} n_i e_i + j\beta.$$

We will choose $u_{\pm} = (1 \pm e_3)/2$, although we already have the idempotent $(1 + e_2)/2$ from *Cl*2. It is obvious that it would be more convenient to choose differently, since we can express elements of *Cl*3 starting from *Cl*2. This choice is just due to the traditional definition of the Pauli matrices in quantum mechanics. The effect is the exchange of the Pauli matrices 2 and 3, which means that physicists should choose the *y*-axis for the direction of the external magnetic field. However, bases and indices are of small importance in geometric algebra; we just need them to communicate with computers, or to interpret experimental data. Hopefully, perhaps computers in the future will know how to calculate without coordinates. Using $e_2 = je_1e_3 = je_1(u_+ - u_-)$, we can write the transformation matrices (two parts)

$$\begin{pmatrix} 1 \\ e_1 \\ e_2 \\ e_3 \end{pmatrix} = \begin{pmatrix} 1 & 0 & 0 & 1 \\ 0 & 1 & 1 & 0 \\ 0 & j & -j & 0 \\ 1 & 0 & 0 & -1 \end{pmatrix} \begin{pmatrix} u_+ \\ e_1u_+ \\ e_1u_- \\ u_- \end{pmatrix}, \quad \begin{pmatrix} e_1e_2 \\ e_3e_1 \\ e_2e_3 \\ j \end{pmatrix} = \begin{pmatrix} j & 0 & 0 & -j \\ 0 & -1 & 1 & 0 \\ 0 & j & j & 0 \\ j & 0 & 0 & j \end{pmatrix} \begin{pmatrix} u_+ \\ e_1u_+ \\ e_1u_- \\ u_- \end{pmatrix}.$$

Defining $e_0 = 1$, we can write a multivector in *Cl*3 in the form $m = \alpha^\mu e_\mu$, $\alpha^\mu \in \mathbb{C}$, $\mu = 0, 1, 2, 3$, where summation over μ is understood. It is not difficult to find the coefficients α^μ (*j* is the "imaginary unit" here).

Note that involutions with e_1 just change the positions of the coefficients in *M* (up to sign), while the transformation $X \to u_+Xu_+$ gives nonzero scalars at positions 1, 4, 7, and 8 (show this), which simplifies calculations. The matrix is

$$[M] = \begin{pmatrix} \alpha + x_3 + jn_3 + j\beta & x_1 + n_2 - jx_2 + jn_1 \\ x_1 - n_2 + jx_2 + jn_1 & \alpha - x_3 - jn_3 + j\beta \end{pmatrix},$$

which gives

$$[e_1] = \begin{pmatrix} 0 & 1 \\ 1 & 0 \end{pmatrix}, \quad [e_2] = \begin{pmatrix} 0 & -j \\ j & 0 \end{pmatrix}, \quad [e_3] = \begin{pmatrix} 1 & 0 \\ 0 & -1 \end{pmatrix},$$

the well-known Pauli matrices (with $i \to j$). With the choice $u_+ = (1 + e_2)/2$, the result will be

$$[e_1] = \begin{pmatrix} 0 & 1 \\ 1 & 0 \end{pmatrix}, \quad [e_2] = \begin{pmatrix} 1 & 0 \\ 0 & -1 \end{pmatrix}, \quad [e_3] = \begin{pmatrix} 0 & j \\ -j & 0 \end{pmatrix}.$$

However, this is just a matter of convenience. A general multivector in *Cl*3 can be expressed as a linear combination of elements of the spectral basis $(u_+, u_-, e_1u_+, e_1u_-)$, although here the coefficients are complex scalars, like $\alpha + j\beta$, α, $\beta \in \mathbb{R}$. If we use the form

$$M = Z + \mathbf{F} = Z + Z_f\mathbf{f}, \quad \mathbf{f}^2 = 1, \quad Z_f = \sqrt{\mathbf{F}^2}, \quad M_\pm = Z \pm Z_f, \quad M\bar{M} = M_+M_- \\ = Z^2 - \mathbf{F}^2,$$

then it follows (check) that $\det[M] = M\bar{M}$, $[M^{-1}] = [M]^{-1}$, and the eigenvalues of $[M]$ are $M_\mp$, while the eigenvectors of $[M]$ are

$$v_\pm = \left(\frac{x_3 + jn_3 \pm Z_f}{x_1 - n_2 + j(x_2 + n_1)}, 1 \right)^T,$$

giving $[M]v_\pm = M_\mp v_\pm$. Recall that for $u_\pm = (1 \pm \mathbf{f})/2$ we have $M = M_+u_+ + M_-u_-$, which gives $Mu_\pm = M_\pm u_\pm$. Note also that we can write

$$v_\pm = \left(\frac{[M]_{11} - Z \pm Z_f}{[M]_{21}}, 1 \right)^T = \left(\frac{[M]_{11} - M_\mp}{[M]_{21}}, 1 \right)^T.$$

There is also a possibility to define an idempotent using $\mathbf{f} = \sqrt{2}e_1 + je_2$, $\mathbf{f}^2 = 1$, $u_\pm = (1 \pm \mathbf{f})/2$, $e_3\mathbf{f} = \sqrt{2}je_2 + e_1$, $e_3\mathbf{f} = -\mathbf{f}e_3$, with the spectral basis $(u_+, u_-, e_3u_+, e_3u_-)$. From

$$M = \alpha + \mathbf{x} + j\mathbf{n} + j\beta = \mathrm{a}u_+ + \mathrm{b}u_- + \mathrm{c}e_3u_+ + \mathrm{d}e_3u_-, \quad \alpha, \beta \in \mathbb{R},$$

we can find complex coefficients a, b, c,and d, for example $\mathrm{a} = \mathrm{a_r} + \mathrm{a_i}j$. The result is (check)

$$\begin{aligned} \mathrm{a_r} &= \alpha + \sqrt{2}x_1 - n_2, \quad \mathrm{a_i} = \beta + x_2 + \sqrt{2}n_1, \\ \mathrm{b_r} &= \alpha - \sqrt{2}x_1 + n_2, \quad \mathrm{b_i} = \beta - x_2 - \sqrt{2}n_1, \\ \mathrm{c_r} &= -x_1 + x_3 + \sqrt{2}n_2, \quad \mathrm{c_i} = -\sqrt{2}x_2 - n_1 + n_3, \\ \mathrm{d_r} &= x_1 + x_3 - \sqrt{2}n_2, \quad \mathrm{d_i} = \sqrt{2}x_2 + n_1 + n_3. \end{aligned}$$

4.3 Zero Divisors and Cancellation of Factors

For nonzero real (or complex) numbers α, β, and γ, we have $\alpha\beta = \alpha\gamma \Rightarrow \beta = \gamma$; however, in some linear spaces there exist products that do not satisfy such a relation. For example, from the relation $a \cdot b = a \cdot c$ it does not generally follow that $b = c$, as for $e_1 \cdot e_2 = e_1 \cdot e_3$, and similarly for the cross product, $e_1 \times e_2 = e_1 \times (e_1 + e_2)$. In geometric algebra, we have for vectors

$$ab = ac \Rightarrow a^2 b = a^2 c \Rightarrow b = c.$$

In addition, we have $ab = ac \Rightarrow a \cdot b + a \wedge b = a \cdot c + a \wedge c$, which means that $a \cdot b = a \cdot c$ and $a \wedge b = a \wedge c$ (grades do not mix). However, knowing that there are zero divisors, such as $(1 + e_1)(1 - e_1) = 0$, we can see immediately that

$$\begin{aligned}(1+e_1)(1-e_1) &= (1+e_1)(1-e_2+e_2-e_1) = 0 \Rightarrow (1+e_1)(1-e_2) \\ &= (1+e_1)(e_1-e_2).\end{aligned}$$

Consequently, we cannot cancel the factor $1 + e_1$. Note that the factor $1 + e_1$ has no inverse, and that $(1 + e_1)/2$ is an idempotent. Consider now the elements $A = (1 + e_1)e_2$ and $B = (1 + e_1)/2$. We have

$$\begin{aligned}AB &= (1+e_1)e_2(1+e_1)/2 = e_2(1-e_1)(1+e_1)/2 = 0, \\ BA &= ((1+e_1)/2)(1+e_1)e_2 = 2\frac{1+e_1}{2}\frac{1+e_1}{2}e_2 = (1+e_1)e_2 \neq 0.\end{aligned}$$

Note that A is a nilpotent, and it has the form of a simple idempotent multiplied by a vector perpendicular to its unit vector ($e_1 \perp e_2$). Using the spectral basis in *Cl*3 (see Sect. 4.2), we can find zero divisors easily. Choosing $u_\pm = (1 \pm e_3)/2$ and complex scalars α_k and γ, we can specify a general multivector as

$$\alpha_1 u_+ + \alpha_2 u_- + \alpha_3 e_1 u_+ + \alpha_4 e_1 u_-,$$

from which, for example, it follows that

$$u_+(\alpha_2 u_- + \alpha_3 e_1 u_+ + \gamma e_2 u_+) = 0, \quad u_-(\alpha_1 u_+ + \alpha_4 e_1 u_- + \gamma e_2 u_-) = 0$$

(check), and so forth (find other examples). To conclude, for nonzero multivectors A, B, and C, from $A(B - C) = 0$ we cannot conclude that $B = C$ is a solution; there is a possibility that A and $B - C$ are zero divisors. However, if they are not zero divisors, then $B = C$ is the solution. Evidently, the spectral basis form of multivectors is suitable for finding whether multivectors are zero divisors. Using the relations from Sect. 4.2, we have

$$M = (1,\ e_1)u_+[M]\begin{pmatrix}1\\ e_1\end{pmatrix},\quad [M] = \begin{pmatrix}\alpha + x_3 + jn_3 + j\beta & x_1 + n_2 - jx_2 + jn_1\\ x_1 - n_2 + jx_2 + jn_1 & \alpha - x_3 - jn_3 + j\beta\end{pmatrix},$$

$$[1+e_1] = \begin{pmatrix}1 & 1\\ 1 & 1\end{pmatrix} \Rightarrow 1 + e_1 = (1,\ e_1)u_+\begin{pmatrix}1 & 1\\ 1 & 1\end{pmatrix}\begin{pmatrix}1\\ e_1\end{pmatrix}$$
$$= u_+ + u_- + e_1u_+ + e_1u_-,$$

which we can write directly, since $u_+ + u_- = 1$. Note that

$$\begin{pmatrix}1 & 1\\ 1 & 1\end{pmatrix} = \begin{pmatrix}1 & 0\\ 0 & 1\end{pmatrix} + \begin{pmatrix}0 & 1\\ 1 & 0\end{pmatrix} = I + \hat{\sigma}_1,$$

where I is the identity matrix. It is usually easy to find the spectral decomposition; for example,

$$1 - e_2 = u_+ + u_- - je_1e_3 = u_+ + u_- - je_1(u_+ - u_-) = u_+ + u_- - je_1u_+ + je_1u_-,$$
$$e_1 - e_2 = e_1u_+ + e_1u_- - je_1u_+ + je_1u_- = (1-j)e_1u_+ + (1+j)e_1u_-.$$

Just keep in mind that $u_+ + u_- = 1$, $u_+ - u_- = e_3$, $u_+u_- = 0$, $e_1u_\pm = u_\mp e_1$, $e_2u_\pm = u_\mp e_2$, etc.

◆ You can now investigate what happened in $(1+e_1)(1-e_2) = (1+e_1)(e_1-e_2)$. ◆ Note also that

$$(1+e_1)(e_1-e_2) = (1+e_1)e_1(e_1-e_2) = (1+e_1)(1-e_1e_2),$$
$$1 - e_1e_2 = 1 - je_3 = u_+ + u_- - j(u_+ - u_-) = (1-j)u_+ + (1+j)u_-.$$

One can create and use the multiplication table for the elements of the spectral basis in such calculations; however, it is better to become familiar with the above calculations (see Sect. 3.8.3).

4.4 Classical Spinors and Matrices

Here we will note some properties of classical spinors. We will discuss properties of spinors in the language of matrices, compared to the language of geometric algebra, which could be enlightening. Felix Klein introduced spinors in 1897, in order to simplify the treatment of the classical spinning top. Further development of the spinor theory was due to Élie Cartan (1913). Spinors are closely related to quaternions; therefore, we can anticipate a nice formulation in geometric algebra. Spinors are important in relativity and in quantum mechanics; however, applications are numerous (see R10). In the language of coordinates, spinors can be represented as complex vectors, complex matrices, etc.

As a start, let us mention a possible geometric interpretation of spinors represented by 2-component complex vectors (in *Cl*3, they are just rotors with dilatation; see Sects. 1.9.13 and 1.14). In Fig. 4.1, we see an image that can help to visualize spinors. A spinor is represented by a vector $\mathbf{r}$ (*flagpole*), defined by the length r, angles (θ, φ), and a *flag* (defined by the angle α). What is not seen in the picture is an overall sign, which is also a

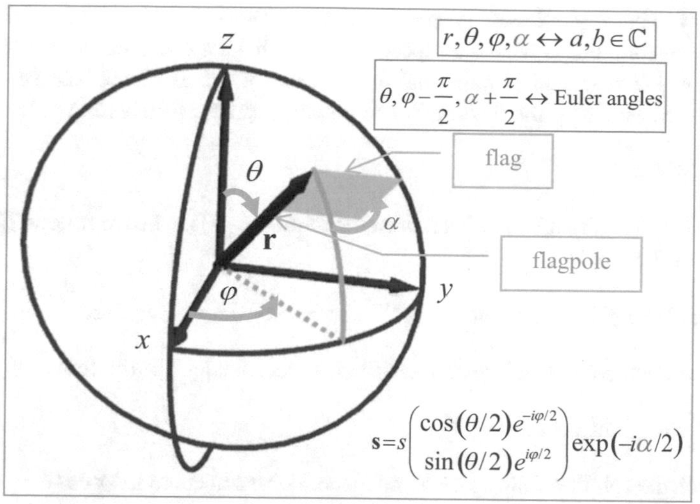

Fig. 4.1 A possible graphical representation of a spinor

part of the definition of a spinor. The sign of a spinor is not connected to the orientation of the flagpole. Spinors with opposite orientations of flagpoles could be orthogonal (in a complex vector space). When a spinor is rotated, its flag rotates, too, as if it were rigidly attached to the flagpole. However, a flag is frequently neglected in applications, as a complex phase $\exp(i\alpha/2)$ (see Sect. 2.10). A rotation about a flagpole changes a flag, that is, changes an overall complex phase. The overall sign of a spinor is tricky; it changes to its opposite after rotation by 2π.

4.4.1 Spinor Formula

It is time to express the spinor from Fig. 4.1 by a formula. If a spinor is specified by the complex numbers a and b (four real parameters), we can represent it by a vector in an abstract complex vector space and write

$$|\psi\rangle = \begin{pmatrix} a \\ b \end{pmatrix},$$

where $|\psi\rangle$ is called a *ket* (due to Dirac). We also define a *bra* vector, as

$$\langle\psi| = (a^*, b^*)$$

(the star denotes the complex conjugate), whence follows the definition of the *scalar product* $\langle\psi|\varphi\rangle$, for example, $\langle\psi|\psi\rangle = aa^* + bb^* \in \mathbb{R}$. This gives us the possibility to normalize a spinor. Here comes a question, namely, how to relate complex vectors and the geometric interpretation from Fig. 4.1. We know that this is possible; it is enough to count real parameters, four from our picture, and four for a complex vector $|\psi\rangle$. ◆ The answer is not obvious, and the curious reader could stop and take some time to find the solution. ◆ The key relation is given using the Pauli matrices. From the point of geometric algebra, this is what we expect; we know that the Pauli matrices are Hermitian matrices, which represent orthonormal unit vectors of the 3D Euclidean vector space. Therefore, we start with the flagpole (**r**) components (in the *expectation value* form)

$$r_k = \langle \psi | \hat{\sigma}_k | \psi \rangle$$

(note that $\hat{\sigma}_k|\psi\rangle$ is also a ket, $\hat{\sigma}_k$ is an operator acting on $|\psi\rangle$), whence follows ($i = \sqrt{-1}$, $x \to 1$, $y \to 2$, $z \to 3$)

$$r_1 = ab^* + ba^*, \quad r_2 = i(ab^* - ba^*), \quad r_3 = aa^* - bb^*, \quad r_k \in \mathbb{R}$$

(check; see Sect. 3.6). A unit vector in spherical coordinates has the form

$$(\sin\theta\cos\varphi, \sin\theta\cos\varphi, \cos\theta),$$

from which the solution follows by comparison. For instance, to get $r\cos\theta = |a|^2 - |b|^2$, we can start from

$$\cos^2\theta = \cos^2(\theta/2) - \sin^2(\theta/2),$$

giving the possibility to define s by $a \equiv s\cos(\theta/2)$, $b \equiv s\sin(\theta/2)$. However, we can also introduce a complex phase and get, generally, $a = s\cos(\theta/2)\exp(i\gamma)$, $b = s\sin(\theta/2)\exp(i\delta)$, which means that our solution (*spinor* in the text) has the form ($\delta - \gamma \equiv \varphi$)

$$\mathbf{s} = s\begin{pmatrix} \cos(\theta/2)e^{-i\varphi/2} \\ \sin(\theta/2)e^{i\varphi/2} \end{pmatrix}\exp(i(\gamma+\delta)/2) = s\begin{pmatrix} \cos(\theta/2)e^{-i\varphi/2} \\ \sin(\theta/2)e^{i\varphi/2} \end{pmatrix}\exp(i\alpha/2),$$

$$\alpha \equiv -(\gamma+\delta)$$

(check this). Here we use the designation **s** (or $|\mathbf{s}\rangle$) instead of $|\psi\rangle$ just to stress the difference between different representations of a vector in an abstract complex vector space. We see that the Pauli matrices were a key to the previous calculation. In the chapter on quantum mechanics (Sect. 2.10), the Pauli matrices are replaced by orthonormal unit vectors. Note that the flagpole length is

$$\langle \mathbf{s}|\mathbf{s}\rangle = s^2 = r = |a|^2 + |b|^2,$$

and we can write the vector components as

$$r_k = \mathbf{s}^\dagger \hat{\sigma}_k \mathbf{s} \equiv \langle \mathbf{s}|\hat{\sigma}_k|\mathbf{s}\rangle,$$

where † means the *Hermitian adjoint* (transposed and complex conjugated).

4.4.2 Exponential of a Matrix

We can define the exponential of a matrix, which is rather simple for the Pauli matrices. From $\hat{\sigma}_k^2 = I$, I is the 2×2 unit matrix, and using the series expansion of the exponential function, it easy to get ($\omega \in \mathbb{R}$)

$$\exp(i\omega\hat{\sigma}_k) = I\cos\omega + i\hat{\sigma}_k\sin\omega,$$
$$\exp(\omega\hat{\sigma}_k) = I\cosh\omega + \hat{\sigma}_k\sinh\omega,$$

which are relations similar to those from geometric algebra; just make use of $I \to 1$, $\hat{\sigma}_k \to e_k$, and $I \to j = e_{123}$. Thus, we have matrices (*spin rotation matrices*)

$$\exp(i(\theta/2)\hat{\sigma}_1) = \begin{pmatrix} \cos(\theta/2) & i\sin(\theta/2) \\ i\sin(\theta/2) & \cos(\theta/2) \end{pmatrix},$$
$$\exp(i(\theta/2)\hat{\sigma}_2) = \begin{pmatrix} \cos(\theta/2) & \sin(\theta/2) \\ -\sin(\theta/2) & \cos(\theta/2) \end{pmatrix},$$
$$\exp(i(\theta/2)\hat{\sigma}_3) = \begin{pmatrix} \exp(i\theta/2) & 0 \\ 0 & \exp(-i\theta/2) \end{pmatrix},$$

or, defining formally

$$\boldsymbol{\sigma} = (\hat{\sigma}_1, \hat{\sigma}_2, \hat{\sigma}_3), \quad \boldsymbol{\sigma}\cdot\mathbf{r} = r_1\hat{\sigma}_1 + r_2\hat{\sigma}_2 + r_3\hat{\sigma}_3$$

(note that $\boldsymbol{\sigma}\cdot\mathbf{r}$ becomes $\mathbf{r}$ in GA), we can find a general matrix $\exp(i(\boldsymbol{\theta}/2)\cdot\boldsymbol{\sigma})$, where $\boldsymbol{\theta} = (\theta_1, \theta_2, \theta_3)$, the three angles that correspond to three rotation axes (x, y, z). ◆ Prove that $(\boldsymbol{\sigma}\cdot\mathbf{r})^2 = r^2$. Check that $\exp(i(\theta/2)\hat{\sigma}_1)\exp(i(\theta/2)\hat{\sigma}_2) \neq \exp(i(\theta/2)(\hat{\sigma}_1 + \hat{\sigma}_2))$. Why? ◆

To calculate matrix exponentials, the reader should consult the literature on linear algebra. Various computer programs could be helpful, for instance *Mathematica*, which has built-in functions, like *MatrixExp*, *MatrixPower*, *Eigensystem*, etc. Here is a typical *Mathematica* input and output:

MatrixExp[θ $JR[3]]//MatrixForm

$$\begin{pmatrix} \mathrm{Cos}[\theta] & -\mathrm{Sin}[\theta] & 0 \\ \mathrm{Sin}[\theta] & \mathrm{Cos}[\theta] & 0 \\ 0 & 0 & 1 \end{pmatrix},$$

where $JR[3] is just a symbol for J_3, the rotation generator (see below), given as a *List*. Calculations in this book have frequently been checked by *Mathematica*. However, some programming was necessary, for example the implementation of *Cl*3 (see L1).

4.4.3 SO(3) and SU(2)

Starting with the spinor

$$\mathbf{s} = s\begin{pmatrix} \cos(\theta/2)e^{-i\varphi/2} \\ \sin(\theta/2)e^{i\varphi/2} \end{pmatrix},$$

we can transform it using spin rotation matrices, like

$$\mathbf{s}' = \exp(i(\varphi_R/2)\hat{\sigma}_3)\mathbf{s}, \quad s = 1,$$

where φ_R is the transformation angle. After some algebra, it follows that

$$s_3' = \langle \mathbf{s}'|\hat{\sigma}_3|\mathbf{s}'\rangle = \langle \mathbf{s}|\exp(-i(\varphi_R/2)\hat{\sigma}_3)\hat{\sigma}_3 \exp(i(\varphi_R/2)\hat{\sigma}_3)|\mathbf{s}\rangle = \cos\theta$$

(check; recall that $\hat{\sigma}_k^\dagger = \hat{\sigma}_k$), and in the same way,

$$s_2' = \sin\theta\sin(\varphi - \varphi_R), \quad s_1' = \sin\theta\cos(\varphi - \varphi_R).$$

However, the same effect can be achieved using the rotation matrix R_3 (in blue):

$$\begin{pmatrix} \cos\varphi_R & \sin\varphi_R & 0 \\ -\sin\varphi_R & \cos\varphi_R & 0 \\ 0 & 0 & 1 \end{pmatrix}\begin{pmatrix} \sin\theta\cos\varphi \\ \sin\theta\sin\varphi \\ \cos\theta \end{pmatrix} = \begin{pmatrix} \sin\theta\cos(\varphi - \varphi_R) \\ \sin\theta\sin(\varphi - \varphi_R) \\ \cos\theta \end{pmatrix}.$$

Note that rotation by R_3 is the same for the two spinor rotation matrices $\pm\exp(-i(\varphi_R/2)\hat{\sigma}_3)$, which means that for every rotation matrix, there are two spinors that can rotate the same way. That is the meaning of the term *double cover*, relating to the spinor group and the rotation group. Unfortunately, there is no space to treat the group theory here. Note also that

$$\exp(i(2\pi/2)\hat{\sigma}_3)\mathbf{s} = \begin{pmatrix} \exp(i\pi) & 0 \\ 0 & \exp(-i\pi) \end{pmatrix} = -\mathbf{s},$$

which is a typical property of spinors.

Matrices of the form $U = \exp(i(\boldsymbol{\theta}/2)\cdot\boldsymbol{\sigma})$ (U is for *unitary*) belong to the representation of the group SU(2), while matrices of the form $R = \exp(\mathbf{J}\cdot\boldsymbol{\theta})$ belong to the representation of the group SO(3). The three matrices (the components of $\mathbf{J}$)

$$J_1 = \begin{pmatrix} 0 & 0 & 0 \\ 0 & 0 & -1 \\ 0 & 1 & 0 \end{pmatrix}, \quad J_2 = \begin{pmatrix} 0 & 0 & 1 \\ 0 & 0 & 0 \\ -1 & 0 & 0 \end{pmatrix}, \quad J_3 = \begin{pmatrix} 0 & -1 & 0 \\ 1 & 0 & 0 \\ 0 & 0 & 0 \end{pmatrix}$$

are the SO(3) rotation *group generators* (check this). These generators are usually defined as iJ_k, but then we must change the formula for R. Prove that ($[a, b] \equiv ab - ba$)

$$[J_i, J_j] = \varepsilon_{ijk} J_k, \quad J_k^3 = -J_k, \quad (\mathbf{J}\cdot\boldsymbol{\theta})^3 = -|\boldsymbol{\theta}|^2 \mathbf{J}\cdot\boldsymbol{\theta}.$$

4.4.4 Spin Rotation Matrices and Vectors

Vectors can be represented using the Pauli matrices

$$\boldsymbol{\sigma}\cdot\mathbf{r} = \begin{pmatrix} r_3 & r_1 - ir_2 \\ r_1 + ir_2 & -r_3 \end{pmatrix}$$

with the determinant $\det(\boldsymbol{\sigma}\cdot\mathbf{r}) = -r^2$ (check). Using the previously defined matrices $U = \exp(i(\boldsymbol{\theta}/2)\cdot\boldsymbol{\sigma})$, we can construct the transformation

$$U' = U\boldsymbol{\sigma}\cdot\mathbf{r}U^{\dagger},$$

for instance,

$$\exp(i(\varphi_R/2)\hat{\sigma}_3)\boldsymbol{\sigma}\cdot\mathbf{r}\exp(-i(\varphi_R/2)\hat{\sigma}_3) = \begin{pmatrix} r_3 & e^{i\varphi_R}(r_1 - ir_2) \\ e^{-i\varphi_R}(r_1 + ir_2) & -r_3 \end{pmatrix}.$$

The vector $\mathbf{r}$ is now

$$(r_1\cos\varphi_R + r_2\sin\varphi_R, -r_1\sin\varphi_R + r_2\cos\varphi_R, r_3),$$

which is just the same effect as for the rotation matrix R_3. What if $\varphi_R = 2\pi$?

4.4.5 Spinors in the Special Theory of Relativity

Starting from the 4-vector (t, x, y, z), we can define

$$X = tI + x\hat{\sigma}_1 + y\hat{\sigma}_2 + z\hat{\sigma}_3 = \begin{pmatrix} t+z & x-iy \\ x+iy & t-z \end{pmatrix},$$

with determinant

$$\det X = t^2 - x^2 - y^2 - z^2 = ds^2,$$

a well-known invariant relativistic interval. In *Cl*3, X would be a paravector. The matrix X is Hermitian, but not traceless, $\operatorname{tr} X = 2t$. If $ds^2 = 0$, we say that the 4-vector (t, x, y, z) is a null 4-vector (*lightlike*). It is important to see that null 4-vectors can be related to the spinors from the previous chapter (the *rank*-1 spinors; general 4-vectors are related to the rank 2 spinors). Let us show this.

Consider a spinor $\mathbf{s}$ (of rank 1), which under a change of inertial system transforms as $\mathbf{s}' = \Lambda\mathbf{s}$, where Λ is a 2×2 matrix to be determined (we can say that this way of transformation defines the spinors of rank 1). The spinors of rank 2 transform as (by definition)

$$X' = \Lambda X \Lambda^{\dagger}.$$

We have seen that it is possible to lower the rank of a rank-1 spinor, just using $\mathbf{s}^{\dagger}\mathbf{s} = \langle \mathbf{s}|\mathbf{s}\rangle$, and thus get a number. The fact that the rank is lowered motivates us to call such a product the *inner product*. However, there is a way to multiply two spinors of rank 1 to get a spinor of rank 2. Specifically, for a spinor **s** we can write

$$\mathbf{s}\mathbf{s}^{\dagger} = |\mathbf{s}\rangle\langle\mathbf{s}| = \begin{pmatrix} a \\ b \end{pmatrix}(a^*, b^*) = \begin{pmatrix} |a|^2 & ab^* \\ ba^* & |b|^2 \end{pmatrix}, \quad \det(\mathbf{s}\mathbf{s}^{\dagger}) = 0.$$

The matrix $\mathbf{s}\mathbf{s}^{\dagger}$ is Hermitian, which is no coincidence; Hermitian matrices generally represent spinors of rank 2. From $\mathbf{s}' = \Lambda\mathbf{s}$ follows (check) $\mathbf{s}'^{\dagger} = \mathbf{s}^{\dagger}\Lambda^{\dagger}$, which means that our spinor of rank 2 transforms as $\mathbf{s}\mathbf{s}^{\dagger} \to \Lambda\mathbf{s}\mathbf{s}^{\dagger}\Lambda^{\dagger}$. Due to $\det(\mathbf{s}\mathbf{s}^{\dagger}) = 0$, we see that the matrix $\mathbf{s}\mathbf{s}^{\dagger}$ can be related to a null 4-vector. ◆ Check that

$$\begin{pmatrix} |a|^2 & ab^* \\ ba^* & |b|^2 \end{pmatrix} = \begin{pmatrix} t+z & x-iy \\ x+iy & t-z \end{pmatrix} \Rightarrow \begin{pmatrix} t \\ x \\ y \\ z \end{pmatrix} = \begin{pmatrix} \left(|a|^2 + |b|^2\right)/2 \\ (ab^* + ba^*)/2 \\ i(ab^* - ba^*)/2 \\ \left(|a|^2 - |b|^2\right)/2 \end{pmatrix}$$

$$= \frac{1}{2}\begin{pmatrix} \langle \mathbf{s}|I|\mathbf{s}\rangle \\ \langle \mathbf{s}|\boldsymbol{\sigma}|\mathbf{s}\rangle \end{pmatrix}. \blacklozenge$$

Generally, we want to relate the spinors X to any invariant interval $ds^2 = \det X$, which means that we have the condition $\det X' = \det(\Lambda X \Lambda^{\dagger}) = \det X$.

4.4.6 Theorems About Determinants

We have three famous theorems about determinants of (square) matrices in linear algebra:

1. $\det(AB) = \det A \det B$,
2. $\det A^T = \det A$,
3. $\det(A^{\dagger}) = (\det A)^*$, $A = A^{\dagger} \Rightarrow \det A \in \mathbb{R}$,

which are not easy to prove (in contrast to geometric algebra; see Sect. 1.17.1). Now we have

$$\det\Lambda(\det\Lambda)^* = 1 \Rightarrow \det\Lambda = \exp(i\lambda), \ \lambda \in \mathbb{R}.$$

4.4.7 SL(2, C)

The complex phase $\exp(i\lambda)$ defines the spinor flag, and we usually use $\lambda = 0$, giving $\det\Lambda = 1$. Such matrices are elements of the group SL(2, C), which preserve the relativistic interval. The fact that we neglect a complex phase in transformations was already discussed in the text. ◆ Consider the matrix (the *spinor Minkowski metric*)

$$\varepsilon = \begin{pmatrix} 0 & 1 \\ -1 & 0 \end{pmatrix}$$

and general element of the group SL(2, C)

$$\Lambda = \begin{pmatrix} a & b \\ c & d \end{pmatrix}, \quad ad - bac = 1.$$

Show that $\Lambda^T \varepsilon \Lambda = \varepsilon$. ◆ Consider now the matrix

$$\Lambda = \exp(\hat{\sigma}_1 \varphi/2) = \begin{pmatrix} \cosh(\varphi/2) & \sinh(\varphi/2) \\ \sinh(\varphi/2) & \cosh(\varphi/2) \end{pmatrix}$$

and the transformation

$$X' = \Lambda X \Lambda^\dagger = \begin{pmatrix} t\cosh\varphi + x\sinh\varphi + z & x\cosh\varphi + t\sinh\varphi - iy \\ x\cosh\varphi + t\sinh\varphi + iy & t\cosh\varphi + x\sinh\varphi - z \end{pmatrix},$$

whence, using tanh$\varphi = v$, cosh$\varphi = \gamma$, and sinh$\varphi = \gamma v$, it follows that

$$X' = \begin{pmatrix} \gamma(t + xv) + z & \gamma(x + tv) - iy \\ \gamma(x + tv) - iy & \gamma(t + xv) - z \end{pmatrix},$$
$$t' = \gamma(t + xv), \ x' = \gamma(x + tv), \ y' = y, \ z' = z.$$

These are the Lorentz transformations, expressed in the spinor formalism. ◆ Find X' in the Clifford basis of *Cl*3. ◆ As in the case of rotations, we can use the 4×4 matrix L (the *Lorentz group*) to get the same transformation

$$\begin{pmatrix} t' \\ x' \\ y' \\ z' \end{pmatrix} = \begin{pmatrix} \cosh\varphi & \sinh\varphi & 0 & 0 \\ \sinh\varphi & \cosh\varphi & 0 & 0 \\ 0 & 0 & 1 & 0 \\ 0 & 0 & 0 & 1 \end{pmatrix} \begin{pmatrix} t \\ x \\ y \\ z \end{pmatrix}.$$

4.4.8 Generators of the Lorentz Group

Generally, defining $\boldsymbol{\varphi} = (\varphi_1, \varphi_2, \varphi_3)$ and $\boldsymbol{\theta} = (\theta_1, \theta_2, \theta_3)$, we can create 2×2 matrices

$$\Lambda = \exp(\boldsymbol{\sigma} \cdot \boldsymbol{\varphi}/2 - i\boldsymbol{\sigma} \cdot \boldsymbol{\theta}/2), \quad \det \Lambda = 1,$$

just as we did in GA (see Sect. 2.7.6). The matrices Λ can Lorentz-transform spinors. ◆ Show that Λ is unitary ($\Lambda\Lambda^\dagger = I$) for pure rotations and Hermitian ($\Lambda = \Lambda^\dagger$) for pure boosts. ◆ Similarly, we can use the generators of the Lorentz group (six of them) and define 4 × 4 matrices $L = \exp(\mathbf{K} \cdot \boldsymbol{\varphi} + \mathbf{S} \cdot \boldsymbol{\theta})$, where $\mathbf{K}$ comprises the boost generators, while $\mathbf{S}$ comprises the rotation generators:

$$S_1 = \begin{pmatrix} 0 & 0 & 0 & 0 \\ 0 & 0 & 0 & 0 \\ 0 & 0 & 0 & -1 \\ 0 & 0 & 1 & 0 \end{pmatrix}, \quad S_2 = \begin{pmatrix} 0 & 0 & 0 & 0 \\ 0 & 0 & 0 & 1 \\ 0 & 0 & 0 & 0 \\ 0 & -1 & 0 & 0 \end{pmatrix}, \quad S_3 = \begin{pmatrix} 0 & 0 & 0 & 0 \\ 0 & 0 & -1 & 0 \\ 0 & 1 & 0 & 0 \\ 0 & 0 & 0 & 0 \end{pmatrix},$$

$$K_1 = \begin{pmatrix} 0 & 1 & 0 & 0 \\ 1 & 0 & 0 & 0 \\ 0 & 0 & 0 & 0 \\ 0 & 0 & 0 & 0 \end{pmatrix}, \quad K_2 = \begin{pmatrix} 0 & 0 & 1 & 0 \\ 0 & 0 & 0 & 0 \\ 1 & 0 & 0 & 0 \\ 0 & 0 & 0 & 0 \end{pmatrix}, \quad K_3 = \begin{pmatrix} 0 & 0 & 0 & 1 \\ 0 & 0 & 0 & 0 \\ 0 & 0 & 0 & 0 \\ 1 & 0 & 0 & 0 \end{pmatrix}.$$

The matrices L can Lorentz-transform four-vectors. Here we also have a double cover related to SL(2, C) and the Lorentz group (why?). ◆ Prove that

$$S_k^3 = -S_k, \quad K_k^3 = K_k, \quad (\mathbf{S} \cdot \boldsymbol{\theta})^3 = -|\boldsymbol{\theta}|^2 \mathbf{S} \cdot \boldsymbol{\theta}, \quad (\mathbf{K} \cdot \boldsymbol{\varphi})^3 = -|\boldsymbol{\varphi}|^2 \mathbf{K} \cdot \boldsymbol{\varphi},$$
$$[S_i, S_j] = \varepsilon_{ijk} S_k, \quad [S_i, K_j] = \varepsilon_{ijk} K_k, \quad [K_i, K_j] = -\varepsilon_{ijk} S_k. ◆$$

Note that the spinor formalism is more compact then the four-vector formalism, and in addition, the rank-1 spinors provide a direct connection with *Cl*3, due to the Pauli matrices.

4.5 Vahlen Matrices and Möbius Transformations

We can formally define matrices with elements different from real or complex numbers, using matrix multiplication for two matrices and a different multiplication for matrix elements. If elements of a matrix are multivectors, then we use the geometric product to multiply elements. Consider now the matrix (see R11)

$$V = \begin{pmatrix} A & B \\ C & D \end{pmatrix},$$

where the matrix elements are invertible multivectors in *Cln* (such elements form a group; the zero element is included). Now we demand some properties of these elements:

1. $AB^\dagger$, $DC^\dagger$, $C^\dagger A$, $B^\dagger D$ are vectors;
2. $\Delta = AD^\dagger - BC^\dagger \neq 0$, $\Delta \in \mathbb{R}$.

Such matrices are called the *Vahlen matrices*, and they are related to the Möbius transformations. ◆ Show that if the diagonal elements of the Vahlen matrices are even, then the off-diagonal elements are odd, and vice versa. Find some Vahlen matrices in *Cl*3. ◆

Appendix

Miroslav Josipović

M. Josipović, *Geometric Multiplication of Vectors*,
Compact Textbooks in Mathematics,
https://doi.org/10.1007/978-3-030-01756-9_5

A toy robot

5.1 The Exponential Function

The exponential function can be defined by the series expansion

$$\exp(x) \equiv \sum_{k=0}^{\infty} \frac{x^k}{k!} = 1 + x + \frac{x^2}{2!} + \frac{x^3}{3!} + \cdots, \quad (*)$$

where it is important to note that for a definite magnitude $|x|$ we have convergence. Therefore, we can define the exponential function of the elements of a geometric algebra. In *Cl*3, we have four important types of elements (numbers): *nilpotents* (dual numbers, square to zero), *imaginary* (square to -1), *hyperimaginary* (square to 1), and *idempotents* (square to itself). Nilpotents are the easiest to deal with: the expansion $(*)$ gives zeros, except for the first two terms, and consequently, we have

$$\exp(\mathbf{N}) = 1 + \mathbf{N}, \quad \mathbf{N}^2 = 0.$$

For idempotents $p^2 = p$, we have

$$\exp(p) = 1 + p \sum_{k=1}^{\infty} \frac{1}{k!} = 1 + p(e - 1) = ep + \bar{p}.$$

From the general form of a nilpotent $p = (1 + \mathbf{f})/2$ in *Cl*3, we get

$$\exp(p) = ep + \bar{p} = \frac{e+1}{2} + \frac{e-1}{2}\mathbf{f}.$$

For imaginary elements $I^2 = -1$, we can take the advantage of the series expansion for the sine and cosine functions

$$\sin(x) \equiv \sum_{k=0}^{\infty} (-1)^k \frac{x^{2k+1}}{(2k+1)!} = x - \frac{x^3}{3!} + \frac{x^5}{5!} - \frac{x^7}{7!} + \cdots,$$

$$\cos(x) \equiv \sum_{k=0}^{\infty} (-1)^k \frac{x^{2k}}{(2k)!} = 1 - \frac{x^2}{2!} + \frac{x^4}{4!} - \frac{x^6}{6!} + \cdots,$$

which gives

$$\exp(\varphi I) \equiv \sum_{k=0}^{\infty} \frac{I^k}{k!} = 1 + \varphi I - \frac{\varphi^2}{2!} - \frac{\varphi^3}{3!} I + \frac{\varphi^4}{4!} - \frac{\varphi^5}{5!} I + \cdots = \cos\varphi + I\sin\varphi.$$

Similarly, for hyperbolic elements $\mathbf{f}^2 = 1$, we use the series expansions

$$\sinh(x) \equiv \sum_{k=0}^{\infty} \frac{x^{2k+1}}{(2k+1)!} = x + \frac{x^3}{3!} + \frac{x^5}{5!} + \frac{x^7}{7!} + \cdots,$$

$$\cosh(x) \equiv \sum_{k=0}^{\infty} \frac{x^{2k}}{(2k)!} = 1 + \frac{x^2}{2!} + \frac{x^4}{4!} + \frac{x^6}{6!} + \cdots,$$

which give

$$\exp(\varphi\mathbf{f}) \equiv \sum_{k=0}^{\infty}\frac{\mathbf{f}^k}{k!} = 1 + \varphi\mathbf{f} + \frac{\varphi^2}{2!} + \frac{\varphi^3}{3!}\mathbf{f} + \frac{\varphi^4}{4!} + \frac{\varphi^5}{5!}\mathbf{f} + \cdots = \cosh\varphi + \mathbf{f}\sinh\varphi.$$

We can look at any term in the trigonometric function expansions and get, say,

$$-\frac{(I\varphi)^3}{3!} = \frac{(\varphi)^3}{3!}I, \qquad -\frac{(I\varphi)^2}{2!} = \frac{\varphi^2}{2!},$$

which suggests the validity of the relations

$$\sinh\varphi = -I\sin(I\varphi), \quad \cosh\varphi = \cos(I\varphi).$$

Due to noncommutativity, we cannot expect that the relation

$$e^A e^B = e^{A+B} \quad (**)$$

is valid generally. Consider commutative elements

$$\begin{aligned} e^{e_1}e^{e_1} &= (\cosh 1 + e_1\sinh 1)^2 = \cosh^2 1 + \sinh^2 1 + 2e_1\sinh 1\cosh 1, \\ e^{e_1+e_1} &= e^{2e_1} = \cosh 2 + e_1\sinh 2 = e^{e_1}e^{e_1}, \end{aligned}$$

where the relation $(**)$ is valid. However, for noncommuting elements we have

$$e^{e_1}e^{e_2} = (\cosh 1 + e_1\sinh 1)(\cosh 1 + e_2\sinh 1),$$

where the bivector e_1e_2 occurs, and

$$\exp(e_1 + e_2) = \exp\left(\sqrt{2}\frac{e_1 + e_2}{\sqrt{2}}\right) = \cosh\sqrt{2} + \frac{e_1 + e_2}{\sqrt{2}}\sinh\sqrt{2},$$

without the bivector. We can state the rule (there are exceptions; see [27])

$$e^A e^B = e^{A+B}, \quad AB = BA.$$

From expansions like $(*)$, we see that for involutions (i) and (analytic) functions (f) we have $\mathrm{i}(\mathrm{f}(M)) = \mathrm{f}(\mathrm{i}(M))$, for example, $\overline{\exp(e_1e_2)} = \exp(\overline{e_1e_2}) = \exp(-e_1e_2)$.

Note also that if the exponent is an element of some subalgebra, then the result belongs to that subalgebra as well. Thus $\exp(e_1e_2)$ belongs to the even part of *Cl*3 (spinors).

From

$$\exp(\theta j\hat{n}) = \cos\theta + j\hat{n}\sin\theta, \quad \theta \to \theta + 2k\pi,$$

we see that an infinite number of elements give the same value of the function. Therefore, there is a problem when we try to define the inverse function (*logarithm*). Usually, we choose to define the logarithm using the element that gives the smallest magnitude (*the principal value*).

5.2 Products as Sums

Here we give some formulas and examples for derived products as sums. The reader should notice differences in the definitions of "contractions," that is, the products that lower the grades of elements.

For homogeneous multivectors (single graded) A_i and B_j, with grades i and j, we define a derived product in the form $\langle A_i B_j \rangle_{g(i,j)}$, where $g(i,j)$ is the function that defines the type of the product. For example, for the outer product we have $g(i,j) = i + j$, $0 \le i + j \le n + 1$, where n is the dimension of the vector space. The summation takes place over the multivectors' grades.

- **outer product**

$$A \wedge B = \sum_{i,j} \left\langle \langle A \rangle_i \langle B \rangle_j \right\rangle_{i+j},$$

$$e_1 \wedge (e_1 e_3) = \left\langle \langle e_1 \rangle_1 \langle e_1 e_3 \rangle_2 \right\rangle_3 = \langle e_1 e_1 e_3 \rangle_3 = \langle e_3 \rangle_3 = 0,$$

$$\alpha \wedge e_1 = \left\langle \langle \alpha \rangle_0 \langle e_1 \rangle_1 \right\rangle_1 = \langle \alpha e_1 \rangle_1 = \alpha e_1.$$

- **left contraction**

$$A \rfloor B = \sum_{i,j} \left\langle \langle A \rangle_i \langle B \rangle_j \right\rangle_{j-i},$$

$$e_1 \rfloor (e_1 e_3) = \left\langle \langle e_1 \rangle_1 \langle e_1 e_3 \rangle_2 \right\rangle_1 = \langle e_1 e_1 e_3 \rangle_1 = \langle e_3 \rangle_1 = e_3,$$

$$\alpha \rfloor e_1 = \left\langle \langle \alpha \rangle_0 \langle e_1 \rangle_1 \right\rangle_1 = \langle \alpha e_1 \rangle_1 = \alpha e_1.$$

- **right contraction**

$$A \lfloor B = \sum_{i,j} \left\langle \langle A \rangle_i \langle B \rangle_j \right\rangle_{i-j},$$

$$e_1 \lfloor (e_1 e_3) = \left\langle \langle e_1 \rangle_1 \langle e_1 e_3 \rangle_2 \right\rangle_{-1} = 0,$$

$$\alpha \lfloor e_1 = \left\langle \langle \alpha \rangle_0 \langle e_1 \rangle_1 \right\rangle_{-1} = 0.$$

- **scalar product**

$$A * B = \sum_{i,j} \left\langle \langle A \rangle_i \langle B \rangle_j \right\rangle_0,$$

$$e_1 * (e_1 e_3) = \left\langle \langle e_1 \rangle_1 \langle e_1 e_3 \rangle_2 \right\rangle_0 = \langle e_3 \rangle_0 = 0,$$

$$\alpha * e_1 = \left\langle \langle \alpha \rangle_0 \langle e_1 \rangle_1 \right\rangle_0 = 0.$$

- **dot product**

$$A \bullet B = \sum_{i,j} \left\langle \langle A \rangle_i \langle B \rangle_j \right\rangle_{|j-i|},$$
$$e_1 \bullet (e_1 e_3) = \langle \langle e_1 \rangle_1 \langle e_1 e_3 \rangle_2 \rangle_1 = \langle e_3 \rangle_1 = e_3,$$
$$\alpha \bullet e_1 = \langle \langle \alpha \rangle_0 \langle e_1 \rangle_1 \rangle_1 = \langle \alpha e_1 \rangle_1 = \alpha e_1.$$

- **inner product**

$$A \cdot B = \sum_{i \neq 0, j \neq 0} \left\langle \langle A \rangle_i \langle B \rangle_j \right\rangle_{|j-i|},$$
$$e_1 \cdot (e_1 e_3) = \langle \langle e_1 \rangle_1 \langle e_1 e_3 \rangle_2 \rangle_1 = \langle e_3 \rangle_1 = e_3,$$
$$\alpha \cdot e_1 = 0 \quad \text{(formally, it is not defined for real numbers).}$$

In implementations, if we choose to make use of these formulas, the summation process could start with the lists of grades of the multivectors (A and B), for example, grades(A) $= (2, 3)$, grades(B) $= (1, 2)$, then we form pairs

$$((2, 1), (2, 2), (3, 1), (3, 2)),$$

choosing these that satisfy the conditions. For instance, for the outer product in $Cl3$, we have the condition $0 \leq i + j \leq 3$, which gives just the pair of grades $(2, 1)$, while for the left contraction in $Cl3$ we have $0 \leq j - i \leq 3$, which gives just the pair of grades $(2, 2)$. This process can reduce the number of geometric multiplications substantially, but it is important to have a fast solution for the function grades( ), which sometimes could be a problem due to small coefficients, like 10^{-11}.

A slightly modified version is to make a predefined list of pairs for each product, applying the conditions to the list of all possible pairs of grades ($(n + 1)^2$ of them). For dimension $n = 8$, we have 81 pairs, which for the outer product reduces to 45, and one could choose to sum over these 45 pairs without making a list of grades. Reducing the number of geometric multiplications becomes very important in higher dimensions. See also Sect. 3.8.4.

5.3 On Idempotents and Spinors

Here we will comment on some relations involving idempotents and spinors, anticipating that it will help the reader in reading the literature.

5.3.1 Idempotents

5.3.1.1 Factorization of Idempotents

First, consider an idempotent p in $Cl3$. We know that the general form of an idempotent is

$$p=(1+\mathbf{n}+j\mathbf{m})/2,\quad \mathbf{m}\cdot\mathbf{n}=0,\quad \mathbf{n}^2-\mathbf{m}^2=1,$$

where we can also write (see Sect. 2.3)

$$p=(1+\hat{\mathbf{n}}\cosh\varphi+j\hat{\mathbf{m}}\sinh\varphi)/2,\quad \hat{\mathbf{m}}^2=\hat{\mathbf{n}}^2=1.$$

In practical applications of idempotents, it is sometimes useful to factor them, so let us see how to do that (Sobczyk, see [42]). Using $\mathbf{n}=|\mathbf{n}|\hat{\mathbf{n}}=\mathrm{n}\hat{\mathbf{n}}$ and the properties of the geometric product, we can write

5

$$p=\mathbf{n}\left(1+\frac{\hat{\mathbf{n}}+j\hat{\mathbf{n}}\mathbf{m}}{\mathrm{n}}\right)/2=\mathbf{n}p_+.$$

The element p_+ is an idempotent, which follows from $\mathbf{m}\cdot\mathbf{n}=0$, meaning that $j\hat{\mathbf{n}}\mathbf{m}=j\mathrm{m}(\hat{\mathbf{n}}\cdot\hat{\mathbf{m}}+\hat{\mathbf{n}}\wedge\hat{\mathbf{m}})=j\hat{\mathbf{n}}\mathbf{m}$ is a vector, and we have the unit vector $\hat{\mathbf{p}}$,

$$\hat{\mathbf{p}}^2=\left(\frac{\hat{\mathbf{n}}+j\hat{\mathbf{n}}\mathbf{m}}{\mathrm{n}}\right)^2=\hat{\mathbf{n}}\frac{1+j\mathbf{m}}{\mathrm{n}}\hat{\mathbf{n}}\frac{1+j\mathbf{m}}{\mathrm{n}}=\hat{\mathbf{n}}\hat{\mathbf{n}}\frac{1-j\mathbf{m}}{\mathrm{n}}\frac{1+j\mathbf{m}}{\mathrm{n}}=\frac{1+\mathrm{m}^2}{\mathrm{n}^2}=1,$$

which means that $p_+=(1+\hat{\mathbf{p}})/2$ is a *simple* idempotent. Note that the components of the unit vector $\hat{\mathbf{p}}$ parallel $(\hat{\mathbf{n}}/\mathrm{n})$ and orthogonal $(j\hat{\mathbf{n}}\hat{\mathbf{m}}\mathrm{m}/\mathrm{n})$ with respect to the unit vector $\hat{\mathbf{n}}$ are obvious. Now it follows that

$$p=p^2=\mathbf{n}p_+\mathbf{n}p_+=\mathrm{n}^2\hat{\mathbf{n}}p_+\hat{\mathbf{n}}p_+\equiv\mathrm{n}^2s_+p_+,$$

where we have defined (note the term $\hat{\mathbf{n}}(j\hat{\mathbf{n}}\hat{\mathbf{m}})\hat{\mathbf{n}}$ and explain its geometric meaning)

$$s_+\equiv\hat{\mathbf{n}}p_+\hat{\mathbf{n}}=\hat{\mathbf{n}}\left(1+\frac{\hat{\mathbf{n}}+j\hat{\mathbf{n}}\mathbf{m}}{\mathrm{n}}\right)\hat{\mathbf{n}}/2=\left(1+\frac{\hat{\mathbf{n}}-j\hat{\mathbf{n}}\mathbf{m}}{\mathrm{n}}\right)/2\equiv(1+\hat{\mathbf{s}})/2,$$

whence it is easy to check that $s_+=(1+\hat{\mathbf{s}})/2$ is a simple idempotent, too. An example may be useful here; therefore, let's specify $\mathbf{n}=\sqrt{2}e_1$, $\mathbf{m}=e_3$, which gives

$$p_+=\left(1+\tfrac{e_1+je_1e_3}{\sqrt{2}}\right)/2=\left(1+\tfrac{e_1+e_2}{\sqrt{2}}\right)/2,\quad s_+=\left(1+\tfrac{e_1-e_2}{\sqrt{2}}\right)/2.$$

5.3.1.2 The Vector Part of Idempotents

Consider now the vector part of the idempotent p:

$$\mathbf{n}+j\mathbf{m}=\hat{\mathbf{n}}\cosh\varphi+j\hat{\mathbf{m}}\sinh\varphi=\hat{\mathbf{n}}(\cosh\varphi+j\hat{\mathbf{n}}\hat{\mathbf{m}}\sinh\varphi).$$

We have $(j\hat{\mathbf{n}}\hat{\mathbf{m}})^2=1$, and therefore,

$$\cosh\varphi+j\hat{\mathbf{n}}\hat{\mathbf{m}}\sinh\varphi=\exp(\varphi j\hat{\mathbf{n}}\hat{\mathbf{m}}),$$

whence, defining $v\equiv\tanh\varphi$, $\gamma\equiv\cosh\varphi$, we have

$$\cosh\varphi + j\hat{\mathbf{n}}\hat{\mathbf{m}}\sinh\varphi = \gamma(1 - \hat{\mathbf{n}} \times \hat{\mathbf{m}}v),$$

where we used the cross product as defined in *Cl*3. Finally, we have

$$\mathbf{n} + j\mathbf{m} = \hat{\mathbf{n}}\gamma(1 - \hat{\mathbf{n}} \times \hat{\mathbf{m}}v),$$

which is an interesting relation; we can interpret it as a boost of the unit vector $\hat{\mathbf{n}}$ in the direction

$$j\hat{\mathbf{n}}\hat{\mathbf{m}} = -\hat{\mathbf{n}} \times \hat{\mathbf{m}},$$

with speed $v = \tanh\varphi$ and proper velocity $\gamma(1 - \hat{\mathbf{n}} \times \hat{\mathbf{m}}v)$.

5.3.1.3 Products of Simple Idempotents

Defining the simple idempotents $x = (1 + \hat{\mathbf{x}})/2$ and $y = (1 + \hat{\mathbf{y}})/2$, we have

$$4xyx = (1 + \hat{\mathbf{x}})(1 + \hat{\mathbf{y}})x = (1 + \hat{\mathbf{x}} + \hat{\mathbf{y}} + \hat{\mathbf{x}}\hat{\mathbf{y}})x = (1 + \hat{\mathbf{x}} + \hat{\mathbf{y}} - \hat{\mathbf{y}}\hat{\mathbf{x}} + 2\hat{\mathbf{x}} \cdot \hat{\mathbf{y}})x.$$

From $\hat{\mathbf{x}}x = x$ and $\hat{\mathbf{y}}x = \hat{\mathbf{y}}\hat{\mathbf{x}}x$, we have finally

$$xyx = \frac{1 + \hat{\mathbf{x}} \cdot \hat{\mathbf{y}}}{2}x,$$

which means that xyx and x are proportional. Starting from the idempotents s_+ and p_+, we can try to find an idempotent of the form Zs_+p_+, where Z is a (commutative) complex scalar:

$$\left(Zs_+p_+\right)^2 = Z^2 s_+p_+s_+p_+ = Z^2\frac{1 + \hat{\mathbf{s}} \cdot \hat{\mathbf{p}}}{2}s_+p_+ = Zs_+p_+,$$

whence follows $Z = 2/(1 + \hat{\mathbf{s}} \cdot \hat{\mathbf{p}})$, $Z \in \mathbb{R}$.

5.3.1.4 Idempotents in Spinor Theories

Idempotents play an important role in spinor theories. For example, consider a rotor defined by Euler angles (see E26; here we have replaced e_1 by e_2),

$$R = \exp(-je_3\phi/2)\exp(-je_2\theta/2)\exp(-je_3\psi/2),$$

and the simple idempotent $u_+ = (1 + e_3)/2 = e_3u_+$. Note that

$$\begin{aligned}
&\exp(-je_3\psi/2)e_3u_+ = \exp(-j\psi/2)u_+,\\
&\exp(-je_2\theta/2)e_3u_+ = \cos(\theta/2)e_3u_+ - je_2e_3u_+\sin(\theta/2) = u_+\cos(\theta/2) + u_-e_1\sin(\theta/2),\\
&\exp(-je_3\phi/2)(u_+\cos(\theta/2) + u_-e_1\sin(\theta/2))\\
&= \exp(-j\phi/2)\cos(\theta/2)u_+ + e_1\exp(j\phi/2)\sin(\theta/2)u_+,
\end{aligned}$$

which gives

$$Ru_+ = \left[e^{-j\phi/2}\cos(\theta/2)u_+ + e_1u_+e^{j\phi/2}\sin(\theta/2)\right]e^{-j\psi/2}.$$

This element contains all the information from the rotor R. From Sect. 4.2.2, we know that matrices of a multivector in the spectral basis of $Cl3$ have in their first column the elements u_+, e_1u_+, and in the second column, the elements u_-, e_1u_-, which means that the matrix of Ru_+ is

5

$$\begin{pmatrix} \exp(-j\phi/2)\cos(\theta/2) & 0 \\ \exp(j\phi/2)\sin(\theta/2) & 0 \end{pmatrix}\exp(-j\psi/2),$$

where the overall phase factor $\exp(-j\psi/2)$ can be neglected. Anyhow, we have a quantity that depends on two complex numbers, which is the basic characteristic of spinors in standard theories. To justify this form further, we can consider a rotation of the unit vector e_3, which is common in quantum mechanics ($u_\pm = (1 \pm e_3)/2$):

$$\begin{aligned} Re_3R^\dagger &= R(u_+ - u_-)R^\dagger = Ru_+R^\dagger - Ru_-R^\dagger = Ru_+u_+R^\dagger - Ru_-u_-R^\dagger \\ &= Ru_+(Ru_+)^\dagger - Ru_-(Ru_-)^\dagger, \end{aligned}$$

where we see how expressions like Ru_+ arise.

Consider now a general element X from $Cl3$ and an element $X_+ = Xu_+$. Now we have

$$X_+u_+ = Xu_+u_+ = Xu_+ = X_+,$$

which means that the subset of elements of the form Xu_+ is closed under the transformation $X \to Xu_+$. We call such a subset of an algebra the *minimal left ideal* (idempotents must be *primitive*; see the literature). The minimal left ideal defines a spinor space. Formally, we can write $S_\pm = Cl3u_\pm$, where $S_\pm$ represent two complementary spinor spaces. The basis of such a spinor space consists of u_+ and e_1u_+, which we can show explicitly. From the general form of a multivector $M = z^\mu e_\mu$, $e_0 = 1$, $z^\mu \in \mathbb{C}$, $\mu = 0, 1, 2, 3$ ($\mathbb{C}$ is due to j), we can write (see problem 4)

$$M = z^0 + z^1e_1 + z^2e_2 + z^3e_3 = z^0 + z^3e_3 + z^1e_1 + z^2je_1e_3,$$

whence follows

$$Mu_+ = \left(z^0 + z^3\right)u_+ + \left(z^1 + z^2j\right)e_1u_+,$$

just the form we need. Note again that the elements of a spinor space are defined by two complex scalars and that their matrices have zeros in the second column:

$$\left(z^0 + z^3\right)\begin{pmatrix} 1 & 0 \\ 0 & 0 \end{pmatrix} + \left(z^1 + z^2i\right)\begin{pmatrix} 0 & 0 \\ 1 & 0 \end{pmatrix} = \begin{pmatrix} z^0 + z^3 & 0 \\ z^1 + z^2i & 0 \end{pmatrix}, \quad j \to i.$$

5.3.2 Bases for Spinorial Matrices

Writing

$$g_1 \equiv \begin{pmatrix} 1 & 0 \\ 0 & 0 \end{pmatrix}, \quad s \equiv \begin{pmatrix} s_1 & 0 \\ s_2 & 0 \end{pmatrix},$$

we have a simple way to express spinorial matrices (like s). It is easy to check that the product of a general complex matrix m and a spin matrix s is a spin matrix. Note that $g_1^2 = g_1$. Then from

$$\hat{\sigma}_3 = \begin{pmatrix} 1 & 0 \\ 0 & -1 \end{pmatrix},$$

it follows that

$$\begin{pmatrix} 1 & 0 \\ 0 & 1 \end{pmatrix} + \begin{pmatrix} 1 & 0 \\ 0 & -1 \end{pmatrix} = \begin{pmatrix} 2 & 0 \\ 0 & 0 \end{pmatrix},$$

which means that the matrix g_1 represents the idempotent $u_+ = (1 + e_3)/2$. Using

$$\hat{\sigma}_1 = \begin{pmatrix} 0 & 1 \\ 1 & 0 \end{pmatrix},$$

we see that the element $e_1 u_+$ is represented by the matrix

$$g_2 = \begin{pmatrix} 0 & 0 \\ 1 & 0 \end{pmatrix},$$

and we have the basis (g_1, g_2) of a linear spinor space in matrix form. Note that g_2^2 is a zero matrix, in accordance with $e_1 u_+ e_1 u_+ = u_- u_+ = 0$. While the basis (g_1, g_2) gives elements of the minimal left ideal over the complex scalars, it is possible to define a basis over the real numbers. The reader can show that such a basis could consist of u_+, $e_2 u_+$, $-e_1 u_-$, and $e_1 e_2 u_+$, with the corresponding matrices f_μ, $\mu = 0, 1, 2, 3$ (find them, using the Pauli matrices).

5.3.3 Observables in Quantum Mechanics

An observable in quantum mechanics is a Hermitian operator, which means that it does not change after transposition and complex conjugation. From the form of the matrix of a general multivector M in $Cl3$,

$$[M] = \begin{pmatrix} \alpha + x_3 + jn_3 + j\beta & x_1 + n_2 - jx_2 + jn_1 \\ x_1 - n_2 + jx_2 + jn_1 & \alpha - x_3 - jn_3 + j\beta \end{pmatrix},$$

we see that the matrix $[M]$ will be Hermitian iff $\beta = n_i = 0$, obtaining the form

$$[M]_H = \begin{pmatrix} \alpha + x_3 & x_1 - jx_2 \\ x_1 + jx_2 & \alpha - x_3 \end{pmatrix}.$$

In terms of the $Cl3$, for such an element we have $M = M^\dagger$ (note again that the character $\dagger$ denotes both the reverse involution in $Cl3$ and the Hermitian adjoint in the matrix formulation), which is a paravector, of the general form $p = p_0 + \mathbf{p} = p_0 + |\mathbf{p}|\hat{\mathbf{p}}$. Defining $u_+ = (1 + \hat{\mathbf{p}})/2$, we can write

$$p = p(u_+ + u_-) = (p_0 + |\mathbf{p}|)u_+ + (p_0 - |\mathbf{p}|)u_-,$$

because of $\hat{\mathbf{p}}u_+ = u_+$. Now it follows that $pu_\pm = (p_0 \pm |\mathbf{p}|)u_\pm$, where $p_0 \pm |\mathbf{p}|$ are *eigenvalues*, while $u_\pm$ are *eigenpotents* (analogous to eigenvectors). Find eigenvalues of $[M]_H$ and compare them with eigenvalues of $pu_\pm$.

5.3.4 The Möbius Strip and Spinors

A Möbius strip can help to visualize the rotational property of spinors. In Fig. 5.1, we have a silver Möbius ring on which the red arrow a starts to move around the ring. The blue arrows represent the moving arrow from 0 to 2π. After a full cycle (2π), the red arrow a becomes the opposite arrow $-a$. The red arrow continues to move (the yellow arrows), and after another full cycle from 2π to 4π, it returns to the starting position.

5.4 Extended Lorentz Transformations. The Speed Limit?

This chapter is **speculative**, with interesting consequences (**new preserved quantities and a change in nature's speed limit**). Those who are faint of heart can take this just as a mathematical exercise.

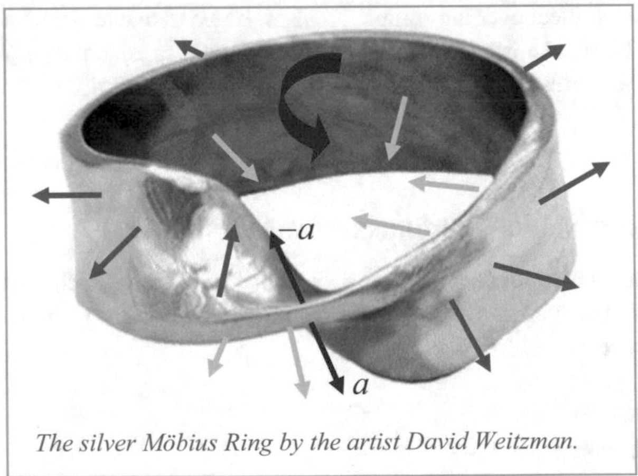

The silver Möbius Ring by the artist David Weitzman.

Fig. 5.1 A Möbius strip helps to visualize the rotational property of spinors

5.4.1 General Bilinear Transformations

Earlier, we defined MA as the square root of

$$M\bar{M} = |M|^2 = t^2 - x^2 + n^2 - b^2 + 2j(tb - \mathbf{x}\cdot\mathbf{n}) \in \mathbb{C},$$

and showed its properties. Now we look for general bilinear transformations $M' = AMB$ that preserve MA (see [15]):

$$|M'|^2 = M'\bar{M}' = AMB\bar{B}\bar{M}\bar{A} = |M|^2|A|^2|B|^2,$$

where we have the possibilities

$$|A|^2 = |B|^2 = \pm 1,$$

which give (see Sect. 2.6)

$$A = e^{Z+\mathbf{F}} \Rightarrow |A|^2 = e^{Z+\mathbf{F}}e^{Z-\mathbf{F}} = e^{2Z} = \pm 1,$$

where we choose the possibility $Z = 0$, although we could consider $Z = j\pi/2$ just as well. Now, the general transformation is given by

$$M' = AMB = \exp(\mathbf{p} + j\mathbf{q})M\exp(\mathbf{r} + j\mathbf{s}),$$

that is, we have 12 parameters from the four vectors in the exponents.

The question is, what motive do we have for considering such transformations. The elements of geometric algebra are linear combinations of the unit blades from the Clifford basis, and each of them actually defines a subspace. If we limit ourselves to the real part of multivectors (paravectors), we put the space of real numbers (grade 0) and the space of vectors (grade 1) in a privileged position. The idea is that we treat all subspaces equally. In fact, this whole structure is based on a new multiplication of vectors; consequently, in manipulating multivectors, we actually manipulate subspaces. For instance, addition of vectors and bivectors is essentially an operation that relates subspaces, and it is important to understand this well. If subspaces are treated equally, then we must consider all possible transformations of subspaces and all possible symmetries; however, they are more than what classical (restricted) Lorentz transformations imply. The reader should be in a position to stop a little and think carefully about this. Recall that symmetries in the flow of time give the law of conservation of energy; translational invariance gives the law of conservation of momentum; and so on. Where do we stop, and why? If we truly accept the naturalness of the new multiplication of vectors, we must accept the consequences of such a multiplication as well, and then they reveal an unusually rich structure of good old 3D Euclidean space. However, the final judgment must be given by experiments (hope).

Considering the invariant MA, expressed in two reference frames, we can compare the real and imaginary parts:

$$t^2 - x^2 + n^2 - b^2 = t'^2 - x'^2 + n'^2 - b'^2,$$
$$tb - \mathbf{x} \cdot \mathbf{n} = t'b' - \mathbf{x}' \cdot \mathbf{n}'.$$

5.4.2 The Real Proper Time

The differential (an infinitely small change) of the multivector $X = t + \mathbf{x} + j\mathbf{n} + jb$ that could represent a general event is

$$dX = dt + d\mathbf{x} + j\,d\mathbf{n} + j\,db,$$

with the MA

$$|dX|^2 = dt^2 - dx^2 + dn^2 - db^2 + 2j(db\,dt - d\mathbf{x} \cdot d\mathbf{n}). \quad (*)$$

Hence, we can try to find conditions for the existence of the **real** proper time. There are many reasons to define a real proper time. For instance, such a definition makes it easy to define a generalized velocity. In the special theory of relativity, a real proper time is defined in the rest frame of a particle. Here, due to additional elements (except the velocity), it will not be sufficient. What we want is

$$|dX|^2 = dt^2 - dx^2 + dn^2 - db^2 + 2j(db\,dt - d\mathbf{x} \cdot d\mathbf{n}) = d\tau^2 \in \mathbb{R}$$

(τ is a *proper time*). The first condition, if we want a real proper time, certainly is the disappearance of the imaginary part of the MA $(*)$ in each system of reference (recall that MA is invariant with respect to our transformations). This means that in all reference frames it must be valid that ($d\dot{b} \equiv h$, $dx/dt \equiv \dot{x}$)

$$dbdt - d\mathbf{x} \cdot d\mathbf{n} = dt^2\left(d\dot{b} - d\dot{\mathbf{x}} \cdot d\dot{\mathbf{n}}\right) = dt^2(h - d\dot{\mathbf{x}} \cdot d\dot{\mathbf{n}}) = 0 \Rightarrow h = d\dot{\mathbf{x}} \cdot d\dot{\mathbf{n}},$$

which implies $h' = d\dot{\mathbf{x}}' \cdot d\dot{\mathbf{n}}'$. If we define $d\dot{\mathbf{x}} \equiv \mathbf{v}$ and $\dot{\mathbf{n}} \equiv \mathbf{w}$, it follows that $h = \mathbf{w} \cdot \mathbf{v}$. The vector $\mathbf{w}$ comes from the bivector part of the multivector; therefore, we expect it to be related to angular momentum-like quantities. Then h could be a flow of such a quantity, as a flow is defined for the flowing of a liquid through a tube, or perhaps, a better interpretation could be a *helicity*, a projection of the spin to the direction of motion.

Bearing in mind the invariance of MA and $d\tau$, $dt \in \mathbb{R}$, we have

$$|dX|^2 = |dX'|^2 = d\tau^2 = dt^2 - dx^2 + dn^2 - db^2 \Rightarrow$$
$$d\tau^2 = dt^2\left(1 - \frac{dx^2}{dt^2} + \frac{dn^2}{dt^2} - \frac{db^2}{dt^2}\right) \equiv \gamma^2\left(1 - v^2 + w^2 - h^2\right) \Rightarrow$$
$$\gamma = 1/\sqrt{1 - v^2 + w^2 - (\mathbf{w} \cdot \mathbf{v})^2} = 1/\sqrt{1 - v^2 + w^2 - w^2v^2\cos^2\alpha} \in \mathbb{R}.$$

5.4.3 The New Limiting Speed(s)

Note that our relativistic factor γ has contributions from all subspaces. In STR, we introduce the rest frame (with the condition $v = 0$), in which we then have $\gamma = 1$; however, here we get $\gamma = 1/\sqrt{1+w^2}$ in that way. Therefore, it would be natural to require the condition $\gamma = 1$ instead of $v = 0$. This condition gives the speed of a "rest frame"

$$-v^2 + w^2 - w^2 v^2 \cos^2\alpha = 0 \Rightarrow v = w/\sqrt{1 + w^2 \cos^2\alpha}.$$

It is not difficult to accept this; perhaps the velocity of a particle cannot be zero. For example, how are we to reconcile the principles of quantum mechanics and the idea of completely peaceful electrons? Including all subspaces and all quantities related to them, it follows that a "rest frame" becomes something like a "center of energy-impulse-angular momentum-helicity frame." The relativistic factor γ is defined as the ratio of two real times, so it must be a real number, which gives the condition (Fig. 5.2)

$$1 - v^2 + w^2 - w^2 v^2 \cos^2\alpha > 0 \Rightarrow v_{\max} < \sqrt{\frac{1+w^2}{1+w^2\cos^2\alpha}}.$$

This is a completely new result: the limiting speed is 1 for $w = 0$ or $\cos\alpha = \pm 1$; otherwise, it is greater than 1. This result is not new in geometric algebra (*Pavšić*, using C-algebras; however, the author got this result independently, in the comment on the article [15]). What could be the physical meaning of this? Consider an electron. It certainly is not a "small ball" (recall the great Ruđer Bošković and his points as the source of a force). To be precise, it has spin, and the spin is a quantity just "like angular momentum," which could be interpreted geometrically (see Sect. 2.10.7). Can we treat the relativistic electron as an Einsteinian relativistic train? Probably not! In the eyes of geometry, it is hard to accept that an electron is just a very small ball with a spin packed into it, like a train with passengers. Relativistic formulas for a train do not depend on the type of cargo in it. However, spin is probably not just a "cargo"; rather, it is a geometric property, and consequently, it should be a part of transformations. If the formulas just

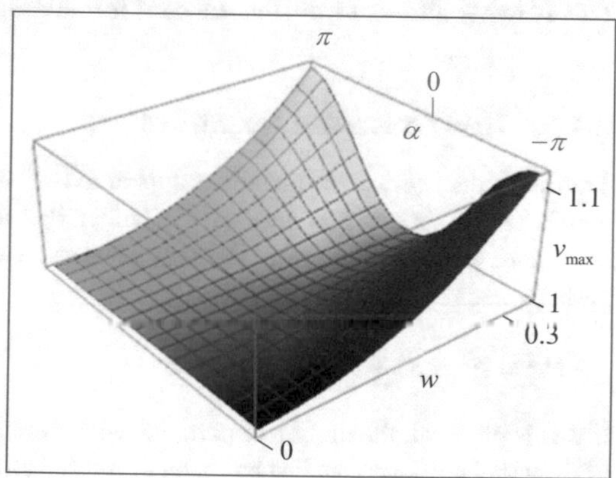

Fig. 5.2 The limit speed in extended Lorentz transformations

derived were applicable to the electron, then its (limiting) speed would depend on the orientation of the spin relative to the velocity vector. Moreover, the speeds for a given energy would depend on the orientation of the spin; consequently, electrons with the same energy, starting at the same moment, are supposed to arrive to the target at different times. Perhaps someone in the future will carry out such an experiment. A positive result would certainly change our current understanding of relativity significantly (this question could be related to the question of the existence of magnetic monopoles, see Sect. 5.4.7). Especially interesting would be to see how electrons behave in quantum tunneling; there are suggestions of some authors that an electron might be moving at speeds exceeding 1. Occasionally, such statements are formulated by introducing complex numbers, creating a whole philosophy, although it is likely a matter of inappropriate mathematics. However, it is hard to be sure, for the time being.

5.4.4 The Generalized Velocity and the Generalized Momentum

Now that we have defined (real and invariant) proper time, we can define the multivector of a generalized velocity

$$V = \frac{dX}{d\tau} = \frac{dt}{d\tau} + \frac{d\mathbf{x}}{dt}\frac{dt}{d\tau} + j\frac{d\mathbf{n}}{dt}\frac{dt}{d\tau} + j\frac{db}{dt}\frac{dt}{d\tau} = \gamma(1 + \mathbf{v} + j\mathbf{w} + jh),$$

with the invariant MA

$$V\bar{V} = 1 \Rightarrow \frac{d(V\bar{V})}{d\tau} = \frac{dV}{d\tau}\bar{V} + V\frac{d\bar{V}}{d\tau} = A\bar{V} + V\bar{A} = 0,$$

which is a kind of expression of the orthogonality of the generalized velocity and the generalized acceleration $A \equiv dV/d\tau$. Note that $A\bar{V} = -\overline{A\bar{V}}$, which means that $A\bar{V}$ is a complex vector. Multiplying the generalized velocity by the (invariant) mass m, we get the generalized momentum

$$P = mV = E + \mathbf{p} + j\mathbf{l} + jH, \quad P\bar{P} = E^2 - p^2 + l^2 - H^2 = m^2.$$

This is rather different from the standard formula for energy–momentum $E^2 - p^2 = m^2$.

5.4.5 New Conserved Quantities

Two additional conserved quantities appear in the energy–momentum formula, the last of which (H) is brand new (see [15]). Under the terms of our derivation, it must be $H = \gamma mh = \mathbf{l} \cdot \mathbf{v}$. Therefore, the new conserved quantity has the form of a *flow* (or, perhaps, *helicity*), and we have finally

$$P\bar{P} = E^2 - p^2 + l^2 - (\mathbf{l} \cdot \mathbf{v})^2 = m^2.$$

If this is physical, the motion of particles with spin should satisfy the law of conservation of flow (helicity), an idea that has been presented already by some authors (unfortunately, I have forgotten the source). For a given energy, the speed of a particle will generally be higher with **w** than without it, which can be deduced from the previous formula: adding a positive term to the negative square of momentum gives the possibility to increase the speed. Let us make sure about this directly:

$$\gamma = E/m = 1/\sqrt{1 - v^2 + w^2 - (\mathbf{w}\cdot\mathbf{v})^2} \Rightarrow$$

$$v = \sqrt{\frac{1 + w^2 - m^2/E^2}{1 + w^2 \cos^2\alpha}} = \sqrt{\frac{1 + (l/E)^2 - m^2/E^2}{1 + (l/E)^2 \cos^2\alpha}} \geq \sqrt{1 - m^2/E^2}.$$

Note that the just discussed formalism reduces to the usual special theory of relativity. It is sufficient to reject the imaginary part of multivectors and keep the real one, that is, paravectors. All "strange" implications just disappear. Of course, there are plenty of possibilities to treat time differently (see [15] for some interesting discussions).

5.4.6 Properties of General Transformations

Consider now some more properties of our general transformations (see [17]). First, note that for $\mathbf{n} = 0$ and $b = 0$ one has the standard Minkowski interval $|X|^2 = |X'|^2 = \tau^2 = t^2 - x^2$. However, generally, we can get the standard Minkowski interval by demanding $\mathbf{n}^2 - b^2 = 0$ and $bt - \mathbf{x}\cdot\mathbf{n} = 0$, whence, using $\dot{\mathbf{x}} = \mathbf{v}$, it follows that $\mathbf{v}\cdot\hat{\mathbf{n}} = \pm 1$. If we interpret $\hat{\mathbf{n}}$ as a spin direction unit vector, then $\mathbf{v}\cdot\hat{\mathbf{n}}$ is the helicity, a projection of the spin unit vector in the direction of motion. We see that an electromagnetic field could be a nice candidate for the condition $\mathbf{v}\cdot\hat{\mathbf{n}} = \pm 1$, since $v = 1$ and $\mathbf{v} \parallel \hat{\mathbf{n}}$. Second, from $n^2 - b^2 = n'^2 - b'^2$ and $tb - \mathbf{x}\cdot\mathbf{n} = t'b' - \mathbf{x}'\cdot\mathbf{n}'$, we can obtain (show this)

$$\begin{aligned} n_1' &= \gamma(n_1 - vb), \\ n_2' &= n_2, \\ n_3' &= n_3, \\ b' &= \gamma(b - vn_1), \end{aligned}$$

which means that we have additional transformations, similar to the standard ones.

5.4.7 Maxwell's Equations Under General Transformations

Now we will comment on Maxwell's equations in relation to our general transformations. We write an electromagnetic field in the form

$$\mathbf{F} = \mathbf{E} + j\mathbf{B},$$

which is a complex vector. We write sources (charges and currents) in the form of a paravector

$$J = \rho - \mathbf{J},$$

as well as the derivative operator

$$\partial = \partial_t + \nabla = \frac{\partial}{\partial t} + e_i \frac{\partial}{\partial x_i},$$

which means that it acts like a paravector in products. The vector part of the derivative operator acts like a vector in products, and therefore, we can write

$$\nabla \mathbf{x} = \nabla \cdot \mathbf{x} + \nabla \wedge \mathbf{x} = \nabla \cdot \mathbf{x} + j\nabla \times \mathbf{x} = div\mathbf{x} + j\,\mathrm{rot}\,\mathbf{x},$$

where the definitions of *divergence* and *curl*, as well as the cross product are introduced, all defined in GA (this is important; see Sect. 2.7.7). Maxwell's equations have a nice, simple form in geometric algebra:

$$\partial \mathbf{F} = J.$$

To see the relationship with the standard Maxwell's equations, we just need to take all grades of the previous relation. Using

$$\partial \mathbf{E} = \partial_t \mathbf{E} + \nabla \cdot \mathbf{E} + j\nabla \times \mathbf{E}, \quad j\partial \mathbf{B} = j\partial_t \mathbf{B} + j\nabla \cdot \mathbf{B} - \nabla \times \mathbf{B},$$

and the facts that $\nabla \cdot \mathbf{x}$ is a real scalar, while $\nabla \times \mathbf{x}$ is a (polar) vector, we have
grade 0: $\nabla \cdot \mathbf{E} = \rho$ (Gauss's law of electricity),
grade 1: $\partial_t \mathbf{E} - \nabla \times \mathbf{B} = -\mathbf{J}$ (Ampère's law),
grade 2: $\partial_t \mathbf{B} + \nabla \times \mathbf{E} = 0$ (Faraday's law),
grade 3: $\nabla \cdot \mathbf{B} = 0$ (Gauss's law of magnetism).

Note the beauty of the representation; **the equations are related to subspaces.**

It is easy to check two facts: first, the geometric product of a paravector with a complex vector is a full multivector (has all grades); second, a general transformation of a paravector (like J) will also result in a full multivector. This means that for bilinear transformations we generally have

$$AJB = \rho' - \mathbf{J}' + j\mathbf{J}'^m + j\rho'^m,$$

where the quantities ρ'^m and $\mathbf{J}'^m$ could be interpreted as a magnetic monopole and a magnetic current. They are not observed experimentally, which means that in electromagnetism we have to transform a paravector to a paravector (that is, AJB must be a paravector), whence, using $p = p^\dagger$ for a paravector p, it follows that

$$(AJB)^\dagger = B^\dagger J^\dagger A^\dagger = B^\dagger J A^\dagger = AJB,$$

where $A = \exp(\mathbf{p} + j\mathbf{q})$ and $B = \exp(\mathbf{r} + j\mathbf{s})$. By comparison, we have a solution

$$\mathbf{r} - j\mathbf{s} = \mathbf{p} + j\mathbf{q}, \quad \mathbf{r} + j\mathbf{s} = \mathbf{p} - j\mathbf{q},$$

giving finally $\mathbf{r} = \mathbf{p}$ and $\mathbf{s} = -\mathbf{q}$, whence our general transformations for electromagnetism are reduced to

$$M \to \exp(\mathbf{p} + j\mathbf{q}) M \exp(\mathbf{p} - j\mathbf{q}) \equiv L M L^{\dagger},$$

which is a common expression for the restricted Lorentz transformations. Note that the eventual existence of magnetic monopoles would demand the general transformations. We could derive similar conclusions for $A\partial B$, since the operator ∂ is a paravector-like quantity.

We can now check a general transformation of the term $\partial \mathbf{F}$,

$$\partial \mathbf{F} \to A \partial \mathbf{F} B = A \partial B \bar{B} \mathbf{F} B,$$

where the term $A\partial B$ requires the restricted Lorentz transformations, which means that the electromagnetic field transforms as

$$\mathbf{F} \to L \mathbf{F} \bar{L}$$

(just use $\bar{B} = L$), and this is an important result. We immediately see that under the boost $\exp(\varphi e_1)$, we have

$$\overline{\exp(\varphi e_1)} = \exp(-\varphi e_1) \Rightarrow E_1' = E_1, \quad B_1' = B_1,$$

while perpendicular components are boosted, in accordance with the EM theory. Note that this follows directly from the commutation properties of the geometric product. This form of transformations of the electromagnetic field follows also from a general analysis of a paravector space (see [9]).

5.5 Visualization of the Electromagnetic Field in Vacuum

To describe the electromagnetic field (EMF) in vacuum, we define a *Faraday*, a name we choose for the object

$$\mathbf{F} = \mathbf{E} + j\mathbf{B}$$

($c = 1$), which (in vacuum) appears to be a nilpotent ($|\mathbf{E}| = |\mathbf{B}|$, $\mathbf{E} \cdot \mathbf{B} = 0$, $\mathbf{F}^2 = 0$). We write Maxwell's equations in vacuum in the form $\partial \mathbf{F} = 0$, where $\partial = \partial_t + \nabla$. Note that we can write

$$\bar{\partial} \partial \mathbf{F} = \left(\partial_t^2 - \nabla^2 \right) \mathbf{F} = 0,$$

which is a wave equation, with the possible solution (see [27])

$$\mathbf{F} = \mathbf{F}_0 \exp(\pm j(\omega t - \mathbf{k} \cdot \mathbf{x})),$$

which for propagation in the e_3 direction and the + sign we can write as

$$\mathbf{F} = \mathbf{F}_0 \exp(j(\omega t - kz)).$$

Due to the commutativity of complex scalars, we can write

$$\mathbf{F} = \mathbf{F}_0 \exp(-jkz) \exp(j\omega t),$$

where $\mathbf{F}_0 \exp(-jkz)$ is a space-dependent nilpotent (since $\mathbf{F}_0$ is a nilpotent), which for $t = 0$ we can display graphically. Note that $j\mathbf{B}$ is a bivector (we can represent it by a parallelogram), while $\mathbf{E}$ is a vector that belongs to the plane of the bivector (*whirl*, Fig. 5.3). On multiplying the spatial part by the time-dependent complex phase $\exp(j\omega t)$, we see that this steady nilpotent begins to rotate about the axis of the wave propagation (see E30). Now we can choose a point in space (say $z = 0$) to see the change in time (Fig. 5.4). Instead of arrows (vectors) that rotate, we should develop an intuition on nilpotents that rotate.

It is straightforward to show that the derivative operator ∂ applied to an exponential function behaves just as we expect: $\exp(Cx) \to C \exp(Cx)$, $C \in \mathbb{C}$. Therefore, from $\partial \mathbf{F} = 0$, for the solution in vacuum we get $(\omega - \mathbf{k})\mathbf{F} = 0$, and on multiplying by $\omega + \mathbf{k}$ from the left, it follows that

$$\left(\omega^2 - |\mathbf{k}|^2\right)\mathbf{F} = 0 \Rightarrow \omega^2 = |\mathbf{k}|^2 \Rightarrow \omega = |\mathbf{k}|,$$

leading to $\left(1 - \hat{\mathbf{k}}\right)\mathbf{F} = 0$, or $\hat{\mathbf{k}}\mathbf{F} = \mathbf{F}$. We already know these relations for nilpotents (see Sect. 2.2). Substituting for $\mathbf{F}$, it follows that

$$\hat{\mathbf{k}}(\mathbf{E} + j\mathbf{B}) = \mathbf{E} + j\mathbf{B},$$

whence, after comparison of the even and odd parts, it follows that $\hat{\mathbf{k}}\mathbf{E} = j\mathbf{B}$, which, using $|\mathbf{E}| = |\mathbf{B}|$, gives $\hat{\mathbf{E}}\hat{\mathbf{B}}\hat{\mathbf{k}} = j$. We see that vectors $\hat{\mathbf{E}}$, $\hat{\mathbf{B}}$, and $\hat{\mathbf{k}}$ compose a right-handed orthonormal frame of vectors, and this is a nontrivial result. ◆ The reader should find a

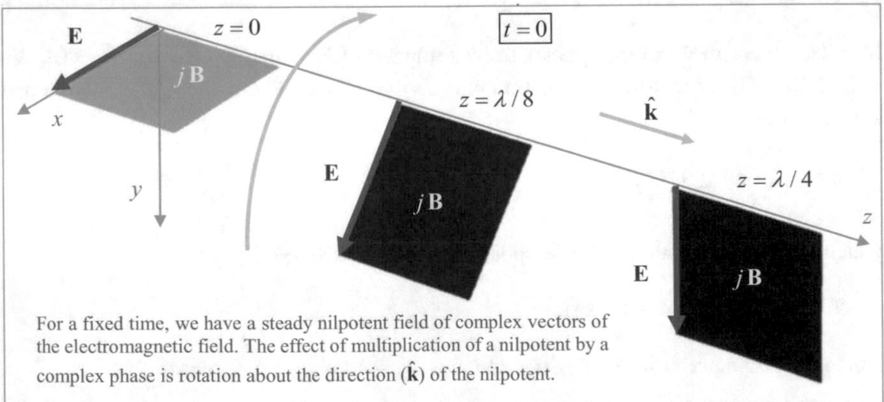
E
z = 0
t = 0
jB
x
y
z = λ / 8
k̂
E
jB
z = λ / 4
z
E
jB
For a fixed time, we have a steady nilpotent field of complex vectors of the electromagnetic field. The effect of multiplication of a nilpotent by a complex phase is rotation about the direction (k̂) of the nilpotent.

Fig. 5.3 The electromagnetic field representation at a fixed moment

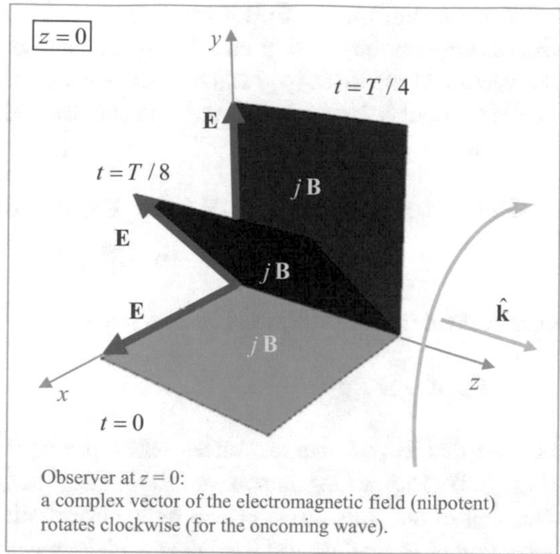

Fig. 5.4 The electromagnetic field representation at a fixed place

possibility for a left-handed orthonormal frame of vectors. (*Hint*: Change the overall sign of the phase of the wave.) ◆

5.6 Paravectors and EM Fields

Consider a paravector $p = p_0 + \mathbf{p}$, a complex vector of an EM field $\mathbf{F} = \mathbf{E} + j\mathbf{B}$, and an element $\varepsilon = \mathbf{F}p\mathbf{F}^\dagger$. From $(\mathbf{F}p\mathbf{F}^\dagger)^\dagger = \mathbf{F}p\mathbf{F}^\dagger$, we see that ε is a paravector, too. Defining the *duality rotation* (see R12) as $\mathbf{F} \to \exp(-j\varphi)\mathbf{F}$, we see that ε is an invariant (due to the commutativity of a complex scalar). This means that grades of ε are also invariant. Let us find them.

From $\mathbf{F}p\mathbf{F}^\dagger = \mathbf{F}p_0\mathbf{F}^\dagger + \mathbf{F}\mathbf{p}\mathbf{F}^\dagger$, we have (the cross product is defined in *Cl*3)

$$\mathbf{F}p_0\mathbf{F}^\dagger = p_0\mathbf{F}\mathbf{F}^\dagger = p_0(\mathbf{E} + j\mathbf{B})(\mathbf{E} - j\mathbf{B}) = p_0\left(E^2 + B^2 + 2\mathbf{E} \times \mathbf{B}\right),$$

where we have a known energy density and the *Poynting vector*. Note that in vacuum, denoting the total energy by ρ and $\mathrm{E} = |\mathbf{E}| = |\mathbf{B}|$, we can write ($\hat{\mathbf{k}}$ is the wave vector)

$$\mathbf{F}p_0\mathbf{F}^\dagger = p_0\rho\left(1 + \frac{2\mathrm{E}^2}{\rho}\hat{\mathbf{E}}\hat{\mathbf{B}}\right) = p_0\rho\left(1 + \hat{\mathbf{E}}\hat{\mathbf{B}}\right) = p_0\rho\left(1 - \hat{\mathbf{k}}\right).$$

For the vector part, we have

$$\mathbf{F}\mathbf{p}\mathbf{F}^\dagger = (\mathbf{E} + j\mathbf{B})\mathbf{p}(\mathbf{E} - j\mathbf{B}) = \mathbf{E}\mathbf{p}\mathbf{E} + \mathbf{B}\mathbf{p}\mathbf{B} + j(\mathbf{B}\mathbf{p}\mathbf{E} - \mathbf{E}\mathbf{p}\mathbf{B}).$$

The real part $\mathbf{EpE} + \mathbf{BpB}$ is just a vector (a kind of reflection of $\mathbf{p}$ on $\mathbf{E}$ and $\mathbf{B}$). We can decompose the vector $\mathbf{p}$ into the component perpendicular to the plane spanned by the vectors of the fields ($\mathbf{p}_\perp$) and the component that belongs to that plane ($\mathbf{p}_\parallel$). ◆ Find explicit formulas for $\mathbf{p}_\perp$ and $\mathbf{p}_\parallel$, using the bivector $\mathbf{E} \wedge \mathbf{B}$. ◆ Then we immediately have the vector

$$\begin{aligned}\mathbf{EpE} + \mathbf{BpB} &= \mathbf{Ep}_\parallel\mathbf{E} + \mathbf{Bp}_\parallel\mathbf{B} + \mathbf{Ep}_\perp\mathbf{E} + \mathbf{Bp}_\perp\mathbf{B}\\ &= \mathbf{Ep}_\parallel\mathbf{E} + \mathbf{Bp}_\parallel\mathbf{B} - \mathbf{p}_\perp\left(E^2 + B^2\right),\end{aligned}$$

connected to the *Maxwell stress tensor*. We can write the term $\mathbf{Ep}_\parallel\mathbf{E} + \mathbf{Bp}_\parallel\mathbf{B}$ in the form

$$E^2\hat{\mathbf{E}}\mathbf{p}_\parallel\hat{\mathbf{E}} + B^2\hat{\mathbf{B}}\mathbf{p}_\parallel\hat{\mathbf{B}}$$

and see that its parts are just scaled reflections of the vector $\mathbf{p}_\parallel$ across the field vectors (Fig. 5.5). The scaling factors are the energy densities of the fields. The geometric meaning of the word *stress* is now quite clear: both fields contribute to the scaling and reflection of $\mathbf{p}_\parallel$, and we just need the parallelogram rule to add the components (here we have omitted to do so). The component $\mathbf{p}_\perp$ is just scaled by the overall energy density.

The complex part is, in fact, just a real number. First, note that $\mathbf{EpB}$ is the reverse of $\mathbf{BpE}$, and $\mathbf{BpE} - \mathbf{EpB}$ is imaginary (grades 2 and 3). Second, only the perpendicular component of $\mathbf{p}$ gives a value different from zero (just write $\mathbf{p}_\parallel = \alpha\mathbf{E} + \beta\mathbf{B}$). Thus, we have

$$\begin{aligned}&j(\mathbf{BpE} - \mathbf{EpB}) = j(\mathbf{Bp}_\perp\mathbf{E} - \mathbf{Ep}_\perp\mathbf{B}) = j\mathbf{p}_\perp(\mathbf{EB} - \mathbf{BE})\\ &= 2j\mathbf{p}_\perp(\mathbf{E} \wedge \mathbf{B}) = 2\mathbf{p}_\perp(j\mathbf{E} \wedge \mathbf{B}),\end{aligned}$$

which is a real number, since $\mathbf{p}_\perp$ and $j\mathbf{E} \wedge \mathbf{B}$ are parallel.

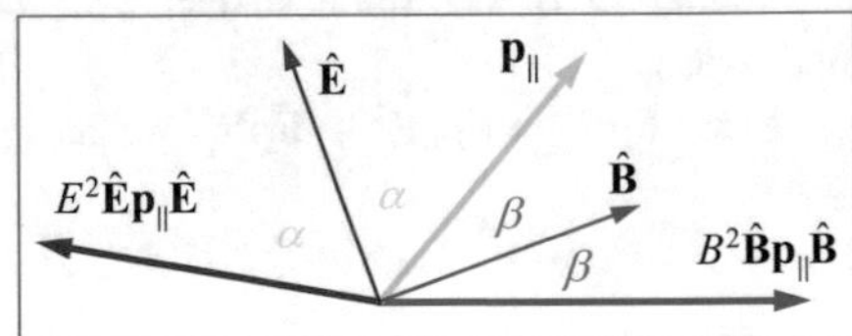

Fig. 5.5 Paravectors and EM field

Note again, we have not introduced coordinates, and this is a great advance of geometric algebra. Actually, the main tool in the calculation was the following simple rule: parallel vectors commute, while orthogonal vectors anticommute. With such a rule, calculation becomes almost child's play. In addition to the ease of calculation, experts in electromagnetism should also appreciate the laconic obviousness of the geometric interpretation.

5.7 Eigenvalues and Eigenelements

In linear algebra, we use the concept of eigenvalues and eigenvectors (*eigensystem*; see Sect. 1.17.4), while in geometric algebra we can introduce the concept of *eigenelements* (not necessarily blades). Generally, we can try to solve the equation

$$MX = \alpha X, \ \ \alpha \in \mathbb{C},$$

where M and X are multivectors from $Cl3$. The complex scalar α is then an eigenvalue, while the solution X is an eigenelement. From the general form $M = Z + \mathbf{F} = Z + Z_f \mathbf{f}$, $\mathbf{f}^2 = 1$, we have

$$M + \bar{M} = 2Z, \quad M\bar{M} = Z^2 - \mathbf{F}^2 \in \mathbb{C}.$$

Now we can write

$$M(M + \bar{M}) = M^2 + M\bar{M} = 2ZM,$$

where M^2 is expressed by M and complex scalars, which can be useful as an iteration formula for higher powers of M. Now we can write

$$\left(M^2 - 2ZM + M\bar{M}\right)X = 0 = \left(\alpha^2 - 2Z\alpha + M\bar{M}\right)X,$$

which gives solutions (see Sect. 4.2.2)

$$\alpha_{\pm} = Z \pm \sqrt{Z^2 - M\bar{M}} = Z \pm \sqrt{\mathbf{F}^2} = Z \pm Z_f = M_{\pm},$$

where we have introduced notation from Sect. 2.4.2. Thus, $M_{\pm}$ are our eigenvalues, and we have to find the eigenelements.

For $\mathbf{F} = 0$, any multivector is an eigenelement with the eigenvalue Z.

If the multivector M is not of the form $Z + \mathbf{N}$, $\mathbf{N}^2 = 0$, we can write

$$\left(Z + Z_f \mathbf{f}\right)X = \left(Z \pm Z_f\right)X, \quad \mathbf{f}^2 = 1,$$

where we see that the "problematic" part is $\mathbf{f}X = \pm X$. However, recalling that $u_{\pm} = (1 \pm \mathbf{f})/2$ are idempotents, we have $\mathbf{f}u_{\pm} = \pm u_{\pm}$, which means that every element from $Cl3$ of the form $X = u_{\pm}Y$ is an eigenelement, where for $X = u_{\pm}$ we say that X is *eigenpotent* (X is idempotent). Explicitly, we have

$$\left(Z + Z_f \mathbf{f}\right)u_{\pm}Y = \left(Z \pm Z_f\right)u_{\pm}Y.$$

In particular, for $M\bar{M} = 0$ (*null multivectors*), we have $(Z + Z_f)(Z - Z_f) = 0$ (see E29) and eigenvalues 0 and $2Z$. Note that such a multivector is proportional to an idempotent, and we can write

$$M = Z + \mathbf{F} = Z \pm Z\mathbf{f} = 2Z(1 \pm \mathbf{f})/2 = 2Zu_{\pm} \Rightarrow MX = 2Zu_r u_s Y, \quad r, s \in \{+, -\},$$

which for $r = s$ gives the eigenvalue $2Z$ and zero otherwise ($u_+u_-Y = u_+u_-u_-Y$, $u_+u_- = 0$).

Another interesting case is for $\mathbf{F}^2 = 0$, meaning that $\mathbf{F} = \mathbf{N}$ is a nilpotent and we have $M\mathbf{N} = (Z + \mathbf{N})\mathbf{N} = Z\mathbf{N}$ or, using the nilpotent direction vector $\hat{k}$, $\left(1 + \hat{k}\right)\mathbf{N} = 2\mathbf{N}$, and defining the projector $P = \left(1 + \hat{k}\right)/2$, we have

$$(Z + \mathbf{N})P\mathbf{N} = ZP\mathbf{N} + \mathbf{N}P\mathbf{N} = ZP\mathbf{N} + \mathbf{N}^2 = ZP\mathbf{N} = Z\mathbf{N}.$$

Thus, we have a unique eigenvalue Z (including $Z = 0$) and an eigenelement $\mathbf{N}$.

In the coordinate form, the geometry of Euclidean 3D space could be enriched by various tensors; however, the simplicity of analysis in the GA language would be unattainable.

5.7.1 Eigensystem of Elements from the Clifford Basis

It is interesting to consider elements from the Clifford basis. For $M = e_k$, we have $u_{\pm} = (1 \pm e_k)/2$ and eigenvalues $\pm Z_f = \pm 1$, which sum to zero. Recall that the sum of eigenvalues of matrices is just their trace. Therefore, we can say that the "trace" of e_k is zero. This is in accordance with the property of the Pauli matrices $\mathrm{tr}\hat{\sigma}_k = 0$. For elements of the form $e_k e_l$, $k \neq l$, we have (check) $\pm Z_f = \pm j$, and the trace is zero. For the Pauli matrices we have $\operatorname{tr}\hat{\sigma}_k\hat{\sigma}_l = 2\delta_{kl}$. For elements of the form $e_k e_l e_m$, we see that they are proportional to j for all distinct indices; otherwise, they are vectors. Since we have a unique eigenvalue j or $-j$, we see that the trace is zero or $\pm 2j$, in accordance with $\operatorname{tr}\hat{\sigma}_k\hat{\sigma}_l\hat{\sigma}_m = 2i\varepsilon_{klm}$, where ε_{klm} is the Levi-Civita symbol. We can consider more complicated products of vectors, like $e_k e_l e_m e_n$; however, we leave that as an exercise. ◆ Find eigenelements for e_k, $e_k e_l$, and $e_k e_l e_m$. ◆

In Sect. 4.1, we discussed some properties of the Pauli matrices; here we can briefly discuss some properties that can help to relate *Cl*3 and matrix formulations (in quantum mechanics, special relativity theory, etc.). First, consider the symmetric and antisymmetric parts of products of the Pauli matrices. We define them as (find their *eigensystem* in *Cl*3)

$$\{\hat{\sigma}_k, \hat{\sigma}_l\} = (\hat{\sigma}_k\hat{\sigma}_l + \hat{\sigma}_l\hat{\sigma}_k)/2 \quad \text{(anticommutator)},$$
$$[\hat{\sigma}_k, \hat{\sigma}_l] = (\hat{\sigma}_k\hat{\sigma}_l - \hat{\sigma}_l\hat{\sigma}_k)/2 \quad \text{(commutator)}.$$

We know that they are related to the inner and outer products, and they are important in matrix formulations of problems. It is not hard to prove the relations

$$\{\hat{\sigma}_k, \hat{\sigma}_l\} = iI\delta_{kl} \quad (I \text{ is the unit } 2\times 2 \text{ matrix}),$$
$$[\hat{\sigma}_k, \hat{\sigma}_l] = i\varepsilon_{klm}\,\hat{\sigma}_m \quad (\text{summation over } m).$$

Now you can see where the i comes from in quantum relations like $[A_x, A_y] = i\hbar A_z$. For the matrix product, we have (write it down in *Cl*3)

$$\hat{\sigma}_k\hat{\sigma}_l = I\,\delta_{kl} + i\varepsilon_{klm}\,\hat{\sigma}_m,$$

which may help to derive various relations. It is common to introduce the formal product of a vector and the Pauli matrices "vector" $\hat{\boldsymbol{\sigma}} \equiv (\hat{\sigma}_1, \hat{\sigma}_2, \hat{\sigma}_3)$ as $\mathbf{a}\cdot\hat{\boldsymbol{\sigma}} \equiv \sum_i \mathrm{a}_i\hat{\sigma}_i$, which in geometric algebra is just the vector $\mathbf{a}$. Here we will give a few formulas, without proof:

$$(\mathbf{a}\cdot\hat{\boldsymbol{\sigma}})(\mathbf{b}\cdot\hat{\boldsymbol{\sigma}}) = \mathbf{a}\cdot\mathbf{b}I + i(\mathbf{a}\times\mathbf{b})\cdot\hat{\boldsymbol{\sigma}},$$
$$\exp(i\theta\hat{\mathbf{n}}\cdot\hat{\boldsymbol{\sigma}}) = I\cos\theta + i\hat{\mathbf{n}}\cdot\hat{\boldsymbol{\sigma}}\sin\theta, \quad \hat{\mathbf{n}}^2 = 1,$$
$$f(x\hat{\mathbf{n}}\cdot\hat{\boldsymbol{\sigma}}) = I\frac{f(x)+f(-x)}{2} + \hat{\mathbf{n}}\cdot\hat{\boldsymbol{\sigma}}\frac{f(x)-f(-x)}{2},$$
$$f(ix\hat{\mathbf{n}}\cdot\hat{\boldsymbol{\sigma}}) = I\frac{f(ix)+f(-ix)}{2} + \hat{\mathbf{n}}\cdot\hat{\boldsymbol{\sigma}}\frac{f(ix)-f(-ix)}{2}.$$

In this form, the formulas are easy to convert to the geometric algebra language and vice versa, but the coordinate approach loses clarity in the expanded form. In addition, formulas in the language of the Pauli matrices are valid in 3D Euclidean vector space only, while *Cl*3 formulas are easy to generalize. Even in *Cl*3, matrices cannot do everything vectors can: just recall rotations or the fact that products of two 2×2 matrices give a 2×2 matrix with a hidden content, while products of vectors give the structure with subspaces and the obvious interpretation. Geometric algebra offers simplicity, generalizations, possibilities for unifications, and we can invoke our powerful geometric intuition. The coordinate approach is like reading a novel in an encrypted language; one must understand the language (in fact, many of them) and try to imagine the content, which often brings frustration. Geometric algebra needs just one language and offers images.

5.8 Duality

Every blade of grade r can be expressed as $A = \|A\|\hat{A}$, where the unit blade $\hat{A}$ can be factored as the geometric product of r orthonormal vectors (say e_i). These unit vectors span an r-dimensional vector subspace (say A), and we can find $n - r$ orthonormal unit vectors (say $f_i, f_i \perp e_j$) that define a new subspace (say A^Δ), which is the complement to A. Now we have n orthonormal vectors that define the unit pseudoscalar $I = \pm e_1 \cdots e_r f_1 \cdots f_{n-r}$ of the whole vector space. The relationship between the subspaces A and A^Δ can be expressed as

$$A^\Delta \equiv A \rfloor I^{-1},$$

which is in accordance with the interpretation of left contraction (see Sect. 1.11), that is, the subspace A^Δ is contained in the space I and is orthogonal to the subspace A. We say that the element A^Δ is the dual to the element A (an orthogonal complement). ◆ Show that $A \wedge A^\Delta \propto I$. ◆ In fact, we can define the duality using any unit blade instead of I, such as $B_r^\Delta = B_r \rfloor A_s{}^{-1}$. However, here we will stay with pseudoscalars. For the vector a, we have

$$a^\Delta = a \rfloor I^{-1} = \left(aI^{-1} - \overleftarrow{I^{-1}} a \right)/2 = aI^{-1},$$

which is true due to the grade involution and commutativity properties of I (see E10). For blades A and B, we have the important relations

$$(A \wedge B)^\Delta = (A \wedge B) \rfloor I^{-1} = A \rfloor \left(B \rfloor I^{-1} \right) = A \rfloor B^\Delta,$$
$$(A \rfloor B)^\Delta = (A \rfloor B) \rfloor I^{-1} = A \wedge \left(B \rfloor I^{-1} \right) = A \wedge B^\Delta,$$

where the second relation is generally valid for $A \subseteq I$, which is true here. Note the symmetry between the products. These formulas are very useful, especially for various simplifications of expressions. We suggest that reader play with these formulas in $Cl3$, in order to get some insight into their geometric content. For example, for $A = e_1$, $B = e_2e_3$, and $C = e_3e_1$, we have

$$(A \wedge B)^\Delta = (e_1 \wedge e_2 \wedge e_3)^\Delta = j^\Delta = e_1 \rfloor (e_2e_3)^\Delta = 1,$$
$$(A \rfloor C)^\Delta = (e_1 \rfloor (e_3e_1))^\Delta = (-e_3)^\Delta = e_1 \wedge (e_3e_1)^\Delta = e_1e_2.$$

◆ Prove that for vectors a, b, and c, the following relations are valid ($\times$ means the cross product):

$$a \cdot b \times c = (a \wedge b \wedge c)^{\Delta},$$
$$a \times (b \times c) = -a \rfloor (b \wedge c). \blacklozenge$$

From the fact that the dual of a blade is a blade follows an interesting consequence, namely the following result.

Theorem 5 *If a vector space is n-dimensional,* the *every* $(n-1)$*-vector is a blade.*

The proof follows from the fact that a vector is 1-blade, which means that its dual (an $(n-1)$-vector) is a blade. We know that the number of unit $(n-1)$-blades in the Clifford basis is equal to n, which means that the dual of any linear combination of unit $(n-1)$-blades gives a vector, and vice versa.

◆ Show that unit $(n-1)$-blades of the Clifford basis (up to sign) can be obtained by the simple trick

$$(e_1 + e_2) \wedge (e_2 + e_3) \wedge \ldots \wedge (e_{n-1} + e_n). \blacklozenge$$

Now we can write the general $(n-1)$-blade as

$$(\alpha_1 e_1 + \alpha_2 e_2) \wedge (\alpha_2 e_2 + \alpha_3 e_3) \wedge \cdots \wedge (\alpha_{n-1} e_{n-1} + \alpha_n e_n) = \sum_{k=1}^{n} b_k B_k,$$

where $B_k = e_1 \cdots e_{k-1} e_{k+1} \cdots e_n$ (note that e_k is missing). To find coefficients $\alpha_i \in \mathbb{R}$, we have to solve a system of n equations; however, we can try to guess the solution, just observing indices. ◆ Show that

$$a_k = \frac{(b_1 \cdots b_{k-1} b_{k+1} \cdots b_n)^{1/(n-1)}}{b_k^{(n-2)/(n-1)}}$$

is a solution. What if $b_k = 0$? Check this solution for *Cl*3 and compare it to the solution from E08.◆

5.9 Permutations of Orthonormal Unit Vectors in *Cl*3

We have two permutations of unit vectors in *Cl*2:

$$(1,2), \quad (2,1),$$

with *signatures* 1 and -1. The transformation matrix of the permutation $(2,1)$ is easy to write down: just use the unit vectors e_2 and e_1 as columns to get

$$Ue_1 = \begin{pmatrix} 0 & 1 \\ 1 & 0 \end{pmatrix} \begin{pmatrix} 1 \\ 0 \end{pmatrix} = \begin{pmatrix} 0 \\ 1 \end{pmatrix} = e_2,$$

with $\det U = -1$. We have an example of an orthogonal transformation, which suggests that we could accomplish such a transformation using reflections in *Cl*2. First, note that

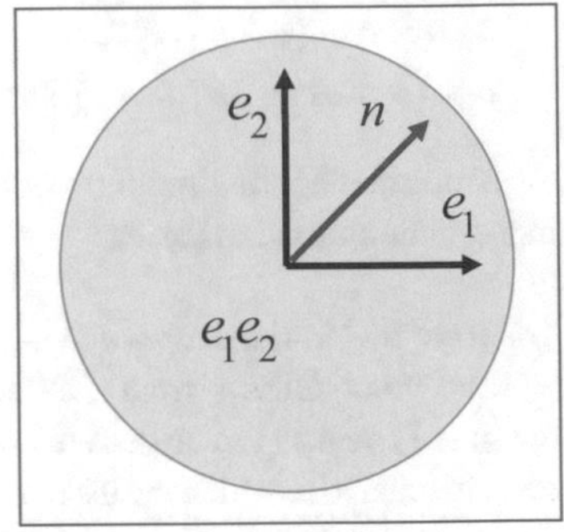

Fig. 5.6 A permutation with negative signature as a reflection

the eigenvectors of U are $(-1, 1)$ and $(1, 1)$, with the real eigenvalues -1 and 1. Defining the unit vector $n = (e_1 + e_2)/\sqrt{2}$, we have

$$ne_1n = e_2, \quad ne_2n = e_1.$$

This is a simple geometric result, and we can achieve a permutation with negative signature using a reflection. The result is obvious from Fig. 5.6.

Now we can try to find similar geometric interpretations for three orthonormal unit vectors in $Cl3$. The permutations of indices with the signature 1 are

$$(1, 2, 3), \quad (2, 3, 1), \quad \text{and} \quad (3, 1, 2),$$

while the permutations of indices with the signature -1 are

$$(1, 3, 2), \quad (2, 1, 3), \quad \text{and} \quad (3, 2, 1).$$

Note that the even permutations $(2, 3, 1)$ and $(3, 1, 2)$ are cyclic permutations, while the odd permutations $(1, 3, 2)$ and $(2, 1, 3)$ have a fixed position. The odd permutation $(3, 2, 1)$ suggests the reverse operation on the pseudoscalar

$$e_1e_2e_3 \to e_3e_2e_1.$$

◆ Check that every permutation applied to the pseudoscalar j gives the signature of the permutation. ◆

As an example of a permutation with negative signature we choose $(1, 3, 2)$, with the transformation matrix (note the columns)

$$U_- = \begin{pmatrix} 1 & 0 & 0 \\ 0 & 0 & 1 \\ 0 & 1 & 0 \end{pmatrix}, \quad \det U_- = -1.$$

The eigenvectors of U_- are

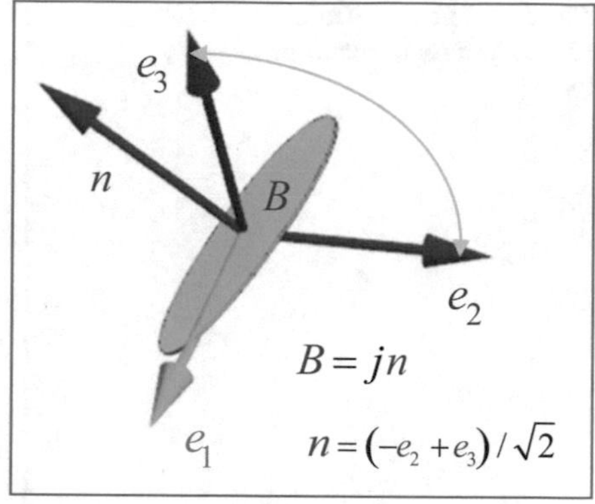

Fig. 5.7 The permutation (1,3,2) as a reflection

$$(0, -1, 1), \quad (0, 1, 1), \quad \text{and} \quad (1, 0, 0),$$

with the (real) eigenvalues -1, 1, and 1. We know that in *Cl*3 we can use planes defined by bivectors (B) and thus reflect vectors (v), like BvB. In the case $(1, 3, 2)$, the vector e_1 is invariant under such a reflection, which means that it belongs to the plane, which must be symmetric with respect to e_2 and e_3. Choosing the eigenvector $n = (-e_2 + e_3)/\sqrt{2}$ and the bivector $B = jn$, we get our permutation (check) as the reflections Be_iB (Fig. 5.7). The geometric meaning is now clear: the vector e_1 is not affected by reflection, while the vectors e_2 and e_3 reflect on B as in a mirror. This is a nice example of a linear transformation expressed using elements of geometric algebra.

For our example of an even permutation, we choose the cyclic permutation $(2, 3, 1)$, with the transformation matrix

$$U_+ = \begin{pmatrix} 0 & 0 & 1 \\ 1 & 0 & 0 \\ 0 & 1 & 0 \end{pmatrix}, \quad \det U_+ = 1,$$

which suggests a rotation as an orthogonal transformation. First, note that the vector

$$n = (e_1 + e_2 + e_3)/\sqrt{3}$$

is not affected by the permutation, and it is easy to check that it is an eigenvector with the eigenvalue 1. Other eigenvectors and eigenvalues are complex, which suggests that such a linear transformation in geometric algebra should have eigenblades with real eigenvalues. Such an eigenblade is obviously the pseudoscalar j; however, there must be an eigenblade of grade two (a bivector). It is clear that our transformation is a rotation with rotation axis n, and there remains the question of the rotation angle. One of the complex eigenvalues is $1/2 + i\sqrt{3}/2$, which suggests the angle $\pm\pi/3$. It appears that the minus sign gives our transformation, while the plus sign gives the permutation $(3, 1, 2)$. Finally, our rotor is

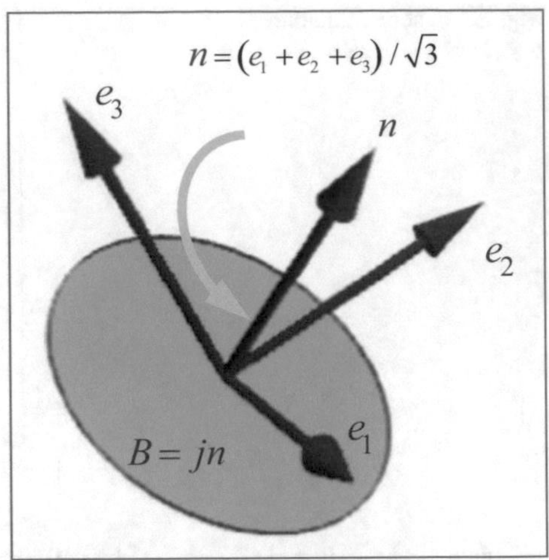

Fig. 5.8 The permutation (2,3,1) as a set of rotations by π/3

$$R = \exp\left(-\frac{\pi}{3}jn\right), \quad RR^{\dagger} = 1,$$

and our permutation appears as a set of rotations $Re_iR^{\dagger}$ (Fig. 5.8). Such transformations may play an important role in quantum mechanics, where the exponent can have a physical interpretation. Note that the bivector *jn* is the eigenblade of the rotor, and it is directly connected to the complex eigenvalues of U_+.

In the chapter on quantum mechanics, we discussed transformations that include the Euler angles, which are much more complicated. The permutations discussed here are a simple and illustrative example. Important to note is that in geometric algebra, transformations can possess a clear geometric content and that we can represent them by elements of the algebra. We suggest that the reader investigate these permutations using the technique explained in the chapter on linear transformations.

Miscellaneous

Miroslav Josipović

M. Josipović, *Geometric Multiplication of Vectors*,
Compact Textbooks in Mathematics,
https://doi.org/10.1007/978-3-030-01756-9_6

6.1 Solutions to Selected Problems

E01

Checking directly, we get $\hat{\sigma}_i^\dagger = \hat{\sigma}_i$. Similarly, it is easy to obtain by a direct multiplication

$$(\hat{\sigma}_2\hat{\sigma}_3)^\dagger = -\hat{\sigma}_2\hat{\sigma}_3 = \hat{\sigma}_3\hat{\sigma}_2 = \begin{pmatrix} 0 & -i \\ -i & 0 \end{pmatrix}.$$

E02

Note that the product of two vectors generally has grades 0 and 2. Multiplying by e_1, from the real scalar part (grade 0) we get a vector. Bivectors from the Clifford basis could be of the form e_1e_k, and they will give vectors, or of the form e_je_k, $j, k \neq 1$, and they will give 3-vectors (grade 3).

E03

$$e_1(e_1e_2) = e_1e_1e_2 = e_2, \quad e_2(e_1e_2) = -e_1,$$

Note that the vector and the bivector anticommute, and this is always true for vectors coplanar to a blade of grade 2. For a 2-blade we can write $a \wedge b = a \wedge (b_\parallel + b_\perp) = a \wedge b_\perp = ab_\perp$, then from $c = \alpha a + \beta b_\perp$, it follows that $aab_\perp = -ab_\perp a$, $b_\perp ab_\perp = -ab_\perp b_\perp$.

E04

$$(e_1 + e_2) \wedge e_2 = e_1 \wedge e_2 + e_2 \wedge e_2 = e_1 \wedge e_2, \quad |e_1 + e_2| = \sqrt{2}, \quad \sin\frac{\pi}{4} = 1/\sqrt{2}.$$

The original version of this chapter was revised. A correction to this chapter can be found at https://doi.org/10.1007/978-3-030-01756-9_7.

E05

$$0 = (\mathbf{a}+\mathbf{b}+\mathbf{c}) \wedge \mathbf{a} = \mathbf{b} \wedge \mathbf{a} + \mathbf{c} \wedge \mathbf{a} \Rightarrow \mathbf{c} \wedge \mathbf{a} = -\mathbf{b} \wedge \mathbf{a} = \mathbf{a} \wedge \mathbf{b}, \quad \text{etc.}$$

E06

From $B = (e_1 + e_2) \wedge e_3 = e_1 \wedge e_3 + e_2 \wedge e_3$ and $B \wedge x = 0$, we have

$$\begin{aligned}
&(e_1 \wedge e_3 + e_2 \wedge e_3) \wedge (a_1 e_1 + a_2 e_2 + a_3 e_3)\\
&= a_2 e_1 \wedge e_3 \wedge e_2 + a_1 e_2 \wedge e_3 \wedge e_1 = -a_2 j + a_1 j = 0 \Rightarrow\\
&a_1 = a_2 \Rightarrow x = a_1(e_1 + e_2) + a_3 e_3;
\end{aligned}$$

however, we could write this directly.

E07

$$(xe_1 + ye_2)e_1 = x + ye_2e_1 = x - yI, \;\; I = e_1e_2;$$

therefore, we have a complex conjugation. The anticommutativity is useful.

E08

If $B_2 = 0$, the result is straightforward. If $B_2 \neq 0$, we have

$$\begin{aligned}
&B_3 e_1 \wedge e_2 - B_2 e_1 \wedge e_3 + B_1 e_2 \wedge e_3\\
&= -B_2 e_1 \wedge \left(e_3 + \frac{B_3}{-B_2} e_2 \right) + B_1 e_2 \wedge \left(e_3 + \frac{B_3}{-B_2} e_2 \right)\\
&= (-B_2 e_1 + B_1 e_2) \wedge \left(\frac{B_3}{-B_2} e_2 + e_3 \right).
\end{aligned}$$

The reader can find other solutions. Note that we can find a vector orthogonal to our bivector:

$$B_3 e_1 \wedge e_2 - B_2 e_1 \wedge e_3 + B_1 e_2 \wedge e_3 = B_1 j e_1 + B_2 j e_2 + B_3 j e_3 = j(B_1 e_1 + B_2 e_2 + B_3 e_3).$$

The reader can check this by calculating the cross product:

$$(-B_2 e_1 + B_1 e_2) \times \left(\frac{B_3}{-B_2} e_2 + e_3 \right).$$

E09

For vectors a and b, we have $B = a \wedge b = a \wedge (b_{\|} + b_{\perp}) = a \wedge b_{\perp} = ab_{\perp}$, meaning that

$$BB^{\dagger} = a^2 b_{\perp}^2 \overset{?}{=} (e_1e_2 + e_3e_4)(e_1e_2 + e_3e_4)^{\dagger} = 2(1 - I),$$

where $I = e_1e_2e_3e_4$, but then we have expressions of different grades.

E10

$$j e_1 = e_1 e_2 e_3 e_1 = -e_1 e_2 e_1 e_3 = e_1 e_1 e_2 e_3 = e_1 j, \quad \text{etc.}$$
$$j^2 = e_1 e_2 e_3 e_1 e_2 e_3 = e_2 e_3 e_2 e_3 = -e_2 e_2 e_3 e_3 = -1.$$

Pseudoscalars have maximum grade; consequently, we can write them as the outer product of n linearly independent vectors, where n is the dimension of the vector space. In such products, we have combinations of indices $(1, 2, \ldots, n)$; however, due to the antisymmetry of the outer product, all terms with two equal indices disappear $(e_1 \wedge e_2 \wedge e_1 = -e_1 \wedge e_1 \wedge e_2 = 0)$.

Generally, we can find conditions under which a unit pseudoscalar commutes with all vectors (and consequently, with all elements of the Clifford basis). It is clear that the unit vector, say e_1, must make n replacements with vectors from the pseudoscalar, so the factor that defines commutativity is $(-1)^{n-1}$ (the vector e_1 commutes with itself), meaning that pseudoscalars of odd dimensions commute with all elements of the Clifford basis. What about the square of unit pseudoscalars? In 4D, we have for indices

$$1,2,3,4 \parallel 1,2,3,4 \xrightarrow{3} 1,1,2,3,4 \parallel 2,3,4 \xrightarrow{2} 2,2,3,4 \parallel 3,4 \xrightarrow{1} 3,3,4 \parallel 4,$$

which gives the general number of replacements $n(n-1)/2$. Now we have the dimensions for the square -1 (in Euclidean vector spaces),

$$2, 3, 6, 7, 10, 11, 14, 15, 18, 19, 22,$$

and the dimensions for the square 1,

$$4, 5, 8, 9, 12, 13, 16, 17, 20, 21, 24.$$

In pseudo-Euclidean vector spaces $\mathfrak{R}^{(p,q)}$, we have to take into account the factor $(-1)^q$. For the commutativity of the pseudoscalar, we need n to be odd, while for the square of the pseudoscalar to be -1 (in Euclidean vector spaces), we need $n(n-1)/2$ to be odd, which means that $n(n-1)/2$ must be odd, from which it follows that

$$(n-1)/2 = 2k + 1 \Rightarrow n = 4k + 3.$$

Consequently, we have periodicity 4, giving dimensions 3, 7, 11, … with pseudoscalars similar to j. For $\mathfrak{R}^{(1,3)}$, the number of replacements is 6, and therefore, the factor $(-1)^q = (-1)^3$ gives negative squares of pseudoscalars.

E11

Obviously, $e_k \wedge e_l \wedge e_m = 0$ for repeated indices. Without repeated indices, we can replace the outer product by the geometric product, and then from $-j = -e_1 e_2 e_3 = e_3 e_2 e_1 \rightarrow 321$, we can write

$$321123 \rightarrow 3223 \rightarrow 33 \Rightarrow 1$$
$$321213 \rightarrow -321123 \Rightarrow -1, \quad \text{etc.}$$

However, note that it is enough to investigate the permutations of indices (1, 2, 3).

E12

The first statement follows from the fact that the outer product of $n + 1$ vectors is zero, where n is the dimension of the vector space (see the text). The second statement follows from the fact that grade k means that we have to select k different indices from n indices of unit vectors, which implies a binomial coefficient.

E13

We have to prove that

$$\sum_{k=0}^{n} \binom{n}{k} = 2^n.$$

From

$$\binom{n}{k} = \frac{n!}{(n-k)!k!},$$

we have

$$\sum_{k=0}^{2} \binom{2}{k} = \frac{2!}{(2-0)!0!} + \frac{2!}{(2-1)!1!} + \frac{2!}{(2-2)!2!} = 1 + 2 + 1 = 2^2.$$

Therefore, we have a starting point for the *induction method*. From the identity (easy to check)

$$\binom{n+1}{k} = \binom{n}{k} + \binom{n}{k-1},$$

we have

$$\sum_{k=0}^{n+1} \binom{n+1}{k} = \sum_{k=0}^{n} \binom{n}{k} + \sum_{k=1}^{n} \binom{n}{k-1} + \binom{n+1}{n+1} = 2^n + 2^n - 1 + 1 = 2 \cdot 2^n$$
$$= 2^{n+1}.$$

E14

Every multivector is a linear combination of elements of the Clifford basis. If we multiply two even basis elements, we have an even number of indices in the product. Every index that is common to both elements will disappear and lower the number of indices by two, giving an even number again. For example (the sign depends on the signature of the vector space),

$$(e_1e_2)(e_1e_3) = -e_1e_1e_2e_3 = -s(e_1)e_2e_3,$$

where $s(e_i)$ is the signature of e_i.

E15

$$nan = n(a_{\|} + a_{\perp})n = (a_{\|} - a_{\perp})nn = a_{\|} - a_{\perp} = 2a_{\|} - a.$$

This is a very important property, which ensures that the "sandwich" form of transformations with versors preserve the grade (vectors transform to vectors). We can follow what happens to a vector in such a transformation ($\mathrm{a}_n = a \cdot n \in \mathbb{R}$ is the component of the vector a parallel to the vector n).

$$\begin{aligned}
\textbf{first reflection}: \quad & a' = -nan = a - 2\mathrm{a}_n n, \\
\textbf{second reflection}: \quad & a'' = -m(a - 2\mathrm{a}_n n)m = -mam + 2\mathrm{a}_n mnm \\
& = a - 2\mathrm{a}_m m + 2\mathrm{a}_n(2\mathrm{n}_m m - n) \\
& = a - 2\mathrm{a}_n n + 2m(2\mathrm{a}_n \mathrm{n}_m - \mathrm{a}_m) = a' + 2m(2\mathrm{a}_n \mathrm{n}_m - \mathrm{a}_m).
\end{aligned}$$

Now we can play with vectors. Let us choose $a = e_1 + e_2 + e_3$, $n = e_1$, $m = e_2$, with $\mathrm{a}_n = \mathrm{a}_m = 1$, and $\mathrm{n}_m = 0$, from which it follows that

$$a' = a - 2e_1 = -e_1 + e_2 + e_3, \qquad a'' = a' - 2e_2 = -e_1 - e_2 + e_3.$$

Note that the reflection of the vector a by e_1 gives the yellow vector (Fig. 6.1) as the mirror image of the vector a in the blue plane e_2e_3 ($\perp e_1$). The green vector is the mirror image of the yellow vector in the green plane e_1e_3 ($\perp e_2$).

Now comes an interesting question: can we find the unit vector u such that $-uau = a''$? It is clear that we can find the plane positioned symmetrically between a and a''. We have

$$\begin{aligned}
& -uau = -e_1 - e_2 + e_3 = a - 2(e_1 + e_2) \Rightarrow \\
& -ua = au - 2(e_1 + e_2)u \Rightarrow \\
& 2(e_1 + e_2)u = au + ua = 2a \cdot u \Rightarrow \\
& (e_1 + e_2) \cdot u = a \cdot u, \quad e_3 \cdot u = 0, \\
& (e_1 + e_2) \wedge u = 0, \quad u \propto (e_1 + e_2),
\end{aligned}$$

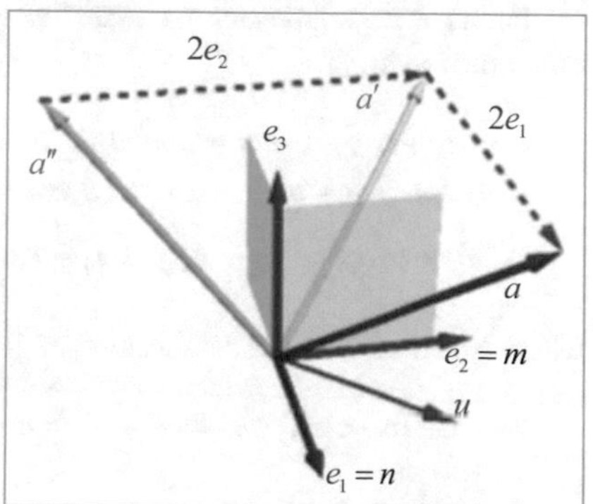

Fig. 6.1 A rotation as two reflections

Table 6.1 The effect of the reflection and the rotation on unit vectors

	e_1	e_2	e_3	j
r	$-e_2$	$-e_1$	e_3	$-j$
R	$-e_1$	$-e_2$	e_3	j

from which we conclude that $u = (e_1 + e_2)/\sqrt{2}$ is a possible solution. ◆ Check that $-uau = a''$. ◆

We can see easily that our two reflections are equivalent to the rotation $R = e_2e_1$, $R^\dagger = e_1e_2$,

$$a'' = -m(-nan)m = RaR^\dagger.$$

Consider the rotation $\mathrm{R}(a) = RaR^\dagger$ and the reflection $\mathrm{r}(a) = -uau$. The Table 6.1 shows the effect of both transformations on the basis unit vectors, from which we see that the two transformations are not equivalent. The main difference is that the rotation does not change the unit pseudoscalar, while the reflection changes its sign. In the chapter on linear transformations (Sect. 1.17), we will see that the determinant of rotations is 1, while reflections have determinant -1.

Now we can show that two reflections in the plane give a rotation by a double angle. Given two reflections $-mam$ and $-nan$, where the unit vector a belongs to the plane spanned by the unit vectors m and n, we have

$$mn = m \cdot n + m \wedge n = \gamma + mb, \quad m \cdot b = 0, \quad b = \hat{b}\sin\theta, \quad \gamma = m \cdot n = \cos\theta$$
$$a^2 = 1, \quad a = a_m + a_b, \quad a_m mb = -mba_m, \quad a_b mb = -mba_b, \quad amb = -mba$$

(a_m and a_b are vectors), from which it follows that

$$\begin{aligned} a' &= mnanm = (\gamma + mb)a(\gamma + bm), \\ a \cdot a' &= \langle a(\gamma + mb)a(\gamma + bm)\rangle = \langle(\gamma a + amb)(\gamma a + abm)\rangle \\ &= \langle \gamma^2 + \gamma bm + \gamma amba + ambabm\rangle = \langle \gamma^2 - 2\gamma mb - mbbm\rangle = \langle \gamma^2 - mbbm\rangle \\ &= \cos^2\theta - \sin^2\theta = \cos(2\theta). \end{aligned}$$

For a general unit vector a, we can find the component that belongs to the plane $m \wedge n$ and orthogonal to it:

$$\begin{aligned} &a = a_\parallel + a_\perp, \quad a_\parallel mb = -mba_\parallel, \quad a_\perp mb = mba_\perp, \\ &a' = mna_\parallel nm + mna_\perp nm = mna_\parallel nm + a_\perp, \\ &a \cdot a' = \cos\varphi = \left(a_\parallel + a_\perp\right) \cdot \left(a'_\parallel + a_\perp\right) = \cos(2\theta) + a_\perp^2, \end{aligned}$$

where (see Sect. 1.11 for the meaning of $\rfloor$, the *left contraction*)

$$b = n - (m \cdot n)m, \quad a_\perp = a - a_\parallel = a - (a\rfloor(mb))bm.$$

E16

We have the formulas

$$a \cdot B = (aB - Ba)/2, \quad a \wedge B = (aB + Ba)/2,$$

which give

$$e_1 \cdot (e_1 e_2) = (e_1 e_1 e_2 - e_1 e_2 e_1)/2 = e_2,$$
$$e_1 \wedge (e_1 e_2) = (e_1 e_1 e_2 + e_1 e_2 e_1)/2 = 0.$$

From the formula

$$e_1 \cdot (\mathbf{a}_1 \wedge \cdots \wedge \mathbf{a}_n) = \sum_{k=1}^{n} (-1)^{k+1} e_1 \cdot \mathbf{a}_k \left(\mathbf{a}_1 \wedge \cdots \wedge \breve{\mathbf{a}}_k \wedge \cdots \wedge \mathbf{a}_n\right),$$

it follows that

$$e_1 \cdot (\mathbf{a} \wedge \mathbf{b}) = e_1 \cdot \mathbf{ab} - e_1 \cdot \mathbf{ba},$$

where the inner product has priority.

Consider now the unit vector $\hat{a}$ and the unit bivector $\hat{B}$. The unit vectors define lines, while bivectors define planes; therefore, we can investigate how their geometric relationship (in 3D Euclidean space) manifests in the geometric product. We will start with the Clifford basis, where our objects have the form

$$\hat{a} = \sum_{k=1}^{3} \mathrm{a}_k e_k, \quad \hat{B} = j \sum_{k=1}^{3} \mathrm{B}_k e_k, \quad \mathrm{a}_k, \quad \mathrm{B}_k \in \mathbb{R}.$$

The product $\hat{a}\hat{B} = \hat{a} \cdot \hat{B} + \hat{a} \wedge \hat{B}$ contains the terms $j(e_k e_l \pm e_l e_k)/2$. Consequently, due to antisymmetry, we have $j(e_k e_l + e_l e_k)/2 = j$, $k = l$, and zero otherwise, whence follows

$$\hat{a} \wedge \hat{B} = j \sum_{k=1}^{3} \mathrm{a}_k \mathrm{B}_k.$$

To find the inner product $j(e_k e_l - e_l e_k)/2 = j e_k e_l$, $k \neq l$, we can use $m = \{1, 2, 3\} \backslash \{k, l\}$ to write

$$j e_k e_l = s e_m \Rightarrow s = j e_k e_l e_m = -\varepsilon_{klm} e_m,$$

where we take the advantage of the result from E11. Now we have

$$\hat{a} \cdot \hat{B} = -\sum_{k \neq l}^{3} \mathrm{a}_k \mathrm{B}_l e_m \varepsilon_{klm}.$$

The inner product is the vector orthogonal to the vector $\hat{a}$, which is easy to prove:

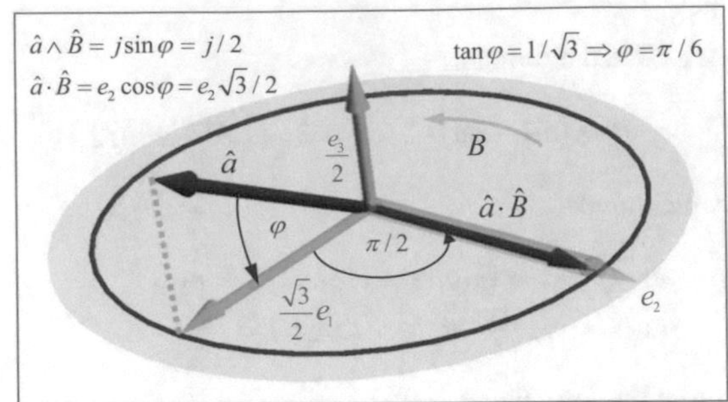

Fig. 6.2 The outer and inner products between a vector and a bivector

$$\left(\frac{aB - Ba}{2}a + a\frac{aB - Ba}{2}\right) = \frac{aBa}{2} - \frac{B}{2} + \frac{B}{2} - \frac{aBa}{2} = 0.$$

For $\hat{a} = \left(\sqrt{3}e_1 + e_3\right)/2$ and $\hat{B} = je_3 = e_1e_2$, we have (Fig. 6.2) $\hat{a}\cdot\hat{B} = e_2\sqrt{3}/2$ and $\hat{a}\wedge\hat{B} = j/2$. From Fig. 6.2, we see that the vector $\hat{a}$ is projected on the plane. Then the projection is rotated by $\pi/2$. If we set $e = \hat{a}\cdot\hat{B}/\left|\hat{a}\cdot\hat{B}\right|$, then there follows the general formula

$$\hat{a}\hat{B} = e\cos\varphi + j\sin\varphi,$$

where φ is the angle between the line and the plane.

Of course, instead of components, we can just write $a = a_{\parallel} + a_{\perp}$ (relative to the plane); then

$$aB = a_{\parallel}\cdot B + a_{\perp}\cdot B + a_{\parallel}\wedge B + a_{\perp}\wedge B = a_{\parallel}\cdot B + a_{\perp}\wedge B, \quad B = x\wedge y,$$

where $a_{\parallel}\cdot B$ is a vector in B perpendicular to $a_{\parallel}$, while $a_{\perp}\wedge B$ is a pseudoscalar (in $Cl3$, a 3-vector generally). The term $a_{\perp}\cdot B$ is zero, since the vector $a_{\perp}$ is perpendicular to all vectors from B, while the term $a_{\parallel}\wedge B$ is zero due to $a_{\parallel} = \alpha x + \beta y, \quad \alpha, \beta \in \mathbb{R}$.

E17

For a detailed treatment, see [12]. First note that in vector spaces with the signature $(p, q, 0)$ all versors are invertible:

$$V^{-1} = V^{\dagger}/\left(VV^{\dagger}\right), \quad VV^{\dagger} \in \mathbb{R},$$

for example,

$$(abc)^{-1} = cba/\left(a^2b^2c^2\right) \Rightarrow (abc)^{-1}(abc) = (abc)(abc)^{-1} = (cbaabc)/\left(a^2b^2c^2\right) = 1.$$

Let us take examples, just to be sure of the terminology here. The element e_1e_2 is a blade ($e_1e_2 = e_1\wedge e_2$), a versor, and a bivector (has the single grade 2). What about $e_1e_2 + e_1e_3$? It is a bivector (has the single grade 2), a versor ($e_1e_2 + e_1e_3 = e_1(e_2 + e_3)$), and a blade ($e_1e_2 + e_1e_3 = e_1\wedge(e_2 + e_3)$). The element $e_1e_2 + e_3e_4$ in 4D is a bivector (has the single grade 2), and it is not a blade (see E09), and therefore, it retains the possibility of being a versor. However, the versor $ab = a\cdot b + a\wedge b$ has grade 2, which gives $a\cdot b = 0$. Therefore, the product ab should be a blade $a \wedge b$, which is impossible. For three noncollinear vectors, we have grade 3, etc.

Let us illustrate this problem. For the versor $ab = a \cdot b + a \wedge b$, we have mixed grades, unless vectors are orthogonal (and we get a blade, bivector) or parallel (and we get a scalar, which is also a blade). However, with a third vector, we have $abc = a \cdot bc + a \wedge bc$ (recall that the geometric product executes last). The bivector $a \wedge b$ defines a plane, so we can write the vector $c = \alpha a + \beta b + c_\perp$ ($c_\perp$ is orthogonal to the plane $a \wedge b$), giving

$$\begin{aligned} &a \wedge b(\alpha a + \beta b + c_\perp) = a \wedge b(\alpha a + \beta b) + a \wedge bc_\perp = a \wedge b(\alpha a + \beta b) + a \wedge b \wedge c_\perp \\ &\quad = ab_\perp(\alpha a + \beta b_\perp + \beta b_\parallel) + a \wedge (b_\parallel + b_\perp) \wedge c_\perp = -\alpha a^2 b_\perp + \beta b_\perp^2 a - \beta a \cdot bb_\perp \\ &+ a \wedge b_\perp \wedge c_\perp \Rightarrow abc = \beta b_\perp^2 a - (\alpha a^2 + \beta a \cdot b) b_\perp + a \cdot bc + a \wedge b_\perp \wedge c_\perp, \end{aligned}$$

where $b_\perp$is perpendicular to a. Consequently, we have grades 1 and 3. For $c_\perp = 0$, we have a vector (blade), while for $c = c_\perp$ ($\alpha = \beta = 0$) and $a \cdot b = 0$, we have a trivector. We see that for orthogonal vectors we have a blade of maximum grade; however, if all vectors belong to some subspace (here it is the plane defined by $a \wedge b$), we can get a blade of grade less than the maximum one. If all vectors are parallel (1D subspace), we have a blade of zero grade (real scalar) for even numbers of vectors or grade 1 for odd numbers of vectors.

If we have a blade, for example $a \wedge b \wedge c$, we can write again

$$a \wedge b \wedge (\alpha a + \beta b + c_\perp) = a \wedge b \wedge c_\perp.$$

However, recall that the shape of a bivector is not important; we have $a \wedge b = a \wedge b_\perp$, and finally

$$a \wedge b \wedge c = a \wedge b_\perp \wedge c_\perp.$$

Note the nice property of k-vectors: $a \wedge b = a \wedge b_\perp$, where $b_\perp$is the rejection to a (1D subspace), from which follows $a \wedge b \wedge c = a \wedge b \wedge c_\perp$, where $c_\perp$ is the rejection to the 2D subspace defined by $a \wedge b$, etc.

In 3D, all of this is quite simple. If a homogeneous (of a single grade) versor has the maximum grade, then it must be a blade $a \wedge b \wedge c = \alpha j$, while if it has grade 2, then it can be written as $x \wedge y$, etc.

E18

$$1, e_1, \ e_2, \ e_3, \ e_1e_2, \ e_1e_3, e_2e_3, \ e_1e_2e_3 \xrightarrow{j} e_1e_2e_3, e_2e_3, e_3e_1, e_1e_2, -e_3, e_2, -e_1, -1.$$

E19

We have the general form in 3D $M = Z + \mathbf{F}$, $\mathbf{F} = x + jn$, $\overleftarrow{\mathbf{F}} = -x + jn$. Consequently,

$$\overleftarrow{(MN)} = \overleftarrow{((Z_1 + \mathbf{F}_1)(Z_2 + \mathbf{F}_2))} = \overleftarrow{(Z_1 Z_2)} + \overleftarrow{Z}_1 \overleftarrow{\mathbf{F}}_2 + \overleftarrow{Z}_2 \overleftarrow{\mathbf{F}}_1 + \overleftarrow{(\mathbf{F}_1 \mathbf{F}_2)},$$

where due to the fact that Z and $Z^* = \overleftarrow{Z}$ both belong to the center of the algebra, the only problematic term is the last one:

$$\overleftarrow{(\mathbf{F}_1\mathbf{F}_2)} = \overleftarrow{(x_1x_2 - n_1n_2 + j(x_1n_2 + n_1x_2))} = x_1x_2 - n_1n_2 - j(x_1n_2 + n_1x_2) = \overleftarrow{\mathbf{F}}_1\overleftarrow{\mathbf{F}}_2.$$

The reverse involution changes the order in versors, and consequently,

$$(\mathbf{F}_1\mathbf{F}_2)^\dagger = (x_1x_2 - n_1n_2 + j(x_1n_2 + n_1x_2))^\dagger = x_2x_1 - n_2n_1 - j(n_2x_1 + x_2n_1) = \mathbf{F}_2^\dagger\mathbf{F}_1^\dagger.$$

Clifford conjugation is a combination of the grade and reverse involutions, so we have

$$\overline{\mathbf{F}_1\mathbf{F}_2} = \left(\overleftarrow{(\mathbf{F}_1\mathbf{F}_2)}\right)^\dagger = \left(\overleftarrow{\mathbf{F}}_1\overleftarrow{\mathbf{F}}_2\right)^\dagger = \left(\overleftarrow{\mathbf{F}}_2\right)^\dagger\left(\overleftarrow{\mathbf{F}}_1\right)^\dagger = \bar{\mathbf{F}}_2\bar{\mathbf{F}}_1.$$

Now we can use involutions to find various parts of a multivector. From the general multivector

$$M = Z + \mathbf{F} = \alpha + x + jn + j\beta,$$

we get

$$\langle M\rangle_+ = \frac{M + \overleftarrow{M}}{2} = \frac{\alpha + x + jn + j\beta}{2} + \frac{\alpha - x + jn - j\beta}{2} = \alpha + jn \quad \text{(even part, \textit{spinors})},$$

$$\langle M\rangle_- = \frac{M - \overleftarrow{M}}{2} = \frac{\alpha + x + jn + j\beta}{2} - \frac{\alpha - x + jn - j\beta}{2} = x + j\beta \quad \text{(odd part)},$$

$$\langle M\rangle_S = \frac{M + \bar{M}}{2} = \frac{\alpha + x + jn + j\beta}{2} + \frac{\alpha - x - jn + j\beta}{2} = \alpha + j\beta \quad \text{(complex scalar)},$$

$$\langle M\rangle_V = \frac{M - \bar{M}}{2} = \frac{\alpha + x + jn + j\beta}{2} - \frac{\alpha - x - jn + j\beta}{2} = x + jn \quad \text{(complex vector)},$$

$$\langle M\rangle_R = \frac{M + M^\dagger}{2} = \frac{\alpha + x + jn + j\beta}{2} + \frac{\alpha + x - jn - j\beta}{2} = \alpha + x \quad \text{(real part, \textit{paravectors})},$$

$$\langle M\rangle_I = \frac{M - M^\dagger}{2} = \frac{\alpha + x + jn + j\beta}{2} - \frac{\alpha + x - jn - j\beta}{2} = jn + j\beta \quad \text{(imaginary part)}.$$

Note that we can write $M^\dagger = Z^* + \mathbf{F}^*$, $\overleftarrow{M} = Z^* - \mathbf{F}^*$, and $\bar{M} = Z - \mathbf{F}$. Use this to show that $MM^\dagger$ is generally real (a paravector), while $M\bar{M}$ is a complex scalar. What about $M\overleftarrow{M}$?

E20

All involutions give the identity. Any two involutions give the third one.

GI: $f(r) = r$;

RI: $1,2,3,4 \xrightarrow{3} 4,1,2,3 \xrightarrow{2} 4,3,1,2 \xrightarrow{1} 4,3,2,1$, from which it follows that

$$f(r) = r(r-1)/2;$$

CI: the combination of both, consequently (Table 6.2),

$$f(r) = r + r(r-1)/2 = r(r+1)/2.$$

Table 6.2 The effect of involutions on the elements from the Clifford basis in *Cl*3

	α	x	jn	$j\beta$
G	+	−	+	−
R	+	+	−	−
C	+	−	−	+
GRC	+	+	+	+

For the reverse involution (RI), we have $f(r)$ even for $r = 4k$ or $r = 4k + 1$, giving $r = 0$, 1| mod 4.

For Clifford involution (CI), we have $f(r)$ even for $r = 0$, 3| mod 4.

To find special multivector types, we can use various relations between involutions; some of them are

$$M = \overleftarrow{M} \Rightarrow \alpha + x + jn + j\beta = \alpha - x + jn - j\beta \Rightarrow x + j\beta = 0 \Rightarrow M = \alpha + jn \quad (even),$$
$$M = \bar{M} \Rightarrow \alpha + x + jn + j\beta = \alpha - x - jn + j\beta \Rightarrow \mathbf{F} = 0 \Rightarrow M = \alpha + j\beta \quad (complex\ scalar,\ Z),$$
$$M = M^{\dagger} \Rightarrow \alpha + x + jn + j\beta = \alpha + x - jn - j\beta \Rightarrow x + j\beta = 0 \Rightarrow M = \alpha + x \quad (paravector).$$

From these relations, we can determine some properties of elements of the algebra, for example, $\overline{M\bar{M}} = \bar{\bar{M}}\bar{M} = M\bar{M}$, which means that $M\bar{M}$ is a complex scalar. Likewise, $MM^{\dagger}$ is a paravector (has grades 0 and 1). We can also write

$$\overleftarrow{M} = \bar{M} \Rightarrow \overleftarrow{\bar{M}} = M \Rightarrow M^{\dagger} = M \quad (paravector),$$
$$\overleftarrow{M} = M^{\dagger} \Rightarrow \overleftarrow{M}^{\dagger} = M \Rightarrow \bar{M} = M \quad (complex\ scalar,\ Z),$$
$$M^{\dagger} = \bar{M} \Rightarrow \overleftarrow{M} = M \quad \text{(even)}.$$

For $MM^{\dagger} = 1$, we call the multivector *unitary*, while for $M\bar{M} = 1$, we call the multivector *unimodular*. For even multivectors, we have $M^{\dagger} = \bar{M}$, which means that such a multivector could be both unitary and unimodular. Typical unitary multivectors are vectors, and consequently, boosts. The Lorentz transformations are represented by unimodular multivectors $L\bar{L} = 1$.

For the versor $V = v_1 v_2 \cdots v_k$, we can apply the fact that the grade involution is an automorphism, so we have

$$\overleftarrow{V} = \overleftarrow{v}_1 \overleftarrow{(v_2 \cdots v_k)} = (-v_1)\overleftarrow{(v_2 \cdots v_k)},$$

and we can proceed with the rest of the versor to get

$$\overleftarrow{V} x V^{\dagger} = (-v_1)(-v_2)\cdots(-v_k) x v_k \cdots v_2 v_1 = (-1)^k v_1 v_2 \cdots v_k x v_k \cdots v_2 v_1.$$

Now we can use the fact that $v_k x v_k$ is a vector, etc. See also Sect. 1.9.15.

E21

From $p = (1 + \mathbf{f})/2$, $\mathbf{f}^2 = 1$, $p^2 = p$, we look for $X = p^{-1}$,

$$Xp = 1 \Rightarrow Xp^2 = p \Rightarrow Xp = 1 = p.$$

E22

$$\begin{aligned} BB^\dagger &= (e_1e_2 + e_2e_3)(e_2e_1 + e_3e_2) = 1 + e_1e_2e_3e_2 + e_2e_3e_2e_1 + 1 \\ &= 1 - e_1e_3 + e_1e_3 + 1 = 2. \end{aligned}$$

We also have

$$\hat{B}e_2 = (e_1e_2 + e_2e_3)e_2/\sqrt{2} = (e_1 - e_3)/\sqrt{2};$$

therefore, we have a rotation in the plane defined by the bivector.

E23

Using a formula from trigonometry, we get one solution ($i \to I$) (Fig. 6.3)

$$x = -\exp(-I\pi/3) = -\cos(\pi/3) + I\sin(\pi/3) = -\frac{1}{2} + I\frac{\sqrt{3}}{2},$$

and

$$\begin{aligned} xe_1 &= \left(-\frac{1}{2} + I\frac{\sqrt{3}}{2}\right)e_1 = -\frac{e_1}{2} - e_2\frac{\sqrt{3}}{2}, \\ x^2 &= \exp(-I2\pi/3) = \cos(2\pi/3) + I\sin(2\pi/3) = -\frac{1}{2} - I\frac{\sqrt{3}}{2}, \\ x^2e_1 &= -\left(\frac{1}{2} + I\frac{\sqrt{3}}{2}\right)e_1 = -\frac{e_1}{2} + e_2\frac{\sqrt{3}}{2}. \end{aligned}$$

Note that

$$e_1 + \left(-\frac{e_1}{2} - e_2\frac{\sqrt{3}}{2}\right) + \left(-\frac{e_1}{2} + e_2\frac{\sqrt{3}}{2}\right) = 0.$$

For the equation $\alpha x^2 + \beta x + 1 = 0$, $\alpha > 0$ (for negative α we can multiply the equation by -1 and start with $-e_1$), we can proceed as follows. With $R = -x\sqrt{\alpha}$, we have

$$R^2 - \frac{\beta}{\sqrt{\alpha}}R + 1 = R^2 - \lambda R + 1 = 0.$$

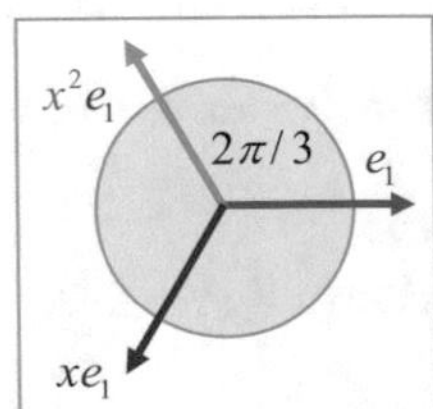

Fig. 6.3 The solution of the quadratic equation

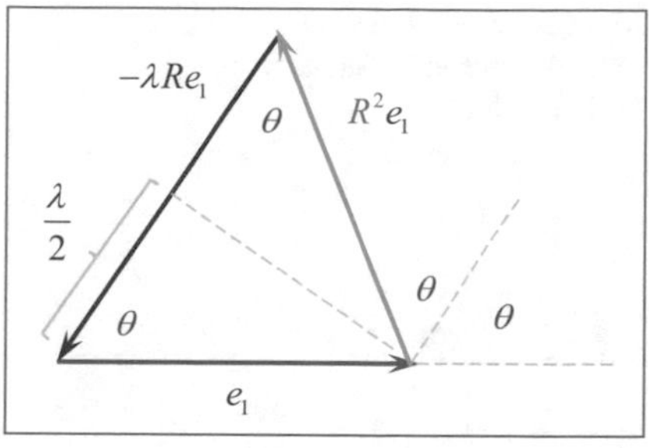

Fig. 6.4 The general solution of a quadratic equation

Now, first take the vector e_1, rotate it by θ, change the sign, and multiply by λ, and second, take the vector e_1 again, rotate it by 2θ, then sum all vectors (Fig. 6.4). The angle follows from $\lambda = 2\cos\theta$. To find the rotor, we define $I = e_2e_1$, giving $R = \exp(I\theta)$ (acts from the left). Our equation becomes

$$
\begin{aligned}
&e_1\cos(2\theta) + Ie_1\sin(2\theta) - e_1\lambda\cos\theta - \lambda Ie_1\sin\theta + e_1 = 0 \Rightarrow \\
&e_1(\cos(2\theta) - \lambda\cos\theta + 1) + e_2(\sin(2\theta) - \lambda\sin\theta) = 0,
\end{aligned}
$$

where the expressions in parentheses are zero. The solution is $x = -\exp(I\theta)/\sqrt{\alpha}$, while the standard solution is

$$
x = \frac{-\beta \pm \sqrt{\beta^2 - 4\alpha}}{2\alpha} = \frac{-\beta \pm \sqrt{D}}{2\alpha} = -\frac{\cos\theta \pm i\sin\theta}{\sqrt{\alpha}}, \quad i = \sqrt{-1}.
$$

This gives $R = \cos\theta \pm i\sin\theta$, which means that for imaginary solutions of the equation, we can draw isosceles triangles; it is enough to make the replacement $i \rightarrow e_2e_1$. To get real solutions, it is enough (check) to make the substitution $\theta \rightarrow i\theta$, giving the identities

$$
\exp(\mp\theta) = \cosh\theta \mp \sinh\theta, \quad e^{-2\theta} - 2e^{-\theta}\cosh\theta + 1 = 0.
$$

Instead of rotations, we have a scaling by the factor $-\exp(-\theta)/\sqrt{\alpha}$.

The discussion above raises questions about other polynomial equations. In [9], the problem of the equation $(R^5 - 1)e_1 = 0$ is discussed, where the solution is connected with the *golden ratio*. Using

$$
R^5 - 1 = (R - 1)\left(1 + R + R^2 + R^3 + R^4\right),
$$

we have

$$
\begin{aligned}
&\left(R^{-2} + R^{-1} + 1 + R + R^2\right)R^2 = 0 \Rightarrow \\
&\left(R^\dagger\right)^2 + R^\dagger + 1 + R + R^2 = 0,
\end{aligned}
$$

where $R^{-1} = R^\dagger$ and R is the rotation by the angle $2\pi/5$.

We can see from Fig. 6.5 that the isosceles red triangle (the so-called *golden triangle*) has the angle $\theta = \pi/5$, which gives

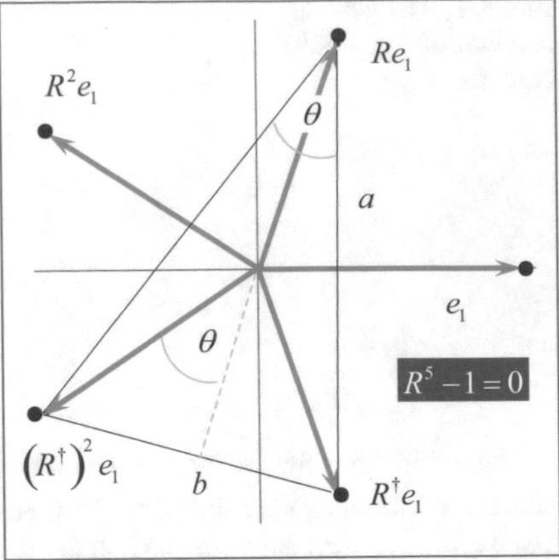

Fig. 6.5 The solution of a special polynomial equation of degree 5

$$\frac{a}{b} = \frac{1}{2\sin(\pi/10)} = \frac{1+\sqrt{5}}{2} = \varphi,$$

where φ is the golden ratio. From $\varphi^{\pm 1} = (\sqrt{5} \pm 1)/2$, it follows that

$$a = 5^{1/4}\sqrt{\varphi}, \quad b = 5^{1/4}/\sqrt{\varphi}.$$

The reader should prove following relations with the golden ratio:

1. $1 + \varphi = \varphi^2$ (*Kepler right triangle*), whence by induction follows $\varphi^{n-1} + \varphi^n = \varphi^{n+1}$;
2. $\varphi = 2\cos(\pi/5)$;
3. Let $p = (1 + u)/2$ be idempotent ($u^2 = 1$) and $g = \varphi p - \varphi^{-1}\bar{p}$. Show that $g\bar{g} = -1$ (*antiunimodular*), $gu\bar{g}u = 1$ (*unimodular*), and $g^{n+1} = g^n + g^{n-1}$, $n \in \mathbb{N}$. Using the *Fibonacci sequence* $F_0 = 0, F_1 = 1, F_{n+1} = F_n + F_{n-1}$, prove that $g^n = F_n g + F_{n-1}$ and $g^n = \varphi^n p - \varphi^{-n}\bar{p}$.
4. $\varphi^n = F_n\varphi + F_{n-1}$, F_n is a *Fibonacci number*;
5. $\lim_{n\to\infty}(F_{n+1}/F_n) = \varphi$;
6. $\sin(\xi \log \varphi) = \xi/2$, $\xi = \sqrt{-1}$ or $\xi = j$;

Note that all equations of the form $(R^p - 1)e_1 = 0$, p is a prime number, can be solved by the method just described, due to the fact that

$$R^p - 1 = (R-1)\left(1 + R + R^2 + \cdots + R^{p-1}\right).$$

E24

For two consecutive rotations, we can write $R_1 = m_1n_1$ and $R_2 = m_2n_2$, whence, because $x^\dagger = x$, where x is a vector, it follows that

$$R = R_2 R_1 = m_2 n_2 m_1 n_1 \Rightarrow RR^\dagger = m_2 n_2 m_1 n_1 n_1 m_1 n_2 m_2 = 1,$$

which is the defining property for rotations. Specifically, in *Cl*3 we have the property that multiplication of two bivectors from the Clifford basis gives a bivector or -1; therefore, R has grades 0 and 2. For a general treatment, see the literature. See also Sect. 3.2.

E25

The first one is obvious, and for the second one, the unit bivector is a linear combination of unit bivectors from the Clifford basis. Consequently, it is enough to prove the relation for a bivector from the basis. For each basis unit vector in *Cl*3, we have exactly two anticommuting unit bivectors, whence

$$\sum_i e_i \hat{B} e_i = -2\hat{B} + \hat{B} = -\hat{B}.$$

E26

Consider the plane defined by the unit bivector $\hat{B}$. Using the rotor R, we can rotate the bivector $\hat{B}' = R\hat{B}R^\dagger$. Let us find the rotor defined by the bivector $\hat{B}'$:

$$\begin{aligned} R' &= \exp\left(-\hat{B}'\alpha/2\right) = \exp\left(-R\hat{B}R^\dagger\alpha/2\right) = \cos(\alpha/2) - R\hat{B}R^\dagger \sin(\alpha/2) \\ &= R\left(\cos(\alpha/2) - \hat{B}\sin(\alpha/2)\right)R^\dagger = R\exp\left(-\hat{B}\alpha/2\right)R^\dagger. \end{aligned}$$

To get the rotor expressed in the Euler angles (see [21]), we start with the rotation about the unit vector e_3, by the angle ϕ. The rotor is

$$R_\phi = \exp(-je_3\phi/2).$$

The vector e_1 is now transformed to $e_1' = R_\phi e_1 R_\phi^\dagger$, which is our new rotation axis, with the rotor

$$R_\theta = \exp\left(-jR_\phi e_1 R_\phi^\dagger \theta/2\right) = R_\phi \exp(-je_1\theta/2) R_\phi^\dagger.$$

Now we have the intermediate rotor

$$R' = R_\theta R_\phi = R_\phi \exp(-je_1\theta/2) R_\phi^\dagger R_\phi = \exp(-je_3\phi/2)\exp(-je_1\theta/2).$$

Note the change of the order. The vector e_3 is now transformed to $e_3' = R' e_3 R'^\dagger$, which is our new, final rotation axis, with the rotor

$$R_\psi = R_\theta R_\phi \exp(-je_3\psi/2) R_\phi^\dagger R_\theta^\dagger.$$

Finally, this gives

$$R = R_\theta R_\phi \exp(-je_3\psi/2) R_\phi^\dagger R_\theta^\dagger R_\theta R_\phi = R_\theta R_\phi \exp(-je_3\psi/2) \Rightarrow$$
$$R = \exp(-je_3\phi/2)\exp(-je_1\theta/2)\exp(-je_3\psi/2).$$

Here, as always, we have a clear geometric interpretation, deeply hidden in a matrix formulation.

E27

Note that the columns of the transformation matrix represent the vectors e_2 and $-e_1$. The secular equation gives

$$\det\left[\begin{pmatrix} 0 & -1 \\ 1 & 0 \end{pmatrix} - \lambda \begin{pmatrix} 1 & 0 \\ 0 & 1 \end{pmatrix}\right] = \det\begin{pmatrix} -\lambda & -1 \\ 1 & -\lambda \end{pmatrix} = \lambda^2 + 1 = 0 \Rightarrow \lambda = \pm i,$$
$$\begin{pmatrix} 0 & -1 \\ 1 & 0 \end{pmatrix}\begin{pmatrix} x \\ y \end{pmatrix} = \begin{pmatrix} -y \\ x \end{pmatrix} = \pm i \begin{pmatrix} x \\ y \end{pmatrix},$$

so we can choose vectors $e_1 \pm ie_2$.

E28

For $\lambda_i \neq \lambda_j$ and $\bar{\mathrm{F}} = \mathrm{F}$, we have $e_i \cdot \bar{\mathrm{F}}(e_j) = \mathrm{F}(e_i) \cdot e_j \Rightarrow \lambda_j e_i \cdot e_j = \lambda_i e_i \cdot e_j$, whence follows $e_i \cdot e_j = 0$.

E29

Note that $M_\pm$ are not affected by the Clifford conjugation; they are complex scalars. From

$$\bar{M} = M_+\bar{u}_+ + M_-\bar{u}_- = M_+u_- + M_-u_+,$$

we have

$$M\bar{M} = (M_+u_+ + M_-u_-)(M_+u_- + M_-u_+) = M_+M_-(u_+ + u_-) = M_+M_-.$$

E30

$$(e_1 + je_2)(\cos\varphi + j\sin\varphi) = e_1\cos\varphi + je_1\sin\varphi + je_2\cos\varphi - e_2\sin\varphi$$
$$= e_1\cos\varphi - e_2\sin\varphi + j(e_1\sin\varphi + e_2\cos\varphi),$$

and we recognize the rotation matrix

$$\begin{pmatrix} \cos\varphi & -\sin\varphi \\ \sin\varphi & \cos\varphi \end{pmatrix}.$$

E31

Consider the term

$$c = \sigma_3 \mathbf{n} + \mathbf{n}\sigma_3 = 2\sigma_3 \cdot \mathbf{n} = 2\cos\theta = \sigma_3 R\sigma_3 R^\dagger + R\sigma_3 R^\dagger \sigma_3 = 2\sigma_3 \cdot \left(R\sigma_3 R^\dagger\right),$$

without dependence on φ, which is possible if R contains a factor dependent on φ that commutes with σ_3. Therefore, we can write

$$R = R_1 \exp(-j\sigma_3\varphi/2), \quad c = 2\sigma_3 \cdot \left(R_1\sigma_3 R_1^\dagger\right).$$

From sin $\theta(\sigma_1 \cos\varphi + \sigma_2 \sin\varphi)$, we see that σ_3 is rotated by θ about σ_2, then by φ about σ_3 (you can make a model and rotate it by hand); therefore, we can write $R_1 = \exp(-j\sigma_2\theta/2)$ and, due to anticommutativity,

$$R = \exp(-j\sigma_2\theta/2)\exp(-j\sigma_3\varphi/2) = \exp(-j\sigma_3\varphi/2)\exp(j\sigma_2\theta/2).$$

E32

Since the reverse involution is an antiautomorphism, we have

$$R_0\psi\sigma_3\psi^\dagger R_0^\dagger = R_0\psi\sigma_3(R_0\psi)^\dagger,$$

whence we see that $\psi \to R_0\psi$ is valid.

E33

For rotors we have

$$R_0 = \exp(-I\varphi/2),$$

so for $\varphi = 2\pi$ we get $R_0 = \exp(-\pi I) = -1$ and $\psi \to R_0\psi = -\psi$. The change of sign of the spinor is due to the 3D geometry, that is, due to rotor properties. This relationship between the spin and geometry is far from obvious in standard quantum mechanics. One can wonder what we might find if we started to dig deeper. In fact, there are numerous surprises, and if you wish to be astonished, please learn geometric algebra.

E34

From the definitions $o \equiv (e + \bar{e})/2$ and $\infty \equiv \bar{e} - e$, we have

$$\begin{aligned}
&o^2 = (e+\bar{e})(e+\bar{e})/4 = (1 + e\bar{e} + \bar{e}e - 1)/4 = 0,\\
&\infty^2 = (\bar{e}-e)(\bar{e}-e) = -1 - \bar{e}e - e\bar{e} + 1 = 0,\\
&o \wedge \infty = \frac{e+\bar{e}}{2} \wedge (\bar{e}-e) = \frac{e\wedge\bar{e}}{2} - \frac{\bar{e}\wedge e}{2} = e\wedge\bar{e},\\
&o \cdot \infty = \frac{e+\bar{e}}{2} \cdot (\bar{e}-e) = -\frac{1}{2} - \frac{1}{2} = -1,\\
&o \cdot o = \frac{e+\bar{e}}{2} \cdot \frac{e+\bar{e}}{2} = \frac{1}{4} - \frac{1}{4} = 0,\\
&\infty \cdot \infty = (\bar{e}-e)\cdot(\bar{e}-e) = -1 + 1 = 0.
\end{aligned}$$

E35

Note that due to the orthogonality of unit vectors, we have $o \cdot \mathbf{p} = \infty \cdot \mathbf{p} = 0$, giving

$$p \cdot p = \left(o + \mathbf{p} + \mathbf{p}^2\infty/2\right) \cdot \left(o + \mathbf{p} + \mathbf{p}^2\infty/2\right) = -\mathbf{p}^2/2 + \mathbf{p}^2 - \mathbf{p}^2/2 = 0,$$

$$p \cdot q = \left(o + \mathbf{p} + \mathbf{p}^2\infty/2\right) \cdot \left(o + \mathbf{q} + \mathbf{q}^2\infty/2\right) = -\mathbf{q}^2/2 + \mathbf{p} \cdot \mathbf{q} - \mathbf{p}^2/2$$
$$= -(\mathbf{p} - \mathbf{q}) \cdot (\mathbf{p} - \mathbf{q})/2.$$

You can also start from $p = \alpha\, o + \beta \mathbf{p} + \gamma\infty$, then find the coefficients.

E36

We have the linear combinations

$$[a] = \sum_{i=1}^{3} \mathrm{a}_i \hat{\sigma}_i \quad \text{and} \quad [b] = \sum_{i=1}^{3} \mathrm{b}_i \hat{\sigma}_i, \quad \mathrm{a}_i, \mathrm{b}_i \in \mathbb{R},$$

and consequently, for the antisymmetric part, we get

$$\frac{[a][b] - [b][a]}{2} = \begin{pmatrix} i(\mathrm{a}_1\mathrm{b}_2 - \mathrm{a}_2\mathrm{b}_1) & \mathrm{a}_3\mathrm{b}_1 - \mathrm{a}_1\mathrm{b}_3 + i(\mathrm{a}_2\mathrm{b}_3 - \mathrm{a}_3\mathrm{b}_2) \\ \mathrm{a}_1 b_3 - \mathrm{a}_3\mathrm{b}_1 + i(\mathrm{a}_2\mathrm{b}_3 - \mathrm{a}_3\mathrm{b}_2) & -i(\mathrm{a}_1\mathrm{b}_2 - \mathrm{a}_2\mathrm{b}_1) \end{pmatrix}.$$

The cross product of vectors a and b is an axial vector,

$$c = a \times b = (\mathrm{a}_2\mathrm{b}_3 - \mathrm{a}_3\mathrm{b}_2, \mathrm{a}_3\mathrm{b}_1 - \mathrm{a}_1\mathrm{b}_3, \mathrm{a}_1\mathrm{b}_2 - \mathrm{a}_2\mathrm{b}_1);$$

therefore, it is straightforward to show that

$$\sum_{k=1}^{3} \mathrm{c}_k \hat{\sigma}_k = -i \frac{[a][b] - [b][a]}{2}.$$

In the quantum mechanics literature, it is customary to write terms like $\boldsymbol{\sigma} \cdot \mathbf{n} \equiv \sum_{i=1}^{3} n_i \hat{\sigma}_i$.

Note, however, that $\boldsymbol{\sigma} \cdot \mathbf{n}$ is just a matrix representation of the vector $\mathbf{n}$. It is straightforward to check that matrix multiplication $(\boldsymbol{\sigma} \cdot \mathbf{n})(\boldsymbol{\sigma} \cdot \mathbf{m})$ gives

$$(\boldsymbol{\sigma} \cdot \mathbf{n})(\boldsymbol{\sigma} \cdot \mathbf{m}) = \mathbf{n} \cdot \mathbf{m} I + \frac{(\boldsymbol{\sigma} \cdot \mathbf{n})(\boldsymbol{\sigma} \cdot \mathbf{m}) - (\boldsymbol{\sigma} \cdot \mathbf{m})(\boldsymbol{\sigma} \cdot \mathbf{n})}{2} = \mathbf{n} \cdot \mathbf{m} I + i(\mathbf{n} \times \mathbf{m}) \cdot \boldsymbol{\sigma}.$$

where I is the 2×2 unit matrix, and the last term is expressed (check) by the cross product. In *Cl*3, this relation is just

$$\mathbf{nm} = \mathbf{n} \cdot \mathbf{m} + \mathbf{n} \wedge \mathbf{m} = \mathbf{n} \cdot \mathbf{m} + j\mathbf{n} \times \mathbf{m},$$

and strange terms, like $\boldsymbol{\sigma} \cdot \mathbf{n}$, are unnecessary. Note that $\boldsymbol{\sigma}$ should be a "vector" whose components are matrices, not to mention that a matrix product $(\boldsymbol{\sigma} \cdot \mathbf{n})(\boldsymbol{\sigma} \cdot \mathbf{m})$ is quite messy when expanded, in contrast to $\mathbf{n} \cdot \mathbf{m} + \mathbf{n} \wedge \mathbf{m}$, where we have a clear geometric picture, and we can draw conclusions without components.

6.2 Problems

1. Let i be the position of an element of the Clifford basis of an n-dimensional vector space. Find the formula for the grade of that element.
2. In $\mathfrak{R}^{2,1}$, multiply e_1e_3 by e_2e_3 using bitwise XOR.
3. Find $MM^{\dagger} - M^{\dagger}M$, where M is a multivector in $Cl3$.
4. Defining $e = e_3$, $P = (1 + e)/2$, and $m = e_1$ (we could choose any other pairs of orthogonal unit vectors), we have $eP = P$, $meP = mP$. For a multivector $M = t + \mathbf{x} + j\mathbf{n} + jb$, we can write $M = z^0 + z^3e + z^1m + jz^2me$, where $z^{\mu} \in \mathbb{C}$ ($z^{\mu} = \langle z^{\mu}\rangle_{\mathfrak{R}} + j\langle z^{\mu}\rangle_{\mathfrak{I}}$). Find z^{μ}. For the elements (c) from the Clifford basis, find cP and $c\bar{P}$. Show that for every multivector M, we can write $MP = \lambda_+P + \lambda_-mP$, λ_+, $\lambda_- \in \mathbb{C}$.
5. For two blades A and B, we can define the *meet* product $A \cap B \equiv A^{\Delta} \cdot B$. For example,

$$e_2e_3 \cap e_1e_3 = (e_2e_3)^{\Delta} \cdot (e_1e_3) = e_1 \cdot (e_1e_3) = e_3.$$

Find $A \cap B$ for elements from the Clifford basis in $Cl3$. Explain.
6. For a vector $\mathbf{a}$ and a unit vector $\hat{\mathbf{n}}$ we can define $\Delta\mathbf{a} = \Delta\theta\hat{\mathbf{n}} \times \mathbf{a}$, where $\Delta\theta$ is a small angle. Decomposing the vector $\mathbf{a}$ into components parallel and orthogonal with respect to $\hat{\mathbf{n}}$, we get $\Delta\mathbf{a} = \Delta\theta\hat{\mathbf{n}} \times \mathbf{a}_{\perp}$. Let $\mathbf{a} = e_1$ and $\hat{\mathbf{n}} = e_3$; then we have $\Delta\mathbf{a} = \Delta\theta e_3 \times e_1 = -e_2\Delta\theta$. Substituting $\hat{\mathbf{n}} \to j\hat{\mathbf{n}}$ and replacing the cross product with the geometric product, we have

$$\Delta\mathbf{a} = \Delta\theta(je_3)e_1 = \Delta\theta(e_1e_2)e_1 = -e_2\Delta\theta.$$

Check that we can write in general $\Delta\mathbf{a} = \Delta\theta\hat{\mathbf{n}} \times \mathbf{a}_{\perp} = \Delta\theta(j\hat{\mathbf{n}})\mathbf{a}_{\perp}$.
7. In EM theory ($c = 1$), we define momentum 4-vectors $p^{\alpha} = (\gamma m, p_1, p_2, p_3)$, a covariant 4-velocity $U_{\alpha} = \gamma(1, -v_1, -v_2, -v_3) = \gamma(1, -\mathbf{v})$, and a contravariant electromagnetic tensor

$$F^{\alpha\beta} = \begin{pmatrix} 0 & -E_1 & -E_2 & -E_3 \\ E_1 & 0 & -B_3 & B_2 \\ E_2 & B_3 & 0 & -B_1 \\ E_3 & -B_2 & B_1 & 0 \end{pmatrix}.$$

Then we can write the Lorentz force in covariant form

$$\frac{dp^{\alpha}}{d\tau} = qF^{\alpha\beta}U_{\beta}. \quad (6.1)$$

In $Cl3$, we use $\mathbf{F} = \mathbf{E} + j\mathbf{B}$, a complex vector, to represent an electromagnetic field. Compare the covariant form (6.1) with

$$\frac{dp}{d\tau} = \dot{p} = \langle q\mathbf{F}u \rangle_{\Re},$$

where $u = \gamma(1 + \mathbf{v})$ and $\langle M \rangle_{\Re} = (M + M^\dagger)/2$.

8. Prove that for vectors in $Cl3$, it is true that $\log(ab) = \log a + \log b + j\theta$.
9. Find $\sqrt{\alpha \pm \beta j}$, $\alpha, \beta \in \mathbb{R}$ in $Cl3$. Discuss different choices of numbers α, β. Can you find all solutions for $\sqrt{\pm 1}$ in $Cl3$?
10. Prove that $\sqrt{M} = \sqrt{M_+ u_+} \pm \sqrt{M_- u_-}$ is generally equivalent to

$$\sqrt{M} = \pm\mathbf{f}\frac{M \pm |M|}{\sqrt{M + \bar{M} \pm 2|M|}}, \quad \mathbf{f}^2 = 1, \quad M = Z + Z_f \mathbf{f}.$$

11. Show that for the Pauli matrices it is true that ([,] denotes the commutator)

$$\left[\hat{\sigma}_i, \hat{\sigma}_j\right] = 2i \ \varepsilon_{ijk} \ \hat{\sigma}_k.$$

12. Find the effect of the transformations $e_i e_j a e_i e_j$ and $e_j e_i a e_i e_j$ on a vector a in $Cl3$.
13. Consider a general multivector M from $Cl3$ and element $p = M^\dagger n M$, n is a unit vector. Show that p is a paravector. Find $p\bar{p}$. Apply your results to $M = R$, where R is a rotor. Is the result a paravector in this case?
14. Consider a general multivector M from $Cl3$ and element $z = \bar{M} n M$, n is a unit vector. Show that z is a complex scalar. Find $zz^\dagger$. Apply your results to $M = B$, where B is a bivector.
15. For two even $Cl3$ multivectors $\psi_1 = \alpha_1 + \beta_1 \hat{B}_1$ and $\psi_2 = \alpha_2 + \beta_2 \hat{B}_2$, where $\hat{B}_i$ are unit bivectors, find $\left\langle \psi_1^\dagger \psi_2 j e_3 \right\rangle j e_3$.
16. Draw 8×8 multiplication tables of the Clifford basis in $Cl3$ for geometric and derived products. Use various methods to calculate products; you will need such a skill. Calculating in your head is welcome. You can also imagine or draw results; geometric intuition is powerful; teach yourself to use it.
17. Analyze the $Cl3$ multivector $1 + m + jn + j$, $m^2 = n^2 = 1$ and its possible parts.
18. Try to implement $Cl3$ on a computer, using any programming language.
19. A general multivector from $Cl3$ can be expressed as $M = \alpha + \mathbf{x} + j\mathbf{n} + j\beta$. Consider two multivectors from 2D Euclidean space $m_i = \alpha_i + x_{i1}e_1 + x_{i2}e_2 + \beta_i I$, $I = e_1 e_2$, $i = 1, 2$. Show that it is possible to express M as $m_1 + jm_2$. Show this also by a direct comparison, using $M = 1 + \sum_i e_i + j\sum_i e_i + j$.
20. Consider a general $Cl2$ multivector $m = \alpha + x_1 e_1 + x_2 e_2 + \beta I$, $I = e_1 e_2$, and idempotents $u_\pm = (1 \pm e_2)/2$. Note that $e_2 u_\pm = \pm u_\pm$, $e_1 u_\pm = u_\mp e_1$, $e_1 u_\pm e_1 = u_\mp$. The elements $(u_+, u_-, e_1 u_+, e_1 u_-)$ form the spectral basis of the algebra, that is, we can write $m = \alpha_1 u_+ + \alpha_2 u_- + \alpha_3 e_1 u_+ + \alpha_4 e_1 u_-$, where $\alpha_i = a_i + b_i I$, $a_i, b_i \in \mathbb{R}$. Show this by a direct comparison, using $m = 1 + e_1 + e_2 + e_1 e_2$. Note that according to problem 19, we can use the same spectral basis for $Cl3$ (see [39], Sect. 4.2).
21. We can define an involution in $Cl3$ as $\mathrm{i} : g \to e_1 g e_1$. Find all $\mathrm{i}(e)$, where e represents elements from the Clifford basis. Find the multivectors M with the property $M = \mathrm{i}(M)$.

22. In the text it is shown how to find functions of multivectors in *Cl*3, using the form of multivectors $M = Z + \mathbf{F} = Z + Z_f\mathbf{f}$, $\mathbf{f}^2 = 1$. Try to implement a similar procedure in *Cl*2. Are there any differences? (*Hint*: See Sect. 4.2)
23. In an n-dimensional vector space choose m vectors and form their outer products P. Prove that P defines a linear space of the dimension $n!/[(n - m)!\, m!]$.
24. A triangle in a plane in 3D is specified by the ordered set of points $\{\mathbf{x}_1, \mathbf{x}_2, \mathbf{x}_3\}$, where $\mathbf{x}_i$ are vectors. Find the oriented area of the triangle. Can you extend this to n points? Imagine a triangulated surface in 3D. Can you find the total area of the surface? (*Hint*: Use the outer product.)
25. From $a \times b \equiv -ja \wedge b$, prove that $a \times b = b \cdot (ja) = -a \cdot (jb)$. What is the geometric interpretation? Note that the old cross product is not a good idea due to the space inversion.
26. Consider a linear transformation F that leaves lengths of vectors and angles between them unchanged. Show that one must have $\bar{\mathrm{F}} = \mathrm{F}^{-1}$ and find detF. Check that rotations and reflections are such transformations. Can you find some other such transformations?
27. In *Cl*3, from $M = Z + \mathbf{F} = Z + Z_f\mathbf{f}$, $\mathbf{f}^2 = 1$, using the spectral decomposition, try to find $\mathbf{f}^{\mathbf{f}}$.
28. Using the spectral decomposition in *Cl*3, apply various functions and their inverses to elements of the Clifford basis. Do the same for paravectors and rotors.
29. Considering the geometric product, can you find all *Cl*3 multivectors for which there is no inverse?
30. In *Cl*3, using both the spectral decomposition and formulas $\log M = \log |M| + \varphi I$, $I = -j\mathbf{f}$, $\varphi = \arctan(|\mathbf{F}|/Z)$, and $|\mathbf{F}| = \sqrt{-\mathbf{F}^2} = jZ_f$, find $\log R$, where R is a rotor.
31. Show that for a *Cl*3 multivector $M = \alpha + \mathbf{x} + j\mathbf{n} + j\beta$, we can derive the norm

$$\|M\| = \alpha^2 + \mathbf{x}^2 + \mathbf{n}^2 + \beta^2 = M * M^\dagger = \langle MM^\dagger \rangle,$$

where for two multivectors we have $\|M_1M_2\| \le \sqrt{2}\|M_1\|\|M_2\|$.
32. Start with the cuboid defined by the vectors e_1, $2e_2$, and $3e_3$, rotating it two ways. First, rotate around e_1 by $\pi/2$, then around e_2 by $\pi/2$. Second, do the same, exchanging e_1 and e_2 (see Figs. 6.6 and 6.7). Find the rotors and the difference between two compound rotors (commutator). What can you say about permutations of the unit vectors indices?

33. For $\mathbf{F} = \mathbf{E} + j\mathbf{B}$ and the main *Cl*3 involutions i, find

$$\mathbf{F}e_\mu \mathrm{i}(\mathbf{F}), \quad e_0 = 1, \quad \mu = 0, 1, 2, 3.$$

34. For a *Cl*3 multivector $M = \sum \mathrm{m}_i c_i$, where the c_i are the eight elements of the Clifford basis, the $\mathrm{m}_i = \mathrm{m}_i(x_1, x_2, x_3) \in \mathbb{R}$ are coefficients that depend on rectangular coordinates, using $e^i = e_i$, we define the derivative operator as

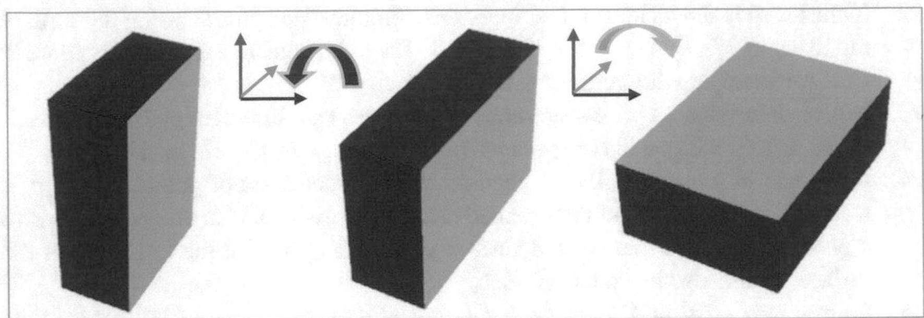

Fig. 6.6 Rotations of a cuboid do not commute

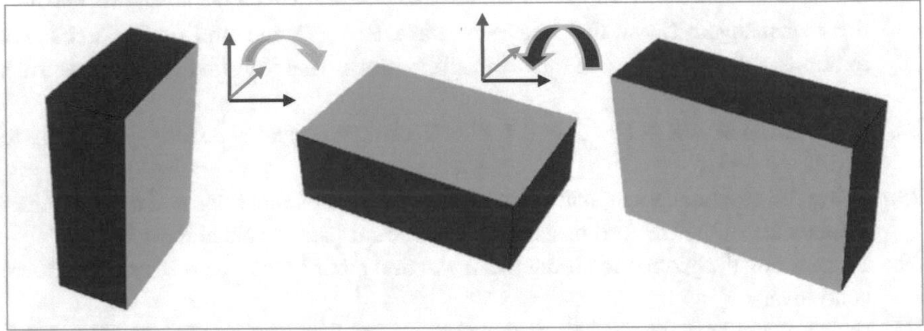

Fig. 6.7 Rotations of a cuboid do not commute

$$\nabla M \equiv \sum_i e^k c_i \partial_k m_i, \quad \partial_k \equiv \frac{\partial}{\partial_{x_k}}, \quad \frac{\partial_{x_l}}{\partial_{x_k}} = \delta_{kl},$$

where $\dfrac{\partial f(x_1, x_2, x_3)}{\partial_{x_k}} = f'_k$ is the partial derivative (we assume the Einstein summation rule over k). We can also define derivatives

$$\nabla \cdot M \equiv \sum_i e^k \cdot c_i \partial_k m_i, \quad \nabla \wedge M \equiv \sum_i e^k \wedge c_i \partial_k m_i,$$

where the operator $\nabla \equiv e^i \partial_i$ (summation) behaves just like a vector. For example, specifying $r = e^i x_i$, we have

$$\nabla r = \sum_i e^k e_i \partial_k x_i = 3 = \nabla \cdot r, \quad \nabla \wedge r = 0.$$

Find $\nabla a \cdot r$, $a = \text{const}$, ∇x_1^k, and ∇r^2. For $E = e^i \mathrm{E}_i(x_1, x_2, x_3)$, find ∇E. For $B = b^i \mathrm{B}_i(x_1, x_2, x_3)$, where the b^i are the unit bivectors from the Clifford basis, find ∇B. What about $\nabla \cdot E$, $\nabla \wedge E$, $\nabla \cdot B$, and $\nabla \wedge B$?

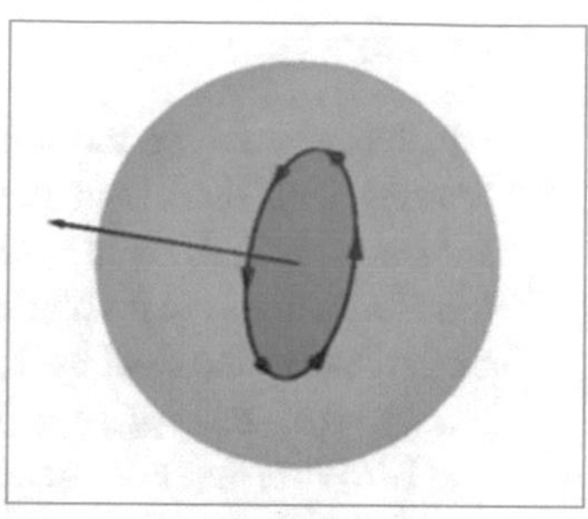

Fig. 6.8 A bivector with arrows

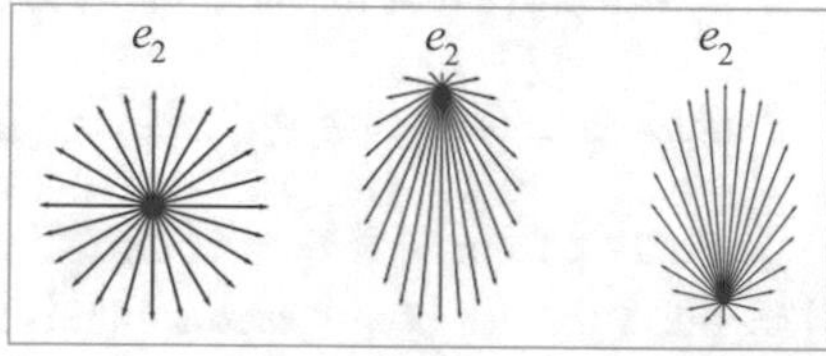

Fig. 6.9 Scaled rotations of a vector

35. Suppose that we want to draw a bivector B in $Cl3$ (green disk) with six little arrows (Fig. 6.8) showing its orientation. How can one do that generally? (*Hint*: Decompose the bivector as the outer product of two vectors, and then use left contraction.)
36. Starting with the unit vector e_2 and using a rotor in the e_1e_2-plane, perform 23 small rotations by $\pi/12$, starting every time from e_2 (series of rotors $\mathrm{R}_i(e_2)$). Then repeat that procedure, but this time multiplying the unit bivector in the exponent of the rotor by a real constant.
    Explain the figures (Fig. 6.9) that are the result of such a game.
37. According to Euler's theorem, to describe rotations in 3D we need three parameters. Consider a rotor in $Cl3$,

    $$R = \alpha + j\mathbf{v} = \exp(j\mathbf{n}\varphi/2),$$

    where $\mathbf{n}$ is the unit vector parallel to the rotation axis, while φ is the rotation angle. It is obvious that $\alpha = \cos(\varphi/2)$ and $e_i \cdot \mathbf{n} \sin(\varphi/2)$ are the *Euler parameters*, which due to $\alpha^2 + v^2 = 1$ reduce to three independent parameters. Find the Euler parameters expressed in terms of the Euler angles. Setting $s_0 = \alpha$ and $\mathbf{s} = \mathbf{n} \sin(\varphi/2)$, show that the rotation matrix is given by

    $$A_{ij} = \delta_{ij} \cos\varphi + 2s_i s_j + 2s_0 \varepsilon_{ijk} s_k,$$

    where δ_{ij} is the Kronecker delta and ε_{ijk} is the Levi-Civita symbol (summation over k is understood). Note that the rotor R is a unit quaternion (defined in $Cl3$). Show that from composition of rotors $R_3 = R_1 R_2$ it follows that

$$\alpha_3 = \alpha_1\alpha_2 - \mathbf{v}_1 \cdot \mathbf{v}_2, \quad \mathbf{v}_3 = \alpha_1\mathbf{v}_2 + \alpha_2\mathbf{v}_1 - \mathbf{v}_1 \times \mathbf{v}_2,$$

which is exactly the formula for multiplication of quaternions (*Hamilton product*).

38. Starting with the list of numbers $l = (1, 2, 3, 4)$, we define the transformations $\mathrm{I}_1 : l = (4, 3, 2, 1)$, $\mathrm{I}_2 : l = (3, 4, 1, 2)$, and $\mathrm{I}_3 : l = (2, 1, 4, 3)$. Show that the defined transformations are involutions. In fact, I_1 is the reverse of the list l, I_2 is the rotation of the list l to the right by 2, while I_3 is the composition $\mathrm{I}_1\mathrm{I}_2$. Now take the spectral basis $B = (u_+, u_-, e_1u_+, e_1u_-)$, $u_\pm = (1 + e_2)/2$, and show that $\mathrm{I}_1(B) = Be_1$, $\mathrm{I}_2(B) = e_1B$, and $\mathrm{I}_3(B) = e_1Be_1$. For a general $Cl3$ multivector $M = z_1u_+ + z_2u_- + z_3e_1u_+ + z_4e_1u_-$, $z_i \in \mathbb{C}$, find $\mathrm{I}_k(M)$. Find a general form of multivectors that satisfy $\mathrm{I}_k(M) = \pm M$.
39. In $Cl3$ prove that for the unit bivectors $B_i = je_i$, one has

$$B_i \otimes B_j = \left(B_iB_j - B_jB_i\right)/2 = -\varepsilon_{ijk}B_k.$$

Find C_{ij}^k from $B_i \otimes B_j = C_{ij}^k B_k$ (summation over k is understood).

40. Show that for homogeneous multivectors (they have a single grade), one has

$$A_rB_s = \langle A_rB_s\rangle_{|r-s|} + \langle A_rB_s\rangle_{|r-s|+2} + \cdots + \langle A_rB_s\rangle_{r+s}.$$

We also define the products

$$A_r \cdot B_s \equiv \langle A_rB_s\rangle_{|r-s|}, \quad A_r \wedge B_s \equiv \langle A_rB_s\rangle_{r+s}.$$

Show that such a definition ensures that the outer product is associative.

41. Prove that for a linear transformation F (in $Cl3$, however, you can generalize), we have

$$\begin{aligned} \mathrm{F}(A_r) \cdot B_s &= \mathrm{F}[A_r \cdot \bar{\mathrm{F}}(B_s)], \quad r \geq s, \\ A_r \cdot \bar{\mathrm{F}}(B_s) &= \bar{\mathrm{F}}[\mathrm{F}(A_r) \cdot B_s], \quad r \leq s. \end{aligned}$$

42. Prove that for vectors a, b, c, d, and a bivector B, one has the formulas

$$\begin{aligned} &(a \wedge b) \cdot (c \wedge d) = a \cdot db \cdot c - a \cdot cb \cdot d, \\ &a \cdot (b \cdot B) = (a \wedge b) \cdot B, \\ &(a \wedge b) \otimes (c \wedge d) = b \cdot ca \wedge d - a \cdot cb \wedge d + a \cdot db \wedge c - b \cdot da \wedge c, \\ &(a \wedge b) \otimes B = (a \cdot B) \wedge b + a \wedge (b \cdot B). \end{aligned}$$

43. Consider two planes in $\mathfrak{R}^3$ and their unit bivectors $\hat{A}$ and $\hat{B}$. If the planes intersect, then we can find a unit vector $\hat{n}$ along their intersection line, and a third plane P, orthogonal to the unit vector $\hat{n}$. We call the angle formed by our planes in the plane P a *dihedral angle* (δ). Prove that $\exp(j\delta\hat{n})$ can be expressed as the geometric product of the unit bivectors $\hat{A}$ and $\hat{B}$. Imagine now a unit sphere and three unit vectors starting at the origin of the sphere that define a spherical triangle on the

sphere. Can you apply the previous results to find the basic relations for a spherical triangle?

44. Consider a general *Cl*3 multivector $M = \alpha + \mathbf{x} + j\mathbf{n} + j\beta$ and the idempotent $u_+ = (1 + e_3)/2$. Show that from given Mu_+, we generally cannot find M. Find the matrix $[Mu_+]$ and interpret your result.
45. Define unit biparavectors $b_{\mu\nu} = \langle e_\mu \bar{e}_\nu \rangle_V$ in *Cl*3 and paravectors $p_{\mu\nu} = \alpha_\mu e_\mu + \alpha_\nu e_\nu$, $\alpha_\mu, \alpha_\nu \in \mathbb{R}$, in the $\mu\nu$-plane, $\mu, \nu = 0, 1, 2, 3$, $e_0 = 1$. Show that biparavectors $b_{\mu\nu}$ rotate paravectors $p_{\mu\nu}$ in the paravector space and find a general formula for $b_{\mu\nu} p_{\mu\nu}$. Find all unit biparavectors and show that only six of them are significantly distinct. Then show that every complex vector is a linear combination of these six unit biparavectors.
46. We defined the exponential function $\exp(I\varphi)$ depending on the square of I:

$$\exp(I\varphi) = \begin{cases} \cos\varphi + I\sin\varphi, & I^2 = -1, \\ \cosh\varphi + I\sinh\varphi, & I^2 = 1, \\ 1 + I\varphi, & I^2 = 0. \end{cases}$$

By a direct calculation, show that these definitions are in accordance with the rule

$$\exp(I\alpha)\exp(I\beta) = \exp(I\alpha + I\beta).$$

Note that the previous formula contains the well-known trigonometric formulas.

47. In the standard formulation of quantum mechanics, we use an abstract complex vector space with vectors $|a\rangle$ (*ket*) or complex conjugated and transposed (adjoint) vectors $|a\rangle^\dagger \equiv \langle a|$ (*bra*). We can define linear combinations over a complex field, like $z_1|a_1\rangle + z_2|a_2\rangle$, $z_1, z_2 \in \mathbb{C}$, representing complex vectors by column lists and adjoint vectors by row lists, for example, for 2-component vectors,

$$|a\rangle \to \begin{pmatrix} a_1 \\ a_2 \end{pmatrix}, \quad \langle a| \to \left(a_1^*,\ a_2^*\right).$$

We also define the inner product of two complex vectors:

$$\langle a|b\rangle \to \left(a_1^*\ a_2^*\right)\begin{pmatrix} b_1 \\ b_2 \end{pmatrix} = a_1^* b_1 + a_2^* b_2.$$

Prove that $\langle a|b\rangle = \langle b|a\rangle^*$. Given a complex matrix M, prove that $\langle a|M|b\rangle = \langle b|M^\dagger|a\rangle^*$, where $M^\dagger$ is the Hermitian conjugate of the matrix M. Prove that $\langle a|M|a\rangle \in \mathbb{R}$ if the matrix M is Hermitian ($M^\dagger = M$). Defining the eigenvectors and eigenvalues of a Hermitian matrix,

$$M|u_i\rangle - \lambda_i|u_i\rangle, \quad i - 1, 2, \quad \lambda_i \in \mathbb{C},$$

prove that if $\lambda_1 \neq \lambda_2$, then $\langle a_1|a_2\rangle = 0$ (orthogonality). Find the eigenvalues and eigenvectors of the Pauli matrices and check the previous relations.

48. Consider a unit ket vector represented by the column

$$|a\rangle \rightarrow \begin{pmatrix} a_1 \\ a_2 \end{pmatrix}, \quad a_1, a_2 \in \mathbb{C}, \quad a_1 (a_1)^* + a_2 (a_2)^* = 1.$$

Show that for every ket vector $|a\rangle$ we can find a spatial unit vector $n = \sum_{i=1}^{3} n_i e_i$ such that

$$\left(\sum_{i=1}^{3} n_i \hat{\sigma}_i \right) \begin{pmatrix} a_1 \\ a_2 \end{pmatrix} = \begin{pmatrix} a_1 \\ a_2 \end{pmatrix},$$

where $\hat{\sigma}_i$ are the Pauli matrices.

49. Using the Pauli matrices, find the matrix representation $[a]$ for the vector a. Then you can "imitate" various transformations from GA, like reflections, rotations, etc. For example, the reflection $-e_i a e_i$ becomes $-\hat{\sigma}_i [a] \hat{\sigma}_i$. Play with transformations to see the effect on $[a]$.
50. Taking advantage of the Clifford conjugation, show that a commutator of two complex vectors is a complex vector, while their anticommutator is a complex scalar.
51. Consider a nilpotent $\mathbf{N}$ and its direction unit vector $\hat{\mathbf{k}}$. Form the rotor $R = R(\hat{\mathbf{k}}, \theta)$ that rotates the nilpotent $\mathbf{N}$ around the vector $\hat{\mathbf{k}}$ by the angle θ and then, using the relation $\hat{\mathbf{k}}\mathbf{N} = -\mathbf{N}\hat{\mathbf{k}} = \mathbf{N}$, show that $R\,\mathbf{N}\,R^{\dagger} = \mathbf{N} \exp(-j\theta)$.
52. In the text, we commented on some properties of the Pauli matrices and showed (Sect. 4.1) that every 2×2 complex matrix can be decomposed over the Pauli matrices with complex coefficients. Now we can use the Clifford basis, defined in the text, to make replacements $1 \rightarrow I$ (2×2 unit matrix) and $e_i \rightarrow \hat{\sigma}_i$. Thus, we get the matrix representation of the Clifford basis. Show that such a matrix representation is the basis for the vector space of 2×2 complex matrices. Expand the multivector $M = \alpha + \mathbf{x} + j\mathbf{n} + j\beta$ in such a basis. What you get?
53. Suppose that we want to rotate a vector w about the unit vector $\hat{e}$ by an angle θ. However, we do not have an implementation of geometric algebra. Can we do that using an advanced calculator, still using results of GA? Show that by decomposing the vector w parallel and perpendicular to $\hat{e}$, we can obtain the rotation as $w' = w_{\parallel} - w_{\perp} \cos\theta + w_{\perp} \times \hat{e} \sin\theta$, which means that we just need the possibility to calculate the standard inner and cross products. Can you figure out more such examples?
54. In *Cl*3, we can define the dual numbers, with the "units" $q_1 = e_1 + je_2$, $q_2 = jq_1$. Show that the only possibility to define the third "unit" q_3 with the property $q_i q_j = 0,\ i, j = 1, 2, 3$ is a linear combination $q_3 = \alpha q_1 + \beta q_2$. Check that the bivectors $B_i = je_i$ have the desired property, however with the outer product, that is, $B_i \wedge B_j = 0,\ i, j = 1, 2, 3$.
55. Consider two elements $|\uparrow\rangle = (1 + e_3)/2$ and $|\downarrow\rangle = e_1 |\uparrow\rangle$ in *Cl*3 (they could represent the two states 1 and 0, for example). We also define the unit vector $\hat{h} = (e_1 + e_3)/\sqrt{2}$, the operators (*raising* and *lowering*) $a_+ = |\uparrow\rangle e_1$ and $a_- = e_1 |\uparrow\rangle$,

as well as a general state $\psi = c_1|\uparrow\rangle + c_0|\downarrow\rangle$, $|c_1|^2 + |c_0|^2 = 1$, $c_1 = \exp(j\alpha)\cos(\theta/2)$, $c_0 = \exp(j\beta)\sin(\theta/2)$, $\alpha, \beta \in \mathbb{R}$.

(a) If $\hat{h} = \exp(X)e_1 = \exp(Y)e_3$, find X and Y.
(b) Find $O|\uparrow\rangle$ and $O|\downarrow\rangle$, if O stands for e_1, $\hat{h}$, and $a_{\pm}$.
(c) Show that $a_{\pm}$ are nilpotents.
(d) Find $[a_+, a_-]$ and $\{a_+, a_-\}$ (*commutator* and *anticommutator*).
(e) Can you express the general state ψ in the form $\psi = f(\phi, \theta, \chi)|\uparrow\rangle$, where (ϕ, θ, χ) are the Euler angles? (*Hint*: $\phi = \alpha - \beta$.)

56. The unit vector n can define a unit sphere in 3D. If we define the *stereographic projection* by

$$x = f(n) \equiv \frac{2}{n + e_3} - e_3 = m - e_3, \quad n \neq -e_3,$$

show that the vector x belongs to the e_1e_2 plane. Find $\hat{m}e_3\hat{m}$, where $\hat{m}$ is a unit vector.

57. Show that $P_{\pm 3} = (1 \pm e_3)/2$, $P_{\pm\alpha} = (1 \pm Ie_4e_5)/2$, and $P_{\pm\beta} = (1 \pm Ie_6e_7)/2$ are idempotents in *Cl*7, where I is the unit pseudoscalar. Define the eight elements $P(k)$ of the form $P_{s_1 3}P_{s_2\alpha}P_{s_3\beta}$, $s_i = \pm$, and show that $\sum_k P(k) = 1$ (the primitive projectors). What can you say about the elements $(1 \pm e_4e_5e_6e_7)/2$?

58. We define the pseudoscalar of the conformal model in $\mathfrak{R}^n$ as $I_{n+1,1} = o \wedge I_n \wedge \infty$, where I_n is the Euclidean unit pseudoscalar. Find $I_{n+1,1}^{-1}$.

59. Using the language of *Cl*2, show that the function of a complex variable $f(z) = z^*$ is not analytic.

60. Show that each of the Pauli matrices contains two independent spinors. Express them in the language of *Cl*3.

61. In the *spacetime algebra* (STA), define the four unit orthonormal vectors γ_μ, $\mu = 0, 1, 2, 3$, with the signature $(1, -1, -1, -1)$. Write down the Clifford basis of STA. Can you identify the even part of the algebra? Define $\sigma_k = \gamma_k\gamma_0$, $k = 1, 2, 3$, and find the multiplication table for σ_k. What do you see?

62. The outer product can be identified as Grassmann's *progressive product*. Grassmann also realized that it is possible to define another, the *regressive product*, which for blades A_r and B_s in the geometric algebra of n-dimensional vector space V_n is defined as

$$A_r \vee B_s \equiv A_r^{\Delta} \rfloor B_s.$$

Show that the grade of the regressive product is $r + s - n$. Show that the "De Morgan" rule $A_r \vee B_s = A_r^{\Delta} \wedge B_s^{\Delta}$ is valid. It is common to define *covectors* (*pseudovectors*) as duals of vectors. Each covector A defines a hyperplane by the nonmetric relation $x \wedge A = 0$. Show that this relation can be written in the form $x \cdot A^{\Delta} = (x \wedge A)I^{-1} = 0$, which demonstrates that **left contraction is a nonmetric operation**. Note that $x \cdot A^{\Delta}$ is a *linear form* on V_n. Show that previous relations are independent of the signature. For $A_r \vee B_s \neq 0$, we can identify $A_r \vee B_s$ as the *meet* (intersection) of subspaces defined by A_r and B_s, while for $A_r \wedge B_s \neq 0$, we identify $A_r \wedge B_s$ as their *join* (union). We can define the cross product of vectors a and b as

$(a \wedge b)^{\Delta}$, independent of dimension or signature. Show that $(a \wedge b)^{\Delta} = b^{\Delta} \vee a^{\Delta}$, which suggests a new meaning of the cross product as an intersection of hyperplanes defined by covectors.

63. Show that for blades A and B, grade(A) $\leq$ grade(B), one has $A \cdot (\bar{\mathrm{F}}B) = \bar{\mathrm{F}}[(\mathrm{F}A) \cdot B]$, where $\bar{F}$ is the adjoint of the linear transformation F, defined as $\langle A\bar{\mathrm{F}}B \rangle \equiv \langle B\mathrm{F}A \rangle$. Show that for $B = I$ (the unit pseudoscalar), we have the relation for the inverse of a linear transformation. Also show that

$$(\det\mathrm{F})\mathrm{F}(A \vee B) = (\mathrm{F}A) \vee (\mathrm{F}B),$$
$$(\det\mathrm{F})\mathrm{F}\left(A^{\Delta} \cdot B\right) = (\mathrm{F}A)^{\Delta} \cdot (\mathrm{F}B).$$

64. Define *covectors* $\underset{\rightarrow i}{e} = e_i I$, where e_i are the vectors of an orthonormal basis, and find $a\underset{\rightarrow}{b}$, $a \rfloor \underset{\rightarrow}{b}$, and $a \wedge \underset{\rightarrow}{b}$ for $a = \sum_i \mathrm{a}_i e_i$ and $\underset{\rightarrow}{b} = \sum_i \mathrm{b}_i \underset{\rightarrow i}{e}$, $\mathrm{a}_i, \mathrm{b}_i \in \mathbb{R}$. Interpret your results.

65. For vectors a and b, find the angle between projections of a and b on the plane defined by the unit 2-blade $\hat{B}$. (This problem is from astronomy.)

66. If A is any invertible element of $Cl3$ and x is a vector, we can define transformations

$$x' \rightarrow AxA^{-1}.$$

Discus conditions on A such that x' is a vector. Show that if x' is a vector, then the transformations are orthogonal and form a group (the Clifford group). We can find corresponding transformations $x' \rightarrow Ox$ in the language of linear algebra that form an orthogonal group. Instead of A, we can use AZ, where Z is a complex scalar, which means that the mapping from the Clifford group onto the orthogonal group is many-to-one. By suitably normalizing elements of the Clifford group, we obtain a subgroup such that the mapping onto the orthogonal group is two-to-one. As an example, consider rotations and note that from $x' \rightarrow RxR^{\dagger}$, $RR^{\dagger} = 1$, we have $R^{-1} = R^{\dagger}$, but also $x' \rightarrow (RZ)x(RZ)^{-1}$. Rotors R are already normalized, and we have a two-to-one mapping of SU(2) onto SO(3) (double cover; see Sect. 1.9.8).

Consider the elements of $Cl3$ such that $A^{-1} = \mathrm{inv}(A)$, where inv is an involution. Discus the possibilities and identify the orthogonal transformations. (Hint: perhaps it is a good idea to work with the spectral decomposition of a multivector $A = A_+ u_+ + A_- u_-$.)

67. The Lorentz force is defined as $\mathbf{F} = q\mathbf{v} \times \mathbf{B}$, where $\mathbf{v}$ is a velocity vector of a charged particle with electric charge q, while $\mathbf{B}$ is a magnetic field vector. Show that if we define the bivector $B = j\mathbf{B}$, then the Lorentz force can be written as $\mathbf{F} = -q\mathbf{v} \rfloor B$. Interpret your result for protons and electrons. (Hint: see Appendix 5.8 and Fig. 1.33.)

6.3 Why Geometric Algebra?

William Kingdon Clifford died young, when geometric algebra (the term he introduced) was just starting to develop and be applied in physics. Grassmann's ingenious work on new geometric objects (like bivectors) was neglected, and Clifford was one of a few who recognized his work and was capable of developing it further. Sadly, people usually do not like new things, especially when they come from schoolteachers, like Grassmann. Then Gibbs comes on the scene, ignores Grassmann's and Clifford's deep insights and develops the vector calculus in 3D, introducing the scalar and the cross products, with extensive use of coordinates. People said, "Wonderful, we multiply vectors and get vectors!" and the era of modern mathematics began. Inevitably, mathematicians and physicists mounted their horses and rode off in many different directions, diminishing the possibility to communicate. They bought the cross product too easily. From time to time, somebody says, "Hello, I have discovered some interesting matrices!" (Pauli), and the other, "Hello, I have discovered some interesting matrices, too!" (Dirac), or, "Hello, I have discovered some interesting objects!" (spinors, many of them), etc. When Einstein published his paper on relativity (1905), people said, "What is this? How are we to calculate?" Unfortunately, nobody recognized that there is a nice way to formulate the new theory in 3D; instead, Minkowski formulated it in 4D, using vectors that have a negative square. What a pity, since all the new theories (Pauli, Dirac, STR) fit in 3D nicely.

Even today, mathematicians generally do not recognize the power of our geometric intuition, and this is hard to understand. Personally, when I saw the idea of a bivector for the first time, I literally jumped from the chair, started to walk around, while a storm of ideas and pictures flooded my mind. I repeated to myself, "What a simple and powerful idea! And how is it that I did not think of it myself?" I could see a bivector in my mind, imagine its actions, and I called it a "magical beast." I speculated that this new stuff would blow away tensors, matrices, differential forms. After studying the subject, I saw that I was right. In fact, after initial difficulties, I was surprised at the magnificent possibilities of unification of mathematical disciplines. What can I say? Today, I am in love with mathematics again, as in the days of my youth. Yes, it is true; my love was dying, it talked too much, in different languages, and that was awful.

To continue, vectors worked well in 3D, coordinates were absorbed in algebras, and geometry was neglected. Some people, Hestenes is the most important, began to alter this situation. Today, the army of geometric algebra users is growing. Why? We can try to give some answers, without pretensions to being complete.

- the possibility of unification of the mathematical language
- makes the most of our geometric intuition
- invertibility (see the text), even for derivative operators
- nontrivial simplifications and generalizations, as for integral theorems, including a complex area
- one tool for all areas in physics and science in general
- powerful implementations on computers, there is no if-then-else (mainly)
- just one implementation for quaternions, complex numbers, dual number, etc., in fact, put here what you want

- operators are elements of the algebra
- object-oriented geometry teaching
- faster software (20% for rotations, according to NASA)
- just one tool for any dimension or signature
- there is no need to learn many different mathematical languages
- possibilities of automatic theorem proving
- automatic code generation
- the possibility to change the architecture of computers; imagine a computer that knows how to calculate with subspaces, with the geometric product incorporated in the processor
- gives new insights into existing theories
- eliminates the ordinary imaginary unit from physics
- learning can start early, through games with oriented objects
- seamlessly integrates the properties of vectors and complex numbers to enable a completely coordinate-free treatment of 2D problems
- reduces "grad, div, curl, and all that" to a single vector derivative that, among other things, combines the standard set of four Maxwell equations into a single equation and provides new methods to solve it
- the GA formulation of spinors facilitates the treatment of rotations and rotational dynamics in both classical and quantum mechanics without coordinates or matrices
- unseen power and simplicity of rotations in any dimension, etc.

Computers are irreplaceable tools today. However, they also have a *coordinate virus*, to paraphrase Hestenes. They prefer numbers, coordinates, matrices. However, try to teach them that $ab + ba$ is a real number, perhaps that $ab - ba$ is an oriented geometric object or, at least, that a is a vector—and please, without cheating with coordinates at some step. One cannot teach a monkey to sing without the possibility of its being able to talk. Even in symbolic languages, like the *Mathematica* language, it is hard to implement geometric algebra just using some set of rules on predefined geometric objects. Computers know how to multiply real numbers; however, it would be nice if they knew how to perform the geometric product. Computers do not think in terms of geometry; they think in terms of real numbers, in fact, just two of them (0 and 1). It would be nice and powerful to have the possibility of saying to a computer, "This is a bivector, and those two vectors are orthogonal." Of course, without specifying a set of rules, such rules should be a part of the processor architecture. Computers were designed at a time of little awareness that elements of algebra can represent geometric objects (even vectors are treated as lists of numbers). It is unfair to compare the implementations of standard linear algebra and geometric algebra on computers that are the product of the algebraic way of thinking. Nevertheless, geometric algebra gives comparable and even better speed results. However, speed should not be the only criterion. For instance, imagine a truly fast program that returns ones and zeros. How is one to interpret the result? Humans have a powerful geometric intuition, and computers should be able to reflect this property of the human mind.

I claim that Grassmann and Whitehead were just one step away from a mathematical system that truly deserves to be regarded as a UNIVERSAL GEOMETRIC ALGEBRA. That system is known in mathematics as Clifford algebra. However, the true universality of Clifford algebra has remained unrecognized, even by mathematicians specializing in the subject. The main reason, I suppose, is that the demonstration of universality requires a wholescale reorganization and redesign of mathematics, integrating into a single mathematical system such superficially disparate systems as quaternion calculus, differential forms and vector analysis. The result is much more than a set of axioms for Clifford algebra along with a few theorems revealing the structure of the algebra. It is a full-blown mathematical language for expressing and elaborating geometric ideas of every sort.

Clifford algebra is commonly regarded as "the algebra of a quadratic form." This, it seems, has been a major barrier to recognizing the geometric universality of the algebra. For it suggests that Clifford algebra is inapplicable to nonmetrical geometry. The main objective of this article is to dispel that misconception conclusively by showing explicitly how nonmetrical geometry can be handled with Clifford algebra. We shall see how Grassmann's progressive and regressive products can be defined within Clifford algebra and employed in a coordinate-free algebraic formulation of projective geometry. This approach has a number of advantages. First, the complementary algebraic and synthetic approaches to projective geometry are brought much closer together, because the primitives in the synthetic formulation correspond directly to algebraic entities and operations. Accordingly, it becomes easier to combine the advantages of both approaches. Second, the algebraic methods of projective and metrical geometry are unified perfectly. This should help integrate the deep ideas of projective geometry with the rest of mathematics. Third, some improvements in linear algebra are suggested to increase its geometrical perspicacity and computational power.
(Hestenes, [28])

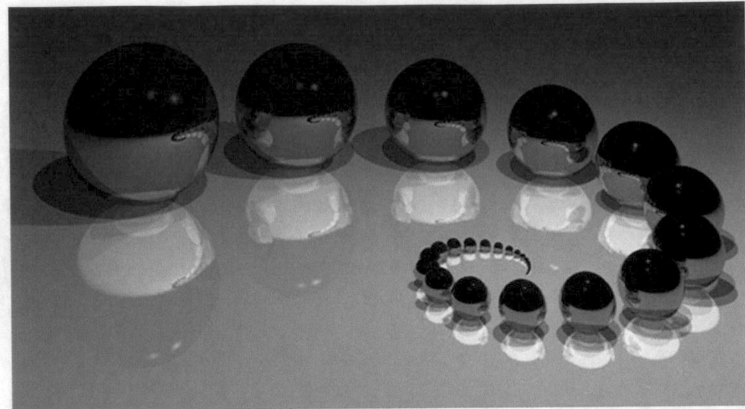

A golden spiral

Ray tracing on spheres by Ken Pratt. In addition to the shape of the image being a golden spiral, the size of the spheres decreases in proportion to the golden ratio, and the colours change in proportion to the golden ratio as well. So it's sort of a 'triple' golden spiral

6.4 Formulas

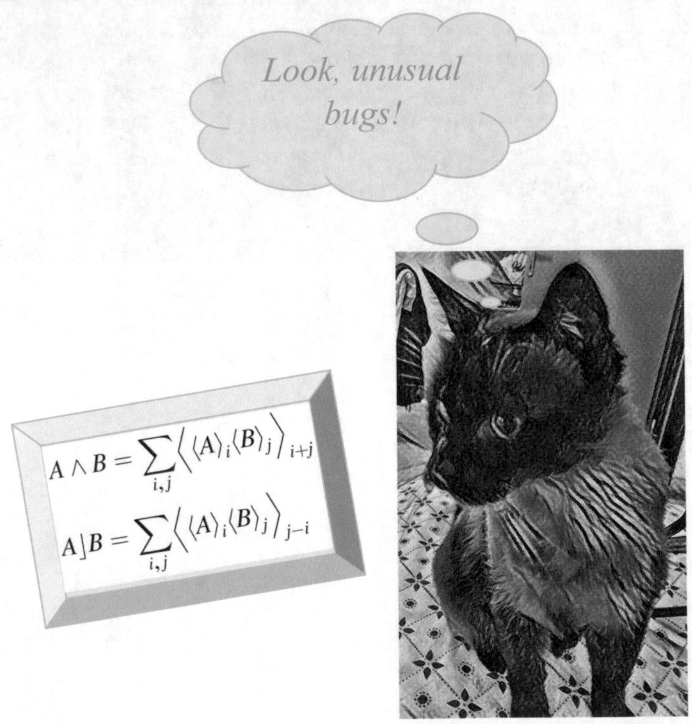

Look, unusual bugs!
$A \wedge B = \sum_{i,j} \left\langle \langle A \rangle_i \langle B \rangle_j \right\rangle_{i+j}$
$A \rfloor B = \sum_{i,j} \left\langle \langle A \rangle_i \langle B \rangle_j \right\rangle_{j-i}$

$$a \cdot b \equiv (ab + ba)/2 = ab_{\|}, \quad b_{\|} = a \cdot ba$$
$$a \wedge b \equiv (ab - ba)/2 = ab_{\perp}, \quad b_{\perp} = b - a \cdot ba$$
$$e_i e_j + e_j e_i = \pm 2\delta_{ij}$$
$$ab = a \cdot b + a \wedge b, \quad a \cdot b = |a||b| \cos\theta, \quad |a \wedge b| = |a||b||\sin\theta|$$
$$ab = |a||b|\hat{a}\hat{b} = |a||b|\big(\hat{a} \cdot \hat{b} + \hat{a} \wedge \hat{b}\big)$$
$$= |a||b|\big(\cos\theta + \hat{B}\sin\theta\big), \ \hat{B} = \frac{\hat{a} \wedge \hat{b}}{\sin\theta}, \ \hat{B}^2 = -1$$
$$ab = |a||b|\exp\big(\hat{B}\theta\big)$$
$$a^{-1} = a/a^2$$
$$A \wedge B = \sum_{i,j} \Big\langle \langle A \rangle_i \langle B \rangle_j \Big\rangle_{i+j} \quad \text{(outer product)}$$
$$A \rfloor B = \sum_{i,j} \Big\langle \langle A \rangle_i \langle B \rangle_j \Big\rangle_{j-i} \quad \text{(left contraction)}$$
$$A \lfloor B = \sum_{i,j} \Big\langle \langle A \rangle_i \langle B \rangle_j \Big\rangle_{i-j} \quad \text{(right contraction)}$$
$$A * B = \sum_{i,j} \Big\langle \langle A \rangle_i \langle B \rangle_j \Big\rangle_0 \quad \text{(scalar product)}$$
$$A \bullet B = \sum_{i,j} \Big\langle \langle A \rangle_i \langle B \rangle_j \Big\rangle_{|j-i|} \quad \text{(dot product)}$$
$$A \cdot B = \sum_{i \neq 0,\, j \neq 0} \Big\langle \langle A \rangle_i \langle B \rangle_j \Big\rangle_{|j-i|} \quad \text{(inner product)}$$
$$A^{\Delta} = A \rfloor I^{-1} \quad \text{(dual)}$$
$$A \vee B = A^{\Delta} \cdot B \quad \text{(regressive product)}$$
$$\mathbf{a} \times \mathbf{b} = -j\mathbf{a} \wedge \mathbf{b} \quad \text{(cross product, a new definition)}$$
$$-j(e_k \wedge e_l \wedge e_m) = \varepsilon_{klm}, \quad \varepsilon_{123} = \varepsilon_{231} = \varepsilon_{312} = -\varepsilon_{213} = -\varepsilon_{132} = -\varepsilon_{321} = 1$$
$$\langle ABC \rangle = \langle CAB \rangle$$
$$a = n^2 a = n(n \cdot a + n \wedge a) = nn \cdot a + nn \wedge a = a_{\|} + a_{\perp}$$
$$\overleftarrow{(MN)} = \overleftarrow{M}\overleftarrow{N} \quad \text{(automorphism)}$$
$$(MN)^{\dagger} = N^{\dagger}M^{\dagger} \quad \text{(antiautomorphism)}$$
$$\overline{MN} = \bar{N}\bar{M} \quad \text{(antiautomorphism)}$$

For a vector a and bivector B we have

$$a \cdot B = (aB - Ba)/2$$
$$a \wedge B = (aB + Ba)/2$$

$$aB = a \cdot B + a \wedge B$$

$$\mathbf{x} \cdot (\mathbf{a}_1 \wedge \cdots \wedge \mathbf{a}_n) = \sum_{k=1}^{n} (-1)^{k+1} \mathbf{x} \cdot \mathbf{a}_k \left(\mathbf{a}_1 \wedge \cdots \wedge \breve{\mathbf{a}}_k \wedge \cdots \wedge \mathbf{a}_n\right)$$

$$R = mn = \exp(-\hat{B}\theta/2), \quad \hat{B} = \frac{n \wedge m}{\sin\theta}$$

$$R = \exp(-je_3\phi/2)\exp(-je_1\theta/2)\exp(-je_3\psi/2)$$

(a rotor expressed through the Euler angles)

$$\mathrm{grade}(A \rfloor B) = \mathrm{grade}(B) - \mathrm{grade}(A)$$

$$\alpha \rfloor B = \alpha B, \quad \alpha \in \mathbb{R}, \quad a \rfloor b \equiv a \cdot b,$$

$$a \rfloor (B \wedge C) = (a \rfloor B) \wedge C + \overleftarrow{B} \wedge (a \rfloor C),$$

$$A \rfloor (B \rfloor C) = (A \wedge B) \rfloor C \quad \text{(universal)}, \quad (A \rfloor B) \rfloor C = A \wedge (B \rfloor C), A \subseteq C \text{ (conditional)}$$

$$x \rfloor M = \left(xM - \overleftarrow{M}x\right)/2, \quad x \wedge M = \left(xM + \overleftarrow{M}x\right)/2, \quad xM = x \rfloor M + x \wedge M$$

$$a \rfloor (A_k I) = (a \wedge A_k) I = a \wedge A_k I, \quad a \wedge (A_k I) = (a \rfloor A_k) I = a \rfloor A_k I,$$

$$A \rfloor B = \left(A \wedge \left(BI^{-1}\right)\right) I$$

$$\mathrm{P}_B(A) = \left(A \rfloor B^{-1}\right) B \quad \text{(projector)}$$

$$x = xBB^{-1} = x \cdot BB^{-1} + x \wedge BB^{-1}, \quad B \text{ is a bivector}$$

$$A \otimes B \equiv (AB - BA)/2 \quad \text{(commutator)}$$

$$(A \otimes B) \otimes C + (C \otimes A) \otimes B + (B \otimes C) \otimes A = 0 \quad \text{(Jacobi identity)}$$

$$BX = B \rfloor X + B \wedge X + B \otimes X, \quad B \text{ is a bivector}$$

6.4.1 Linear Transformations

$$\mathrm{F}(I) \equiv I \det \mathrm{F}, \quad \det \mathrm{F} \in \mathbb{R}$$

$$\bar{\mathrm{F}}(e_j) = e^i e_j \cdot \mathrm{F}(e_i)$$

$$\mathrm{F}^{-1}(A) = I\bar{\mathrm{F}}\left(I^{-1}A\right)(\det \mathrm{F})^{-1}, \quad \bar{\mathrm{F}}^{-1}(A) = I\mathrm{F}\left(I^{-1}A\right)(\det \mathrm{F})^{-1}$$

6.4.2 *Cl*3

$$\mathbf{a} \wedge \mathbf{b} = \begin{vmatrix} je_1 & je_2 & je_3 \\ a_1 & a_2 & a_3 \\ b_1 & b_2 & b_3 \end{vmatrix}$$

$$\mathbf{a}\wedge\mathbf{b}\wedge\mathbf{c}=\begin{vmatrix} a_1 & a_2 & a_3 \\ b_1 & b_2 & b_3 \\ c_1 & c_2 & c_3 \end{vmatrix} j$$

$$M=Z+\mathbf{F},\quad Z=t+bj,\quad \mathbf{F}=\mathbf{x}+j\mathbf{n},\quad t,b\in\mathbb{R}$$

$$M\bar{M}=(Z+\mathbf{F})(Z-\mathbf{F})=Z^2-\mathbf{F}^2=\bar{M}M$$

$$\mathbf{F}^2=x^2-n^2+j(\mathbf{xn}+\mathbf{nx})=x^2-n^2+2j\mathbf{x}\cdot\mathbf{n}$$

$$M^{-1}=\bar{M}/M\bar{M}$$

$$\mathbf{N}=\mathbf{x}+j\mathbf{n},\quad x=n,\ \mathbf{x}\cdot\mathbf{n}=0,\quad \mathbf{N}^2=0\quad \text{(nilpotent)}$$

$$\mathbf{f}=\mathbf{n}\cosh\varphi+j\mathbf{m}\sinh\varphi,\quad \mathbf{n}^2=\mathbf{m}^2=1,\quad \mathbf{n}\perp\mathbf{m},\quad \mathbf{f}^2=1$$

$$u_\pm=(1\pm\mathbf{f})/2,\quad u_++u_-=1,\quad u_+-u_-=\mathbf{f},\quad u_+u_-=u_-u_+=0,$$

$$u_\pm^2=u_\pm,\quad \bar{u}_+=u_-$$

$$\mathbf{F}^2\neq 0\Rightarrow M=Z+\mathbf{F}=Z+\sqrt{\mathbf{F}^2}\mathbf{f}\equiv Z+Z_f\mathbf{f},\ \mathbf{f}^2=1,\quad M_\pm=Z\pm Z_f$$

$$M=M_+u_++M_-u_-,\quad \mathrm{f}(M)=\mathrm{f}(M_+)u_++\mathrm{f}(M_-)u_-$$

$$\log M=\log|M|+\varphi\mathbf{I},\quad \varphi=\arctan(|\mathbf{F}|/Z),\quad |M|=\sqrt{M\bar{M}}$$

$$\mathbf{I}=\mathbf{n}\sinh\varphi+j\mathbf{m}\cosh\varphi,\quad \mathbf{n}^2=\mathbf{m}^2=1,\quad \mathbf{n}\perp\mathbf{m},\quad \mathbf{I}^2=-1$$

$$\varphi=argM\equiv\arctan(|\mathbf{F}|/Z),\quad \cos\varphi=Z/|M|,\quad \sin\varphi=|\mathbf{F}|/|M|,$$

$$M=Z+\mathbf{F}=|M|(\cos\varphi+\mathbf{I}\sin\varphi)$$

$$M^n=|M|^n[\cos(n\varphi)+\mathbf{I}\sin(n\varphi)]$$

$$M=Z+\mathbf{F}=Z+Z_f\mathbf{f}=\rho\left(\frac{Z}{\rho}+\frac{Z_f}{\rho}\mathbf{f}\right)$$

$$=\rho(\cosh\varphi+\mathbf{f}\sinh\varphi),\quad \rho=\sqrt{M\bar{M}}=\sqrt{Z^2-Z_f^2}$$

$$\vartheta\equiv\tanh\varphi\Rightarrow M=\rho(\cosh\varphi+\mathbf{f}\sinh\varphi)=\rho\gamma(1+\vartheta\mathbf{f}),\quad \gamma^{-1}=\sqrt{1-\vartheta^2}$$

$$u=\gamma(1+\vartheta\mathbf{f}),\ u\bar{u}=1\quad (\textit{proper velocity})$$

$$M=\rho\gamma(1+\vartheta\mathbf{f})=k_+u_++k_-u_-\Rightarrow k_\pm=\rho\gamma(1\pm\vartheta)=\rho k^{\pm1}$$

$$k=\sqrt{(1+\vartheta)/(1-\vartheta)},\quad \varphi=\log k$$

6.4.3 The Special Theory of Relativity

$$p\bar{p}=(t+\mathbf{x})(t-\mathbf{x})=t^2-x^2\in\mathbb{R}$$

$$t^2-x^2=t^2(1-v^2)=\tau^2\Rightarrow t^2/\tau^2=1/(1-v^2)=\gamma^2$$

$$u\equiv\gamma(1+\mathbf{v})$$

$$\gamma = \cosh\varphi, \quad \gamma v = \sinh\varphi, \quad u = \cosh\varphi + \hat{\mathbf{v}}\sinh\varphi = \cosh\varphi(1 + \hat{\mathbf{v}}\tanh\varphi)$$
$$= \exp(\varphi\hat{\mathbf{v}})$$
$$\gamma(1 + v\hat{\mathbf{v}}) = k_+u_+ + k_-u_- \Rightarrow k_\pm = \gamma(1 \pm v) = \cosh\varphi \pm \sinh\varphi, \quad \varphi \equiv \log k$$
$$k^{\pm 1} = \cosh\varphi \pm \sinh\varphi, \quad k = \sqrt{(1+v)/(1-v)}, \quad u = ku_+ + k^{-1}u_-$$

6.4.4 Lorentz Transformations

$$L = BR, \quad L\bar{L} = 1, \quad \text{where}$$
$$B = \cosh(\varphi/2) + \hat{\mathbf{v}}\sinh(\varphi/2) = \exp(\varphi\hat{\mathbf{v}}/2),$$
$$R = \cos(\theta/2) - j\hat{\mathbf{w}}\sin(\theta/2) = \exp(-j\hat{\mathbf{w}}\theta/2)$$
$$p' = LpL^\dagger = BRpR^\dagger B^\dagger = BRpR^\dagger B$$
$$B = \sqrt{u} = \frac{u+1}{\sqrt{2(\gamma+1)}}$$

6.4.5 Electromagnetic Field

$$\mathbf{F} = \mathbf{E} + j\mathbf{B}, \quad c = 1$$
$$\mathbf{B} = B_1e_1 + B_2e_2 + B_3e_3, \quad j\mathbf{B} = j(B_1e_1 + B_2e_2 + B_3e_3)$$
$$= B_3e_1e_2 + B_2e_3e_1 + B_1e_2e_3$$
$$\mathbf{F}^2 = E^2 - c^2B^2 + 2jc\mathbf{E}\cdot\mathbf{B}$$
$$E = B, \mathbf{E}\perp\mathbf{B}, \ \mathbf{F} = \mathbf{E} + j\mathbf{B} \Rightarrow \mathbf{F}^2 = 0$$
$$\mathbf{F}\mathbf{F}^\dagger = E^2 + c^2B^2 + 2c\mathbf{E}\times\mathbf{B}$$
$$\mathbf{F}_0e^{\pm jkx}e^{j\omega t}, \quad \mathbf{F}_0 = \mathbf{E}_0 + j\mathbf{B}_0 \quad \text{(EM wave in vacuum)}$$
$$\dot{p} = m\dot{u} = e\langle\mathbf{F}u\rangle_R \quad \text{(Lorentz force)}$$
$$\partial\,\mathbf{F} = \mathbf{J} \quad \text{(Maxwell equations)}$$

6.4.6 Quantum Mechanics

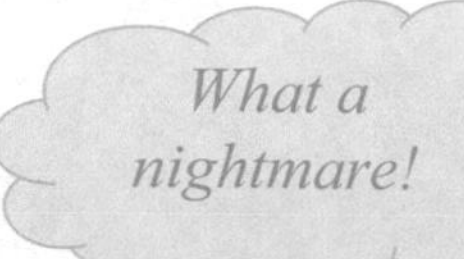

$$|\psi\rangle = \begin{pmatrix} a^0 + ia^3 \\ -a^2 + ia^1 \end{pmatrix} \leftrightarrow \psi = a^0 + a^k j\sigma_k \quad \text{(spinor)}$$

$$\hat{\sigma}_k|\psi\rangle \leftrightarrow \sigma_k\psi\sigma_3 \quad \text{(operators)}$$

$$i|\psi\rangle \leftrightarrow \psi j\sigma_3$$

$$\mathbf{s} \equiv \frac{1}{2}\hbar\psi\sigma_3\psi^\dagger \quad \text{(spin vector)}$$

$$\langle\psi|\hat{s}_k|\psi\rangle = \frac{1}{2}\hbar\langle\sigma_k\psi\sigma_3\psi^\dagger\rangle = \sigma_k \cdot \mathbf{s} \quad \text{(expected value)}$$

$$\rho = \psi\psi^\dagger, \quad R = \rho^{-1/2}\psi \text{ (rotor)}, \quad \mathbf{s} = \frac{1}{2}\hbar\rho R\sigma_3 R^\dagger \quad \text{(spin vector)}$$

6.4.7 Conformal Model (*Hestenes*)

$$e,\ e^2 = 1, \quad \bar{e},\ \bar{e}^2 = -1, \quad \{e, e_1, e_2, e_3, \bar{e}\}$$

$$o = (e + \bar{e})/2, \quad \infty = \bar{e} - e, \quad \{o, e_1, e_2, e_3, \infty\}$$

$$p = o + \mathbf{p} + \mathbf{p}^2\infty/2, \quad p \cdot p = 0, \quad p \cdot q \propto (\mathbf{p} - \mathbf{q}) \cdot (\mathbf{p} - \mathbf{q})$$

Correction to: Geometric Multiplication of Vectors

Miroslav Josipović

M. Josipović, *Geometric Multiplication of Vectors*,
Compact Textbooks in Mathematics,
https://doi.org/10.1007/978-3-030-01756-9_7

Correction to: M. Josipović, *Geometric Multiplication of Vectors*, Compact Textbooks in Mathematics, https://doi.org/10.1007/978-3-030-01756-9

This book was inadvertently published without updating the following corrections. An error in the production process unfortunately led to publication of this chapter prematurely, before incorporation of the final corrections. The version supplied here has been corrected and approved by the author.

Reference Labels:

A.n Appendix n has been removed from the list

Corrections:

Page 25, 1.6.6 Idempotents, line 4, "E21: Show that the trivial idempotent is the only one with an inverse." has been changed to "◆ E21: Show that the trivial idempotent is the only one with an inverse.◆"

The updated online versions of these chapters can be found at
https://doi.org/10.1007/978-3-030-01756-9_1
https://doi.org/10.1007/978-3-030-01756-9_2
https://doi.org/10.1007/978-3-030-01756-9_3
https://doi.org/10.1007/978-3-030-01756-9_4
https://doi.org/10.1007/978-3-030-01756-9_6
https://doi.org/10.1007/978-3-030-01756-9

1.6.7 Zero Divisors, line 7, "Find $(\alpha u_+ + \beta u_-)^n$, $\alpha, \beta \in \mathbb{R}$, $n \in \mathbb{N}$" has been changed to "◆ Find $(\alpha u_+ + \beta u_-)^n$, $\alpha, \beta \in \mathbb{R}$, $n \in \mathbb{N}$.◆"

Page 29, 1.8.1 Merging Multiplication Tables, error in multiplication table below 1st paragraph. The correction presentation is given below.

$\cdot$	e_1	e_2	e_3
e_1	1	0	0
e_2	0	1	0
e_3	0	0	1

$\oplus \; j\otimes$

$\times$	e_1	e_2	e_3
e_1	0	e_3	$-e_2$
e_2	$-e_3$	0	e_1
e_3	e_2	$-e_1$	0

$\rightarrow$

GP	e_1	e_2	e_3
e_1	1	je_3	$-je_2$
e_2	$-je_3$	1	je_1
e_3	je_2	$-je_1$	1

Page 49, line 9, "blades in Euclidean vector" has been changed to "blades in nondegenerate vector"

Page 60, 1.14 Spinors, line 3, "In geometric algebra, such transformations…" has been changed to "In geometric algebra, such elements…"

Page 82, lines 1 and 2, "transformations for small velocities…" has been changed to "transformations for the infinite speed of light"

Page 82, 2.3 Idempotents and Hyperbolic Structure, line 1, "For $\mathbf{F}^2 = x^2 - n^2 = 1$" has been changed to "For $\mathbf{F}^2 = 1$".

Page 113, 2.10.2 Spinors in *Cl3*,

Line 8, "while from (Sect. 2.4)" has been changed to "while from (2.4)"

Line 8 and 9, after "observable have components $(0, 0, \pm 1)$.)" new sentence "Note that ψ is just a quaternion (in GA)." has been added.

Page 136, formula below the line "with the commutation property" has been changed to $e_1 z = z^* e_1$.

Page 144, line 6, "(as for matrices)" has been changed to "(as for the Pauli matrices)"

Page 144, line 18, "(see Sect. 4.1)" has been changed to "(see Sect. 4.1.1)"

How to Proceed Further?

Miroslav Josipović

M. Josipović, *Geometric Multiplication of Vectors*,
Compact Textbooks in Mathematics,
https://doi.org/10.1007/978-3-030-01756-9

This question does not have the same answer for everyone; however, we can give a few ideas. Apart from studying the literature, it would be good to apply your knowledge to specific problems. You can also write to the author and point out errors in this book. However, several guidelines can be of value in general.

Geometric algebra should begin to live in high schools, and we need to invest considerable effort to achieve that goal. If you want to take part, you can join the group *Pre-University Geometric Algebra* on Linkedin (the owner is James Smith). Games with oriented objects could be introduced at an early age.

It would be a good idea to learn about *spacetime algebra* (STA). A good place to begin is the book *Space-Time Algebra*, by David Hestenes, Birkhäuser, 2015. We will outline two reasons for learning this powerful algebra. First, many authors use this algebra, and second, there is a beautiful formulation of the general theory of relativity in a flat space and in the language of this algebra (see Ref. [20]).

Geometric calculus is an inevitable and difficult bite, and one of the best sources for this topic is Ref. [25].

Thank you for reading and good luck!

Quotes

Miroslav Josipović

M. Josipović, *Geometric Multiplication of Vectors*,
Compact Textbooks in Mathematics,
https://doi.org/10.1007/978-3-030-01756-9

There are two versions of math in the lives of many Americans: the strange and boring subject that they encountered in classrooms and an interesting set of ideas that is the math of the world, and is curiously different and surprisingly engaging. Our task is to introduce this second version to today's students, get them excited about math, and prepare them for the future.

(Jo Boaler in *What's Math Got to Do with It?*, Penguin 2008)

We all agree that your theory is crazy, but is it crazy enough?

(Niels Bohr)

A mathematician is a blind man in a dark room looking for a black cat which isn't there.

(Charles Darwin)

It is a striking (and not commonplace) fact that Clifford algebras and their representations play an important role in many fundamental aspects of differential geometry. These algebras emerge repeatedly at the very core of an astonishing variety of problems in geometry and topology. Even in discussing Riemannian geometry, the formalism of Clifford multiplication will be used in place of the more conventional exterior tensor calculus. The Clifford multiplication is strictly richer than exterior multiplication; it reflects the inner symmetries and basic identities of the Riemannian structure. The effort invested in becoming comfortable with this algebraic formalism is well worthwhile.

(H. Lawson Jr. and M.L. Michelson, 1989)

The geometric algebra is a conceptually appealing and mathematically powerful formalism. If you want to understand rotations, Lorentz transformations, spin-1/2 particles, and supersymmetry, and you want to do actual calculations elegantly and (relatively) easily, then the geometric algebra is the thing to learn.

(Andrew J.S. Hamilton)

It is my experience that proofs involving matrices can be shortened by 50% if one throws the matrices out.

(Emil Artin in Geometric Algebra, p. 14)

It appears as one of the fundamental principles of Nature that the equations expressing basic laws should be invariant under the widest possible group of transformations.

(Dirac, P.A.M. 1973, Proceedings of the Royal Society of London Series A, 333, 403)

Credits

Miroslav Josipović

M. Josipović, *Geometric Multiplication of Vectors*,
Compact Textbooks in Mathematics,
https://doi.org/10.1007/978-3-030-01756-9

Figure	Credits
Figure in Sect. 1.6.9	The image *adapted* with kind permission of Kamenko Ćulap, the man in the adapted image (photo by Držislav Korade)
Fig. 1.29, Reflection on water	*Image reprinted with kind permission from* https://www.flickr.com/photos/stephi2006/12725183683/
Fig. 1.29, Total reflection in optical fibers	*The image adapted with kind permission from* https://www.flickr.com/photos/borshop/14869592682/
Fig. 1.29, Andromeda galaxy rotates	*The image adapted with kind permission from* https://apod.nasa.gov/apod/ap150830.html Image Credit & Copyright: Robert Gendler
Fig. 1.29, Neutron stars rotate fast	The image *adapted* with kind permission from https://www.flickr.com/photos/kevinmgill/14773475650/
Figure in Sect. 1.11.4	The image *adapted* with kind permission of Eckhard Hitzer
Fig. 1.36	The image of binaries adapted with kind permission from https://www.flickr.com/photos/gsfc/15403844862/ credit: NASA's Goddard Space Flight Center/S. Wiessinger
G. Sobczyk, Sect. 1.15.1	The image *adapted* with kind permission of Garret Sobczyk
W. Baylis, Sect. 2.7.5	The image *adapted* with kind permission of William E. Baylis
Chap. 3, Robot is playing a trumpet	The image *adapted* with kind permission from https://www.flickr.com/photos/marufish/3522355556/
Chap. 4, An old train	The image *adapted* with kind permission from https://www.flickr.com/photos/91807507@N03/14372621646/
Chap. 4, A modern train	The image *adapted* with kind permission from https://www.flickr.com/photos/foolish_adler/2380994170/in/photostream/
W. Pauli, Sect. 4.1.1	The image *reprinted* with kind permission from CERN
Chap. 5, A toy robot	The image *adapted* with kind permission from https://www.flickr.com/photos/perijove/7298486740/
Fig. 5.1	The used Möbius Ring by the artist David Weitzman (in image)
Argon laser test, Sect. 5.6	The image *adapted* with kind permission from https://www.flickr.com/photos/mightyohm/3938307729/

(continued)

Figure	Credits
Golden spiral, Sect. 6.3	The image *adapted* with kind permission of Ken Pratt http://kenpratt.net/portfolio/raytracer/
A cat, Sect. 6.4	The image *adapted* with kind permission of Sanja Josipović, original photo of the cat Luna by Vesna Josipović

The other figures are made by the author, using Wolfram *Mathematica* and the Package that can be found at L1

Glossary

A

algebra of physical space (APS) *Cl*3, Pauli algebra

antiautomorphism $\mathrm{f}(MN) = \mathrm{f}(N)\mathrm{f}(M)$

automorphism $\mathrm{f}(MN) = \mathrm{f}(M)\mathrm{f}(N)$

B

binary star a system of two stars bound by their mutual gravitational forces

Bondi factor (coefficient) $k = \sqrt{(1+v)/(1-v)}$; v is the velocity of an inertial reference frame

C

center of an algebra subset of the algebra whose elements commute with all elements of the algebra

comoving frame a frame of reference that is not attached to a moving object and can be treated as an inertial reference frame in which the moving object has the velocity zero at some instant of time

contraction an operation that generally lowers the grade (or rank) of an element of an algebra

D

dual number number of the form $\alpha + \beta h$, $h^2 = 0$, $\alpha, \beta \in \mathbb{R}$

E

expectation value the probabilistic *expected value* of the result of measurement of an experiment

Euclidean vector space a vector space in which the square of nonzero vectors is a positive real number

F

Faraday in *Cl*3, a common name for a complex vector of an electromagnetic field, $\mathbf{F} = \mathrm{E} + j\mathbf{B}$

G

geometric calculus a calculus in geometric algebra

Gibbs products standard scalar and cross products

group a set of objects with a binary operation (closure) that is associative and has a unit element and inverses

M. Josipović, *Geometric Multiplication of Vectors*,
Compact Textbooks in Mathematics,
https://doi.org/10.1007/978-3-030-01756-9

H

Hermitian adjoint (of a matrix) transposition followed by complex conjugation
Hermitian matrix a matrix equal to the Hermitian adjoint, like the Pauli matrices
hyperbolic number (split complex, perplex; see double number)

I

ideal (left) of an algebra for an algebra A, it is a subset L of the algebra with the property $a \in A, l \in L \Rightarrow al \in L$

J

Jacobi identity a property of a binary operation that describes how the order of evaluation (the placement of parentheses in a multiple product) affects the result of the operation: $a \otimes (b \otimes c) + b \otimes (c \otimes a) + c \otimes (a \otimes b) = 0$

K

Kronecker delta a special symbol, δ_{ij}, equal to 1 for $i = j$, zero otherwise

L

left ideal see *ideal*
Levi-Civita symbol ε_{ijk}, represents a collection of numbers; defined from the *sign of a permutation* of the natural numbers 1, 2, 3 (it can be defined with any number of indices)

M

Minkowski vector space a vector space with signature $(n-1, 1)$

N

null-vector $a^2 = 0, \quad a \neq 0$ ($a \cdot a = 0$ in CGA model)

O

observable in quantum mechanics, a quantity that can be measured
operator in linear algebra, an object that operates on vectors, usually represented by matrices (in geometric algebra, "operators" are elements of the algebra)

P

paravector a sum of a real number and a vector, $\alpha + \mathbf{x}$, $\alpha \in \mathbb{R}$

Q

quantum information theory a quantum mechanical information theory contained in quantum systems

quaternion discovered by Hamilton, a number of the form $q^0 + q^i I_i$ (summation), $i = 1, 2, 3$, $I_i^2 = -1$, $I_i I_j = -I_j I_i$, $I_1 I_2 = I_3$, etc. (cyclic), $I_1 I_2 I_3 = -1$

R

reduced mass for a two-body problem we define $\mu = m_1 m_2 / (m_1 + m_2)$

S

spacetime algebra (STA) a geometric algebra based on $\mathfrak{R}(1,3)$

spinor elements of a (complex) vector space that can be associated with Euclidean space; in *Cl*3, they are elements of the even part of the algebra

standard model the theory of particle physics, describing three of the four known fundamental forces in the universe (the electromagnetic, weak, and strong interactions)

symmetry a transformation that leaves a property of a system unchanged

T

Taylor expansion (series) a representation of a function as an infinite sum of terms that are calculated from the values of the function's derivatives at a single point

tensor a geometric object that describes linear relations between geometric vectors, scalars, and other tensors

U

unimodular $M\bar{M} = 1$, where $\bar{M}$ is the Clifford conjugation of the multivector M

unipodal number $z_1 + z_2 h$, $z_1, z_2 \in \mathbb{C}$, $h^2 = 1$

unitary $MM^\dagger = 1$, where $M^\dagger$ is the reverse involution of the multivector M (a complex square matrix U is unitary if its conjugate transpose $U^\dagger$ is also its inverse)

V

vector space a collection of objects called vectors, objects with special transformation rules, which may be added together and multiplied ("scaled") by numbers, called scalars; usually, scalars are real (a real vector space) or complex (a complex vector space)

W

wedge the character $\wedge$, common in GA to designate the *outer product*

whirl a complex vector with the property $\mathbf{F}^2 \in \mathbb{R}$ (author's suggestion)

X

x-rays a form of electromagnetic radiation, Röntgen rays (there are indications that Tesla discovered x-rays before Röntgen)

Literature

1. C.J. Baez, *Division algebras and quantum theory*. arXiv:1101.5690v3 (2011)
2. W.E. Baylis et al., *Motion of charges in crossed and equal E and B fields*. AJP **62**(10), 899–902 (1993)
3. W.E. Baylis, *Geometry of Paravector Space with Applications to Relativistic Physics* (Kluwer Academic, Boston, MA, 2004)
4. W.E. Baylis, S. Hadi, *Rotations in n dimensions as spherical vectors*, in *Applications of Geometric Algebra in Computer Science and Engineering*, ed. by L. Dorst, C. Doran, J. Lasenby, (Springer, New York, 2002), pp. 79–90
5. W.E. Baylis, *Classical eigenspinors and the Dirac equation*. Phys. Rev. A **45**, 7 (1992)
6. W.E. Baylis, Ch. 18: *Eigenspinors in electrodynamics*, in *Clifford (Geometric) Algebras with Applications to Physics, Mathematics, and Engineering*, ed. by W. E. Baylis, (Birkhäuser, Boston, MA, 1996)
7. W.E. Baylis, Ch. 19: *Eigenspinors in quantum theory*, in *Clifford (Geometric) Algebras with Applications to Physics, Mathematics, and Engineering*, ed. by W. E. Baylis, (Birkhäuser, Boston, MA, 1996)
8. W.E. Baylis, Ch. 20: *Eigenspinors in curved spacetime*, in *Clifford (Geometric) Algebras with Applications to Physics, Mathematics, and Engineering*, ed. by W. E. Baylis, (Birkhäuser, Boston, MA, 1996)
9. W.E. Baylis, *Electrodynamics: A Modern Geometric Approach*, Progress in Mathematical Physics, vol 17 (Birkhäuser, Boston, MA, 2002)
10. W.E. Baylis, G. Sobczyk, *Relativity in Clifford's geometric algebras of space and spacetime*. Int. J. Theor. Phys. **43**, 2061 (2004)
11. W.E. Baylis, *Surprising symmetries in relativistic charge dynamics*. arXiv:physics/0410197v1 (2004)
12. T. Bouma, G. Memowich, *Invertible homogeneous versors are blades*. (2001), http://www.science.uva.nl/ga/publications
13. A. Bromborsky, *An introduction to geometric algebra and calculus*. (2014), http://www2.montgomerycollege.edu/departments/planet/planet/Numerical_Relativity/bookGA.pdf
14. J. Chappell, et al., *Geometric algebra for electrical and electronic engineers*. (IEEE, 2014)
15. J.M. Chappell, J.G. Hartnett, N. Iannella, D. Abbott, A. Iqbal, *Exploring the origin of Minkowski spacetime*. arXiv:1501.04857v3 (2015)
16. J.M. Chappell, A. Iqbal, L.J. Gunn, D. Abbott, *Functions of multivector variables*. arXiv:1409.6252v1 (2014)
17. J.M. Chappell, A. Iqbal, N. Iannella, D. Abbott, *Generalizing the Lorentz transformations*. arXiv:1611.02564v1 (2016)
18. J.M. Chappell, A. Iqbal, N. Iannella, D. Abbott, *Revisiting special relativity: a natural algebraic alternative to Minkowski spacetime*. PLoS One **7**(12), e51756 (2012)
19. C. Doran, *Geometric algebra and its application to mathematical physics*, Thesis, 1994
20. C. Doran, A. Lasenby, S. Gull, *Gravity as a gauge theory in the spacetime algebra*, in *Fundamental Theories of Physics*, ed. by F. Brackx, R. Delanghe, H. Serras, (Springer, Dordrecht, 1993)
21. C. Doran, A. Lasenby, *Geometric Algebra for Physicists* (Cambridge University Press, Cambridge, 2003)
22. C. Doran et al., *Spacetime algebra and electron physics*, in *Advances in Imaging and Electron Physics*, vol. 95, (Academic Press, San Diego, CA, 1996), pp. 271–386
23. L. Dorst, D. Fontijne, S. Mann, *Geometric algebra for computer science*. (2007), http://www.geometricalgebra.net

M. Josipović, *Geometric Multiplication of Vectors*,
Compact Textbooks in Mathematics,
https://doi.org/10.1007/978-3-030-01756-9

24. L. Dorst, *The inner products of geometric algebra*, in *Applications of Geometric Algebra in Computer Science and Engineering*, ed. by L. Dorst, C. Doran, J. Lasenby, (Birkhäuser, Boston, MA, 2002)
25. D. Hestenes, G. Sobczyk, *Clifford Algebra to Geometric Calculus* (Reidel, Dordiecht, 1984)
26. D. Hestenes, *Mathematical viruses*, in *Clifford Algebras and Their Applications in Mathematical Physics*, ed. by A. Micali et al., (Kluwer Academic, Dordrecht, 1992)
27. D. Hestenes, *New Foundations for Classical Mechanics* (Kluwer Academic, Dordrecht, 1999)
28. D. Hestenes, *Universal geometric algebra*. Simon Stevin Q. J. Pure Appl. Math. **62**(3–4) (1988)
29. E. Hitzer, J. Helmstetter, R. Ablamowicz, *Square roots of –1 in real Clifford algebras*. arXiv:1204.4576v2 (2012)
30. E. Hitzer, *Angles between subspaces computed in Clifford algebra*. arXiv:1306.1825v1 (2013)
31. B. Jancewicz, *Multivectors and Clifford Algebra in Electrodynamics* (World Scientific, Singapore, 1988)
32. G.L. Jones, *The Pauli algebra approach to relativity*, Thesis, Univ. of Windsor, 1994
33. A. Lasenby, C. Doran, S. Gull, *Gravity, gauge theories and geometric algebra*. arXiv:gr-qc/0405033 (2004)
34. A. Macdonald, *Linear and geometric algebra*. (2010), http://faculty.luther.edu/~macdonal/laga/
35. A. Macdonald, *Vector and geometric calculus*. (2012), http://faculty.luther.edu/~macdonal/vagc/
36. N.D. Mermin, *Relativity without light*. Am. J. Phys. **52**, 119 (1984)
37. O.A. Mornev, *Idempotents and nilpotents of Clifford algebra (in Russian)*, Гиперкомплексные числа в геометрии и физике. **2**(12), том 6 (2009)
38. V. de Sabbata, B.K. Datta, *Geometric Algebra and Applications in Physics* (Taylor & Francis, London, 2007)
39. G. Sobczyk, *Geometric matrix algebra*. Linear Algebra Appl. **429**, 1163–1173 (2008)
40. G. Sobczyk, *New Foundations in Mathematics: The Geometric Concept of Number* (Birkhäuser, Boston, MA, 2013)
41. G. Sobczyk, *Special relativity in complex vector algebra*. arXiv:0710.0084v1 (2007)
42. G. Sobczyk, *Vector analysis of spinors*. (2015), http://www.garretstar.com
43. G. Sobczyk, *The missing spectral basis in algebra and number theory*. Am. Math. Mon. **108**(4), 336–346 (2001)
44. G. Trayling, W. Baylis, *The Cl7 approach to the standard model*. Prog Math Phys **34**, 547–558 (2011)
45. J. Vince, *Geometric Algebra for Computer Graphics* (Springer, London, 2008)

Some More Interesting Texts

E. Chisolm, *Geometric algebra*. arXiv:1205.5935v1 (2012)

J.S.R. Chisholm, A.K. Common, *Clifford Algebras and Their Applications in Mathematical Physics* (Reidel Publishing Company, Boston, MA, 1985)

D. Hestenes, *Celestial mechanics with geometric algebra*. Celest. Mech. **30**, 151–170 (1983)

D. Hildenbrand, *Foundations of Geometric Algebra Computing* (Springer, Berlin, 2013)

E. Hitzer, *Introduction to Clifford's geometric algebra*. (2012). arXiv:1306.1660v1

E. Hitzer, *Multivector differential calculus*. arXiv:1306.2278v1 (2013)

N. Hitzer, *Kuroe: Applications of Clifford's Geometric Algebra*. arXiv:1305.5663 (2013)

K. Kanatani, *Understanding Geometric Algebra* (CRC Press, Hoboken, NJ, 2015)

D.E. Kosokowsky, *Dirac theory in the Pauli algebra*, Thesis, Univ. of Windsor, 1991

A. Macdonald, *A survey of geometric algebra and geometric calculus*. http://faculty.luther.edu/~macdonal/GA&GC.pdf, (2016)

E. Meinrenken, *Clifford algebras and Lie groups*. http://isites.harvard.edu/fs/docs/icb.topic1048774.files/clif_mein.pdf, (2005)

C. Perwass, *Geometric Algebra with Applications in Engineering* (Springer, Berlin, 2009)
J. Vaz Jr., *The Clifford algebra of physical space and Dirac theory*. Eur. J. Phys. **37** (5) (2016)
J. Wei, *Quantum mechanics in the real Pauli algebra*, Thesis, Univ. of Windsor, 1991
Y. Yao, *New relativistic solutions for classical charges in an electromagnetic field*, Thesis, Univ. of Windsor, 1998

References and Links to Specific Subjects

References

R1: D. Hestenes, *Spin and uncertainty in the interpretation of quantum mechanics*. Am. J. Phys. **47**, 399 (1979)
R2: F. Witte, *Classical physics with geometric algebra*. (2007)
R3: B. Jancewicz, *A system of units compatible with geometry*. CEJP **2**(1), 204–219 (2004)
R4: C.J.L. Doran et al., *Spacetime algebra and electron physics*. Adv. Imaging Electron Phys **95**, 271–386 (1996)
R5: H.S.M. Coxeter, S.L. Greitzer, *Geometry revisited*, MAA. (1967)
R6: A.D. Gardiner, C.J. Bradley, *Plane Euclidean Geometry: Theory and Problems* (UKMT, Leeds, UK, 2005)
R7: R.J. Wareham, J. Lasenby, *Generating fractals using geometric algebra*. AACA **21**, 647–659 (2011)
R8: D. Fontijne, *Efficient Implementation of Geometric Algebra* (Morgan Kaufmann, Amsterdam, 2007)
R9: G. Aragón-Camarasa, et al., *Clifford algebra with mathematica*. arXiv:0810.2412v1 (2008)
R10: A.M. Steane, *An introduction to spinors*. arXiv:1312.3824v1 (2013)
R11: L. Pertti, *Clifford Algebras and Spinors* (Cambridge University Press, Cambridge, 2001)
R12: J.D. Jackson, *Classical Electrodynamics* (Wiley, New York, 1999)

Links to Supplementary Materials

L1: http://extras.springer.com/2019/978-3-030-01755-2
L2: https://ocw.mit.edu/resources/res-8-001-applied-geometric-algebra-spring-2009/

3D Geometry

http://geocalc.clas.asu.edu/GA_Primer/GA_Primer/index.html

Some Web Resources

https://gaupdate.wordpress.com
http://geocalc.clas.asu.edu/html/Evolution.html (the diagram at the end of the book is taken from this page and adapted by the author with the kind permission of David Hestenes)
https://staff.science.uva.nl/l.dorst/clifford/

Software

Cinderella. https://www.cinderella.de/tiki-index.php
M Clifford. Mathematica package. https://arxiv.org/abs/0810.2412, 2018
http://www.fata.unam.mx/investigacion/departamentos/nanotec/aragon/software/clifford.m
CLIFFORD. http://math.tntech.edu/rafal/
Clifford algebra for CAS Maxima. https://github.com/dprodanov/clifford
Clifford Multivector Toolbox for MATLAB. http://clifford-multivector-toolbox.sourceforge.net/
CLUCalc/CLUViz. http://www.clucalc.info/
GA20 and GA30 (Formerly called *pauliGA*). https://github.com/peeterjoot/gapauli
Gaalet. https://sourceforge.net/projects/gaalet/
Gaalop. http://www.gaalop.de/
GABLE. https://staff.fnwi.uva.nl/l.dorst/GABLE/index.html
Gaigen. https://sourceforge.net/projects/g25/
GAlgebra. https://github.com/brombo/galgebra
GA Sandbox. https://sourceforge.net/projects/gasandbox/
GAViewer. http://www.geometricalgebra.net/downloads.html; this nice tool is recommended with the text. You can manipulate the images to some extent
GluCat. https://sourceforge.net/projects/glucat/
GMac. https://gacomputing.info/gmac-info/
SpaceGroupVisualizer. http://spacegroup.info/
GrassmannAlgebra package. https://sites.google.com/site/grassmannalgebra/thegrassmannalgebrapackage
Versor. http://versor.mat.ucsb.edu/
Some important details can be found at https://gacomputing.info/ga-software/
https://ga-explorer.netlify.com/index.php/ga-online-resources/

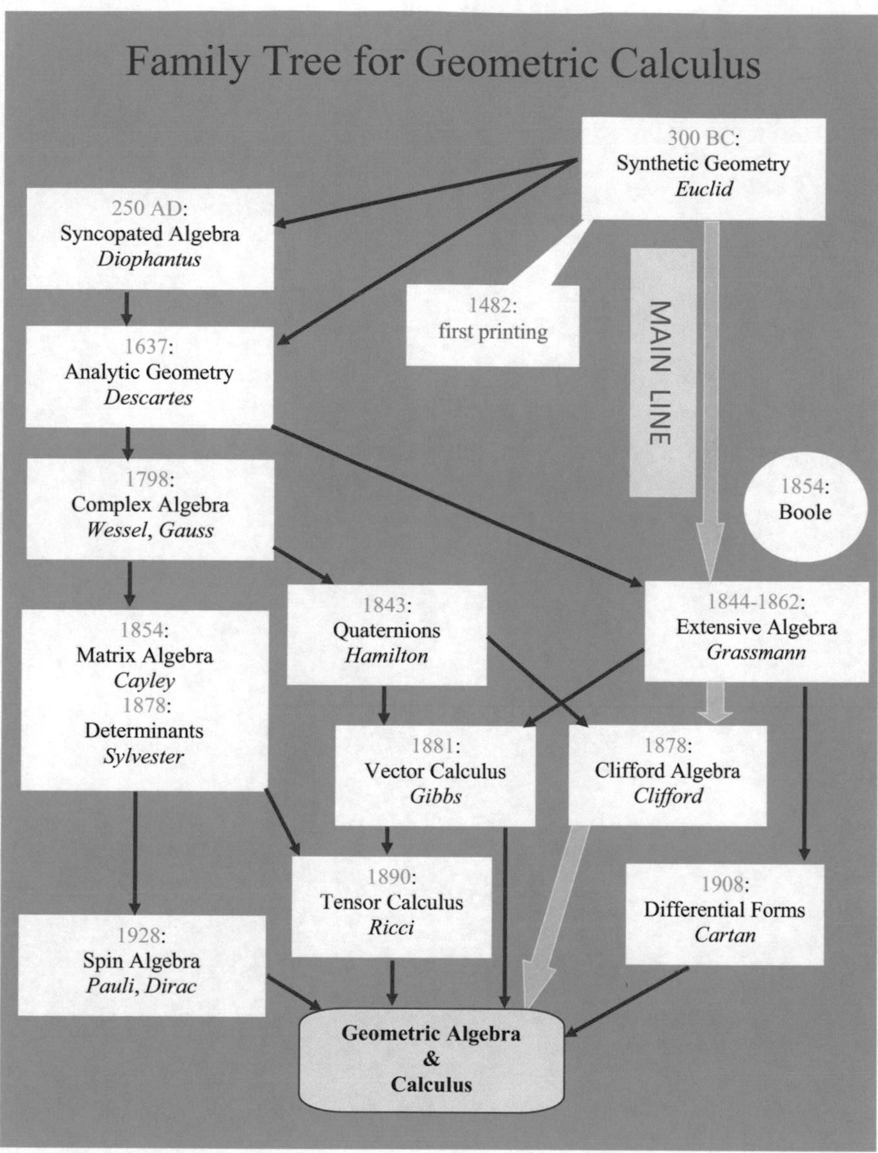

Family Tree for Geometric Calculus, The image *adapted* with kind permission of David Hestenes.
http://geocalc.clas.asu.edu/html/Evolution.html

Printed in the United States
By Bookmasters